Structural Integrity

.

Structural Integrity

Theory and Experiment

edited by

E. S. FOLIAS

University of Utah
Salt Lake City, Utah, U.S.A.

Reprinted from *International Journal of Fracture*, Vol. 39, Nos. 1–3 (1989)

Kluwer Academic Publishers

Dordrecht / Boston / London

ISBN-13: 978-94-010-6906-9   e-ISBN-13: 978-94-009-0927-4
DOI: 10.1007/978-94-009-0927-4

Published by Kluwer Academic Publishers,
P.O. Box 17, 3300 AA Dordrecht, The Netherlands.

Kluwer Academic Publishers incorporates
the publishing programmes of
D. Reidel, Martinus Nijhoff, Dr W. Junk and MTP Press.

Sold and distributed in the U.S.A. and Canada
by Kluwer Academic Publishers,
101 Philip Drive, Norwell, MA 02061, U.S.A.

In all other countries, sold and distributed
by Kluwer Academic Publishers Group,
P.O. Box 322, 3300 AH Dordrecht, The Netherlands.

TABLE OF CONTENTS

Some of the attendees at the University of Utah meeting
honoring Professor M. L. Williams on the occasion of his 65th anniversary.

Preface

It is propitious that the 25th year of publication of the *International Journal of Fracture* should coincide with the opportunity to support the recognition by the Society of Engineering Science of the 65th birthday of the Founding Editor-in-Chief, Professor M.L. Williams.

At its 24th Annual Meeting at the University of Utah in September 1987, the Organizing Committee of the Society chose to honor Professor Williams by reserving several sessions for contributions from his colleagues working in the mechanics of fracture and associated mechanical property-structure relationships. We are therefore pleased to welcome Professor E.S. Folias, Professor of Mechanical Engineering, University of Utah, and a member of the Editorial Committee as the Guest Editor for the first three issues of Volume 39 of the Journal. We have also noted that the University of Utah was one of the three homes for the Journal, sandwiched between the original home at the California Institute of Technology in 1965 and its current location at the University of Pittsburgh.

In observing these dual anniversaries, the publishers not only enthusiastically support the presentation of these special papers, but also wish to extend to Professor Williams our own best wishes on his personal anniversary, and to thank him and all the authors, reviewers, and particularly M.C. Williams, J.L. Swedlow and the Regional Editors for their respective contributions as we observe this 25th milestone.

F.W.B. van Eysinga
President
Dordrecht, The Netherlands   *Kluwer Academic Publishers Group*
January 1989

Professor M. L. Williams

International Journal of Fracture 39: ix–xiv (1989)
© Kluwer Academic Publishers, Dordrecht

Foreword

It has been a pleasure for a few of us in various capacities and with different knowledge bases to join in organizing and presenting a tribute to our colleague and friend on the occasion of his 65th birthday. Professor Williams has contributed his talents over a diverse range of activity spanning the academic, industrial, and public service arenas while maintaining an enviable sense of humor and sense of proportion. In this volume we focus mainly upon his technical work and merely note for the record that the observance in Salt Lake City was suitably accompanied by a very non-technical evening session enlivened by appropriate anecdotes from the past which were contributed by the many colleagues and friends of the "roastee."

Max Lea Williams, Jr. was born on 22 February 1922 in Aspinwall, Pennsylvania, a suburb of Pittsburgh. He spent his early years in that vicinity and ultimately attended the Carnegie Institute of Technology, now Carnegie Mellon University, from which he graduated in 1942, having received a degree in mechanical engineering (aeronautics option). During his college days he was quite active in extra curricular activities, one of them interestingly enough being editor of the *Carnegie Technical*, the monthly engineering school publication — perhaps portending his editorship of the International Journal of Fracture.

Upon graduation he was called to active duty with the Air Force and served approximately four years, first as a squadron engineering officer in the 15th Air Force, and then as a flight test engineer at Wright Field in Dayton, Ohio. The practical field maintenance experience coupled with the subsequent opportunity to flight test a wide variety of Air Force experimental aircraft at the Dayton center for technical research, development, and evaluation had a strong influence on his later career. He flew, for example, in the Northrop Flying Wing and the C-74 Douglas Globemaster, which was a forerunner of the present day large cargo aircraft. This personal experience with rapidly changing designs, many influenced by the emergence of supersonic flight which had not been usually presented in pre-war aeronautics courses, motivated his return to post-graduate studies. He therefore entered the graduate program at the Gugenheim Aeronautical Laboratory of the California Institute of Technology (GALCIT) in 1946, originally only for a master's degree, but his residence there eventually extended to 20 years.

He joined the faculty in 1948 as a part-time lecturer in flight testing and systems engineering and graduated in 1950 with a doctorate in aeronautics. His work was based upon a major research program of Professor E.E. Sechler's dealing with one of the early research investigations into the stress analysis of swept aircraft and missile wings [1]. With the missile wing idealized as a swept plate with reentrant corners, the study led naturally to research into the stress state at angular corners and ultimately produced an extensive series of papers on stress singularities, the best known being the two on stresses in the vicinity of angular corners of thin plates subjected to bending [2] and stretching [3].

From this work came the limit case when the angular corner was reduced to a zero opening angle, or crack, which was the first of many contributions by Dr. Williams to the mechanics of fracture. His 1957 paper [4] in the *Journal of Applied Mechanics* developed the now called

"Williams series" for analyzing symmetrically and anti-symmetrically loaded plates. The latter paper also predicted the characteristic "bug-eye" photoelastic fringe pattern to be expected at crack tips as well as the angular orientation about the tip at which various maximum and minimum critical stresses and strain energies would occur, e.g. the maximum principal tensile stress would occur at ± 60 degrees either side of the direction of the original crack when subjected to symmetrical loading. Also, the two principal stresses at the crack tip were equal, thus suggesting a decreased amount of yielding and a higher degree of applicability of the elastic solution near the (locally yielded) crack tip. Another interesting outgrowth was the direct interpretation of the stress intensity factor in terms of the local (elliptical) radius of curvature at the tip [5]. Eigenfunction expansions, or Williams series, were also to prove useful in numerical analysis; first in collocation analysis of fracture geometries and second in finite element analysis using the singular crack tip elements.

While much of his analytical work at the time related to fracture, Dr. Williams also published a pair of papers stimulated by concern in the aircraft industry for blast and thermal loading in skin panels [6, 7]. By considering (infinite) thin plate strips loaded by varying pressure and arbitrary temperature distribution through the thickness he deduced a straight-forward exact solution for this geometrical limit case of the von Karman non-linear large deflection equations, supplemented with the necessary thermal terms derived from a variational formulation of the minimum potential energy. Interestingly enough, his direct solution, presented in graphical, parametric form for easy design use, circumvented the cumbersome development of the special case treated by Timoshenko in Chapter I of his classical text, *Theory of Plates and Shells* (1940).

In 1959, Dr. Williams published another classic paper, also dealing with stress singularities, which produced a local crack tips singularity of the (sometime) oscillating type between dissimilar media [8]. His result arose from a geophysics seminar presented at Caltech and subsequently sparked personal interest in the mechanics of adhesive fracture and different bonding orientations [9], as well as stimulating many extensions in the literature, including the astute observation of A.H. England that if elastic displacement oscillations were to actually occur the faces would interact! Practically speaking the extent of the oscillating effect is confined to the tip region, probably within any yielded region.

The foregoing analytical papers at Caltech were complemented by a parallel series of experimental works, frequently utilizing a modified Ellis million frame per second camera constructed at GALCIT [10]. Blast loading around underground structures and their shock isolation was modelled photoelastically by Arenz [11] and high speed crack propagation in steel photographed by W.M. Beebe, M.E. Jessey, H.W. Liu, and S.R. Valluri [12]. At almost the same time, and in support of analytical work on solid propellant rocket motor design, Dr. Williams applied a photoelastic testing device used by A.J. Durelli to publish, with D.D. Ordahl, the first series of parametric stress concentration factors for pressurized solid rocket grains [13, 14] and followed up with a similar series for thermally loaded grains using a Weibull deformeter and exploiting the Biot–Muskhelishvili displacement dislocation principle.

By the late 1950's, his research group was becoming well established in the fracture field and was unique in its interest in fracture phenomena regardless of the material. At about this time, the USSR launched Sputnik and stimulated enhanced international interest in rocketry. Inasmuch as solid propellant rocket grains are composed of filled, rubbery, viscoelastic materials in which cracks could cause catastrophic fracture, this aeronautical design generated considerable special interest which had to start with a knowledge of mechanical properties and materials characterization before fracture behavior could be assessed in an orderly way. The GALCIT interest in solid rocket structural integrity attracted industrial sponsorship which, in 1961, led to the popular and well circulated GALCIT 61-5 report [15] co-authored with P.J. Blatz and R.A. Schapery. This period was one of intense activity and the contributions had a strong influence on the early direction of structural integrity analysis in the rocket industry because the personal research of Dr. Williams, his Research Fellows, and senior graduate students was reinforced by their vigorous participation in various joint industry-government-academic working committees, steering groups, e.g. JANAF Solid Propellant Information Agency, as well as their being frequent lecturers at many industrial short courses. Just last year, Dr. Williams and Professor J.E. Fitzgerald, a former colleague at Utah and now at the Georgia Institute of Technology, were honored by the American Institute of Aeronautics and Astronautics (AIAA) for their achievements in rocket structural integrity. One of the last major papers given by Dr. Williams from his Caltech position was an invited presentation to the AIAA in 1966 on practical engineering stress analysis methods of linear viscoelastic materials [16]. This summary and survey paper covered material characterization, stress analysis and experimental methods as known at the time.

Space does not permit a commentary in depth on the entire scope of the GALCIT group's activity, particularly outside the rocket area described above, but one gains an appreciation of the breath of interest from the subject matter of several of the dissertations: photoelastic material characterization by R.J. Arenz, fracture micro-mechanisms by W.G. Knauss, large deformation phenomena by W.L. Ko and G.H. Lindsey, fundamental thermodynamic interpretation of fracture by R.A. Schapery, numerical analysis techniques in the work of J.L. Swedlow, that of E.S. Folias on the effect of single and double curvature on fracture initiation, and work on dynamic fracture by D.D. Ang, which complemented the experimental work of W.M. Beebe, and in 1959 related closely to the Broberg running crack problem of the same time period. In addition, there were also many contributions from national and international visitors, such as the joint papers with Professor Takeshi Kunio of Keio University on solid rocket design [17, 18].

As a final comment regarding Professor Williams' 20 year period at Caltech, it is perhaps not surprising in view of the group's fracture activity, its collegiality, and its diversity that, in conjunction with Dr. Williams and Dr. Swedlow being Founder Members of the International Conference on Fracture (ICF), Caltech was a natural choice and Professor Williams an obvious person to accept Professor Takeo Yokobori's challenge to initiate the first technical journal dealing solely with the mechanics of fracture. As mentioned earlier, the first issue of the *International Journal of Fracture* (*Mechanics*), a quarterly at the time, came off the press, from an antecedent of the present publisher, in March 1965.

It was in 1965 that Professor Williams somewhat reluctantly, we believe, left the research culture of Caltech to become Dean of the College of Engineering at the University of Utah. With increased administrative duties, his research role, while still technical, became more that of a supporter and collaborator, particularly to Professor K.L. DeVries and his students. On the public service side, Professor Williams became increasingly active and held several committee memberships, most notably on the Biomaterial Research Advisory

Committee of the National Institute for Dental Research, the NASA Research and Technology Committee on Chemical Rocket Propulsion and the National Science Foundation Advisory Committee for Engineering. In 1972 he was honored as a Sigma Xi National lecturer on the Mechanics of Fracture.

The collaboration with Professor DeVries was rewarding as well as productive, mainly because of their mutual interest and reinforcing capabilities in mechanics and physics. The joint work began shortly after the First ICF in Sendai, Japan (1965) when Professor Williams, to assist communication in English, was invited by Dr. S.N. Zhurkov, now an Honorary Editor of the Journal, to read his paper on an electron spin resonance method for deducing from microscopic measurements the macroscopic behavior of certain polymers. This paper and the ensuing discussions between Dr. Zhurkov and members of his research group in Leningrad led the Williams–DeVries collaboration along several different paths. Simultaneously, following the prior interest in polymers and material characterization developed through the viscoelastic behavior of rocket fuel material, a stimulating collaboration emerged with Dr. F.N. Kelley, then of the Air Force Rocket Propulsion Laboratory, and Professor F.R. Eirich, of Brooklyn Polytechnic Institute, based upon the NATO Advanced Study Institute organized by Professor Williams in 1967 to join polymer chemists and engineers. While this five-way set of acquaintances seldom resulted in five-way collaborations, it frequently sparked bi-lateral and tri-lateral efforts — with changing partnerships.

The NATO meeting led Kelley and Williams to "invent" an Interaction Matrix which gave a protocol for assessing property-structure relationships in polymers [19, 20], e.g. relaxation modulus to chain stiffness, cross-link density, etc. Eirich and Williams then co-authored a 1972 National Science Foundation Workshop Monograph on *Polymer Engineering and Its Relevance to National Materials Development* [21], followed by another in 1979 on *Mechanics-Structures Interactions with Implications for Material Fracture* [22]. On direct research collaboration, Professor Williams first devoted his time to assisting in the interpretation of the spin resonance work being conducted by Professor DeVries and his students, especially that of D.K. Roylance, E.R. Simonson, K. Leheru and B.A. Lloyd, with the view toward engineering applications. Second, there evolved with a separate segment of the DeVries research group a specific thrust in adhesive fracture, primarily from the continuum viewpoint, but utilizing the microstructural characteristics and polymer chain properties wherever possible in understanding the process. From the mechanics point of view, the major question often asked at Utah was: How does this (microstructure) research help to understand, control or influence the specific adhesive fracture energy, i.e. viscoelastic fracture toughness? Professor Williams reviewed some of his concerns and possibilities for cohesive and adhesive fracture at the 5th U.S. National Congress of Applied Mechanics [23].

This second portion of the DeVries–Williams collaboration produced the pressurized "blister-test" [24], developed in conjunction with W.B. Jones, as an extension of a point load experiment reported by Malyshev and Salganik in the *International Journal of Fracture* (1965) to assess the adhesive fracture criterion for softer polymeric materials as well as to provide an isolated controlled environment within the blister during pressurization so chemical environmental influences at the debonding crack tip could be assessed. Numerical analysis as well analytical, was contributed in the dissertation by G.P. Anderson for several variations of the blister test, as well as developing a conical plug experiment whereby the ratio of normal and shear stress could be varied by combining pull-out force and twist. These experiments lead to the first investigations we know of to suggest that new fracture surface area might have three modes with a constant fracture toughness — instead of three modes of fracture toughness and one reference area. G.P. Anderson and S.J. Bennett, members of the DeVries group, collaborated to make the first estimate of possible (viscoelastic) time dependence of the adhesive fracture energy [25] and raised the interesting question of the possible proportionality between the fracture energy and relaxation modulus. Much of the research in adhesion contributed by Anderson, Bennett and DeVries was assembled in their 1977, Academic Press book, *Analysis and Testing of Adhesive Bonds.*

Aside from a few exceptions, Dr. Williams' active, personal research came to an end when he moved to the University of Pittsburgh in 1973 as Dean of the School of Engineering. By then his technical effort was concentrated at the advisory level to government, where he became increasingly involved. During his early years at Pitt he was a member of the National Materials Advisory Board and chairman of its Council on Materials, Structures, and Design, then chairman of the Structural Mechanics Sub-committee for Automotive Research of the Presidential Office of Science and Technology Policy, an occasional associate member of the Defense Science Board, an advisor to the Department of Defense and the Department of State. In 1975, he received the National Adhesives Award from the American Society of Testing and Materials, and later served a four year term on the Penn State Advisory Council for its Advanced Research Laboratory.

We have not failed to note however, that upon achieving Dean Emeritus status in 1985, essentially terminating his administrative duties at the University of Pittsburgh, he was appointed Distinguished Service Professor of Engineering and returned to the role of teacher. While he still continues his public service role, most recently having been appointed to a four year term on the U.S. Air Force Scientific Advisory Board (1985–1989), he served for two years after his retirement from administration as the General Lew Allen Distinguished Visiting Professor of Aeronautics at the Air Force Institute of Technology. He thus returned to his old territory in Dayton, Ohio, where he then spent a third year in Dayton as Science Advisor to the Commander of the Acquisition Logistics Center before returning to the Pitt campus for the 1988–1989 academic year.

Professor Williams has had a very full 65 years. Our description of it has of necessity been subjective, but we hope representative. At this time we take the opportunity to thank the authors who have contributed to this M.L. Williams "Zeitschrift-65," and his friends and colleagues who helped celebrate this Salt Lake City event in person, as well as many more who were there in spirit. So for the community,

Happy Birthday Greetings,

K.L. DeVries
E.S. Folias
F.N. Kelley
W.G. Knauss
R.A. Schapery
J.L. Swedlow

Editorial Committee

References to work of M.L. Williams

1. "An Initial Approach to the Overall Structural Problems of Swept Wings Under Static Loads," co-authors E.E. Sechler and Y.C. Fung, *Journal of Aeronautical Sciences*, 17; 10, 639–646 October (1950).
2. "Surface Stress Singularities Resulting from Various Boundary Conditions in Angular Corners of Plates Under Bending," Proceedings of First National Congress of Applied Mechanics, June (1951).
3. "Stress Singularities Resulting from Various Boundary Conditions in Angular Corners of Plates in Extension," *Journal of Applied Mechanics*, 19:4, 526–528 December (1952). See also "Discussion," *Journal of Applied Mechanics*, 20:4, 590 December (1953).
4. "On the Stress Distribution at the Base of a Stationary Crack," *Journal of Applied Mechanics*, 25:1, 109–114 March (1957).
5. "Discussion of: 'Analysis of Stresses and Strains near the End of a Crack Traversing a Plate' (JAM, September 1957, by G.R. Irwin)," *Journal of Applied Mechanics* 25:2, 299–302 June (1958).
6. "Large Deflection Analysis for a Plate Strip Subjected to Normal Pressure and Heating," *Journal of Applied Mechanics*, 22:4, 458–464 December (1955).
7. "Further Large Deflection Analysis for a Plate Strip Subjected to Normal Pressure and Heating," *Journal of Applied Mechanics*, 25:2, 251–258 June (1958).
8. "The Stresses Around a Fault or Crack in Dissimilar Media," *Bulletin of Seismological Society of America*, 49:2, 199–204 April (1959).
9. "Crack Point Stress Singularities at a Bi-Material Interface," co-author A.R. Zak, *Journal of Applied Mechanics*, Series E, 30:1, 142–143 March (1963).
10. "Some Exploratory Photoelastic Studies in Stress Wave Propagation", co-authors M.E. Jessey and R.R. Parmerter, *Proceedings of Society of Experimental Stress Analysis*, 17:2, 121–134 October (1960).
11. "A Photoelastic Technique for Ground Shock Investigation," co-author R.J. Arenz, Proceedings of Fifth AFBM-STL Symposium, Ballistic Missile and Space Technology, IV, 137–152, Academic Press, New York, August (1960).
12. "Development and Application of a High-Speed Camera for Metallographic and Crack Propagation Studies," co-authors W.M. Beebe, M.E. Jessey, H.W. Liu and S.R. Valluri, ARL 64-49 Aerospace Research Laboratories, USAF, Wright-Patterson AFB, Ohio, April (1964).
13. "Some Preliminary Photoelastic Design Data for Stress in Rocket Grains," co-author D.D. Ordahl, *Jet Propulsion*, 27:6, 657–662 June (1957).
14. "Some Thermal Stress Design Data for Rocket Grains," *ARS Journal*, 29:4, 260–267 April (1959).
15. "Fundamental Studies Relating to Systems Analysis of Solid Propellants," co-authors P.J. Blatz and R.A. Schapery, Final report, GALCIT 101 Thiokol Chemical Corp., AD-256-905 (GALCIT SM 61-5) February (1961).
16. "The Structural Analysis of Viscoelastic Materials," *AIAA Journal* 2:5, 785–808, May (1964).
17. "Mechanical Behavior of Viscoelastic Materials," co-author T. Kunio, *Journal of Japan Society for Mechanical Engineering*, 68:552, Showa 40 (1965).
18. "An Analysis of Adhesive Debonding in Case-Bonded Solid Propellant Rocket Motors," co-author T. Kunio, Proceedings of the Eighth International Symposium on Science and Technology of Space, Tokyo, Japan, p. 217–222, August (1969).
19. "The Relation Between Engineering Stress Analysis and Molecular Parameters in Polymeric Materials," co-author F.N. Kelley, Proceedings of Fifth International Congress of Rheology, 3: 185–202, Tokyo, Japan (1970).
20. "The Engineering of Polymers for Mechanical Behavior," co-author F.J. Kelley, *Rubber Chemistry and Technology*, 42:4, 1175–1185, September (1969).
21. "Polymer Engineering and its Relevance to National Materials Development," Edited by F.R. Eirich and M.L. Williams, A National Science Foundation Workshop, Washington, D.C., University of Utah (UTECH DO73-013), December (1972).
22. Mechanics − Structure Interactions with Implications for Material Fracture, M.L. Williams (ed.), National Science Foundation Workshop May 1979, University of Pittsburgh, 1980.
23. "Stress Singularities, Adhesion, and Fracture," Proceedings of Fifth U.S. National Congress of Applied Mechanics, pp. 451–464, June (1966).
24. "The Continuum Interpretation for Fracture and Adhesion," *Journal of Applied Polymer Science*, 13: 29–40 (1969).
25. "The Time Dependence of Surface Energy in Cohesive Fracture," co-authors, S.J. Bennett and G.P. Anderson, *Journal of Applied Polymer Science*, 14: 735–745 (1970).

International Journal of Fracture 39: 1–13 (1989)
© Kluwer Academic Publishers, Dordrecht

The near-tip field at high crack velocities

K.B. BROBERG
Department of Solid Mechanics, Lund Institute of Technology, Box 118, S-221 00 Lund, Sweden

Received 20 September 1987; accepted 1 April 1988

Abstract. Several velocity regions, distinctly different as regards crack edge propagation characteristics, can be distinguished. The simplest case is mode III with only the subsonic and the supersonic regions. For modes I and II four different regions can be recognized. When analyzing the near-tip field at leading and trailing edges it is found that some velocity regions are forbidden. The most important field characteristic is the energy flow to or from the edge. A clear difference exists between modes I and II: for mode I the whole region between Rayleigh and irrotational wave velocities is forbidden, for mode II only the subsonic super-Rayleigh region.

In attempts to provoke crack edge propagation at a velocity in a forbidden region, the result appears to be edge propagation at velocities alternating between velocities in non-forbidden regions, above and below the attempted velocity.

A study of the stresses ahead of the edge region of a mode II crack expanding in both directions indicates that the edge might accelerate (by a jump) spontaneously from a sub-Rayleigh to an intersonic velocity.

1. Introduction

In a series of papers [1–3] Williams studied stress singularities at crack edges and notches. He showed, among other things, that the near-edge situation can be described without considering other parts of the body. Williams's interest was focussed on static cases, but for crack edges the methods used can be applied also to dynamic cases.

In the present paper very rapid crack growth will be studied. Only small scale yielding will be considered, and linear, istropic elasticity is assumed. This implies that *autonomy* [4] prevails, i.e., the stress field near the crack edge is uniquely specified through the mode and the velocity of propagation. In modes I and II four distinct crack edge velocity regions are recognized, whereas mode III exhibits only two regions; see Fig. 1.

For modes II and III a distinction should be made between *leading* and *trailing* edges, cf. [5]. At a leading edge the process is *tearing apart* (decohesion), at a trailing edge it is *healing*. Healing does not necessarily imply return to the original condition, but only that the sliding motion has been arrested and that renewed sliding motion requires addition of energy.

For mode I a similar distinction is hardly needed or appropriate.

Since the process of tearing apart requires energy, a necessary condition for propagation of a leading edge is that energy is supplied, i.e., that there is a positive energy flow from the outer field to the crack edge. The energy flow per unit of crack edge advance and edge length will be called the energy flow for simplicity.

The crack propagation characteristics for leading edges at subsonic velocities are well known. A systematic survey is given by, e.g., Achenbach and Bazant [6]. For mode III one finds that there is an energy flow from the outer field to a leading crack edge for all subsonic velocities. For modes I and II the situation is different: a theoretical analysis results in

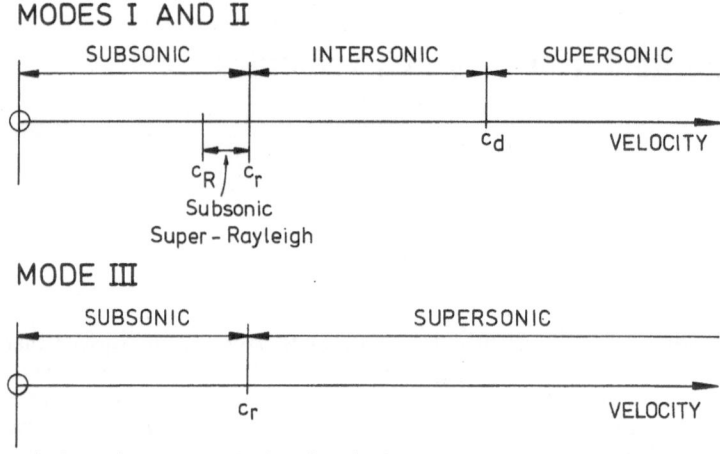

Fig. 1. Crack edge velocity regions. c_r = velocity of equivoluminal waves, c_d = velocity of irrotational waves and c_R = velocity of Rayleigh waves.

negative energy flow at subsonic super-Rayleigh velocities [7] to a leading edge, which consequently cannot be propagated at such velocities.

For a trailing edge the situation is different. The process consists of healing, which does not require energy supply. Rather one can expect the healing process to be associated with some (though very small) energy release.

Crack propagation in the intersonic region has been studied by Freund [8], Burridge et al. [9] and Broberg [10–12].

For simplicity friction stresses are not included in the treatment of modes II and III. However, since only linear problems are studied, the case of constant friction stress along the crack surfaces can be included simply by superposition.

The energy flow argument appears to imply that certain crack edge velocity regions are forbidden. What happens if one tries to force a crack to propagate at a velocity in a forbidden region? Such attempts can, for instance, consist of wedging or of moving outer loads. Such cases will be discussed.

2. Subsonic velocities

2.1. Mode III

The near-tip field at leading and trailing edges can be illustrated by studying a mode III crack of constant length $2b$, moving at constant velocity V, in a large body, subjected to a remote shear stress τ_∞; see Fig. 2. Near the leading edge the displacement is

$$w \approx \pm \frac{\sqrt{2}\pi\tau_\infty b}{G} \left(\frac{r}{b}\right)^{1/2} [(1 - \gamma^2 \sin^2\phi)^{1/2} - \cos\phi]^{1/2}, \tag{1}$$

where (r, ϕ) are polar coordinates with origin at the crack edge, G the modulus of rigidity and $\gamma = V/c_r$, where c_r is the propagation velocity of equivoluminal waves. The energy flow

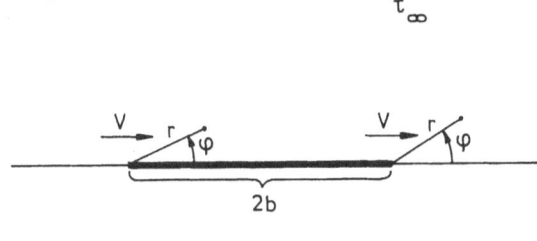

Fig. 2. Unidirectional mode III slip propagation.

to the crack edge is

$$\frac{\mathrm{d}W}{\mathrm{d}S} = \frac{2\pi\tau_\infty^2 b}{G}(1 - \gamma^2)^{1/2} \tag{2}$$

showing that the necessary condition for propagation of a leading edge in the subsonic region is fulfilled.

Near the trailing edge, assuming zero energy release to the outer field, one obtains:

$$w \approx \pm\frac{\pi\tau_\infty b}{G}\left[1 - \frac{\sqrt{2}}{3\pi}\left(\frac{r}{b}\right)^{3/2} g(\phi)\right], \tag{3}$$

where

$$g(\phi) = \cos\phi[(1 - \gamma^2\sin^2\phi)^{1/2} + \cos\phi]^{1/2} - (1 - \gamma^2)^{1/2}[(1 - \gamma^2\sin^2\phi)^{1/2} - \cos\phi]^{1/2}.$$

Equation (3) shows that there is a displacement offset at a trailing edge. Since autonomy prevails, (1) and (3) give the near-tip displacement field near leading or trailing edges in general, apart from a displacement offset and an amplitude factor.

2.2. Modes I and II

The energy flow to a leading mode I or II edge is

$$\frac{\mathrm{d}W}{\mathrm{d}S} = \frac{K^2}{4(1 - k^2)G}f(\beta), \tag{4}$$

where K is the stress intensity factor, $k = c_r/c_d$, where c_d is the propagation velocity of irrotational waves,

$$f(\beta) = \begin{cases} \dfrac{2k^2(1 - k^2)\beta^2(1 - \beta^2)^{1/2}}{R(\beta)} & \text{for mode I} \\[2ex] \dfrac{2k(1 - k^2)\beta^2(k^2 - \beta^2)^{1/2}}{R(\beta)} & \text{for mode II}, \end{cases}$$

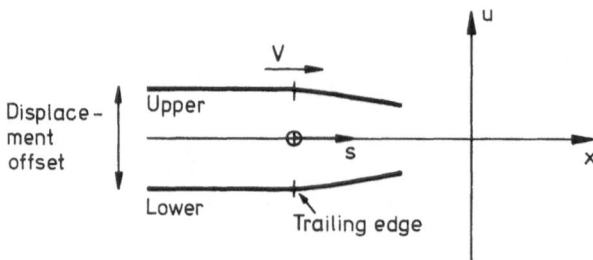

Fig. 3. Displacement u of upper and lower surface of a trailing mode II edge.

and where

$$R(\beta) = 4k^3(1 - \beta^2)^{1/2}(k^2 - \beta^2)^{1/2} - (2k^2 - \beta^2)^2$$

is the Rayleigh function, which possesses a simple zero for $\beta = c_R/c_d$, where c_R is the Rayleigh wave velocity.

One observes that the energy flow to the edge is positive only for $V < c_R$. Therefore, a mode I or II leading edge cannot be propagated at a constant (or smoothly varying) velocity in the subsonic super-Rayleigh region ($c_R < V < c_r$). What about a trailing mode II edge?

Regard a trailing mode II edge, moving at constant velocity, V, in positive direction along the x-axis in a large body, see Fig. 3. Assuming that the energy release from the edge is zero one obtains for $y = 0$ and the neighbourhood of the trailing edge:

$$\frac{\partial u}{\partial x} = -\frac{C\beta^2(k^2 - \beta^2)}{R(\beta)} s^{1/2}, \tag{5}$$

where u is the displacement in positive x-direction of the upper crack surface, C is an undetermined constant, and s is the distance from the edge. Equation (5) can be found from [5], where the displacement field is also given.

Equation (5) shows that transition from a sub-Rayleigh to a super-Rayleigh velocity would hardly be possible, since $\partial u/\partial x$ would change sign. This would imply that $\partial u/\partial x$ vanishes somewhere between the trailing and the leading edge, which, in turn, implies that the sliding motion has come to a stop. Propagation at constant (or smoothly varying) velocity of a trailing mode II edge therefore does not appear to be possible in the subsonic, super-Rayleigh region.

3. Intersonic velocities

3.1. Mode II

For the intersonic case, $c_r < V < c_d$, the idealization to a point-size dissipative region turns out to be incompatible with the requirement of non-zero energy dissipation. A comparatively simple way to overcome this difficulty is to assume a Barenblatt type process region [4, 5, 10]. Then one obtains the following expression for shear stress, displacement gradient

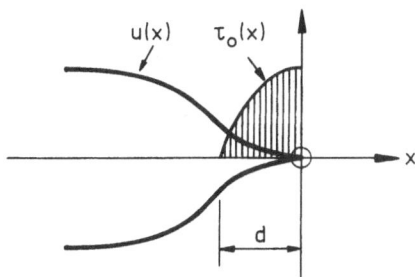

Fig. 4. Crack surface displacement $u(x)$ and process region shear stress $\tau_0(x)$ near a mode II leading edge.

and energy flow to a leading mode II edge [10]; see Fig. 4:

$$\tau_{xy} = \tau_0(0) - \frac{1}{\pi} \sin \pi\gamma \cdot x^{1-\gamma} \int_0^d \frac{\tau_0(0) - \tau_0(-s)}{s^{1-\gamma}(s+x)} \, ds \quad \text{for } y = 0, \quad x > 0 \tag{6}$$

$$\frac{\partial u}{\partial x} = \pm \frac{f_1(\beta)}{\pi G} \left\{ \sin \pi\gamma \cdot |x|^{1-\gamma} \int_0^d \frac{\tau_0(0) - \tau_0(-s)}{s^{1-\gamma}(s+x)} \, ds + \pi \cos \pi\gamma [\tau_0(0) - \tau_0(x)] \right\}$$

$$\text{for } y = 0, \quad x < 0 \tag{7}$$

$$\frac{dW}{dS} = -2 \frac{f_1(\beta)}{\pi G} \sin \pi\gamma \int_0^d \int_0^d \frac{x^{1-\gamma}}{s^{1-\gamma}} \tau_0(-x) \frac{\tau_0(-s) - \tau_0(-x)}{s - x} \, ds \, dx, \tag{8}$$

where d is the length of the process region,

$$\gamma = \pi^{-1} \tan^{-1} \frac{4k^3(1 - \beta^2)^{1/2}(\beta^2 - k^2)^{1/2}}{(2k^2 - \beta^2)^2}, \quad 0 < \gamma \leqslant 1/2$$

$$f_1(\beta) = \frac{\beta^2}{4k^2(1 - \beta^2)^{1/2}} \sin \pi\gamma$$

and $\tau_0(x)$ is the shear stress along the process region.

The function $\tau_0(x)$ is probably dependent on the edge velocity. Stress continuity and smooth closing of the process region are assumed and the trailing edge is assumed to be energy-neutral.

Physical reasons require that $\tau_0(x)$ possesses a maximum at $x = 0$. Then dW/dS is positive, i.e., energy flows towards the leading edge at intersonic propagation velocities! Mode II crack propagation at constant (or smoothly varying) velocity in this region should therefore, in principle, be possible.

For the trailing edge similar expressions are obtained, assuming for a moment that there exists a process region (of healing) of finite length. The difference is mainly that γ should be exchanged by $1 - \gamma$. The energy flow to the edge is negative, which shows that propagation of a trailing mode II edge at a constant or smoothly varying velocity in the intersonic region is possible. The assumption of an energy-neutral trailing edge (which corresponds to

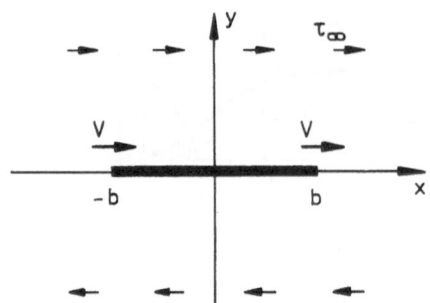

Fig. 5. Unidirectional mode II slip propagation.

vanishing process region) results in displacement gradients compatible with propagation of a leading edge, followed by a trailing edge at any velocity in the intersonic region.

This can be seen from the example of a sliding region of length 2b, travelling with constant velocity V in a large body, subjected to a remote shear stress $\tau_{xy} = \tau_\infty$; see Fig. 5. Shear stress and displacement gradient along the x-axis are (with the coordinate system moving with the crack):

$$
\tau_{xy} = -\frac{x}{|x|} \cdot \frac{\sin \pi\gamma}{\pi} \cdot |b - x|^{1-\gamma}|b + x|^\gamma \int_{-b}^{b} \frac{\tau_0(s)\,ds}{(b - s)^{1-\gamma}(b + s)^\gamma(s - x)} \quad \text{for} \quad |x| > b
$$

(9)

$$
\frac{\partial u}{\partial x} = \mp \frac{f_1(\beta)}{\pi G} \left[\sin \pi\gamma \cdot (b - x)^{1-\gamma}(b + x)^\gamma \oint_{-b}^{b} \frac{\tau_0(s)ds}{(b - s)^{1-\gamma}(b + s)^\gamma(s - x)} \right.
$$

$$
\left. - \pi \cos \pi\gamma \cdot \tau_0(x) \right] \quad \text{for} \quad |x| < b.
$$

(10)

The reason why a finite size process region had to be assumed has to do with the nature of the singularity that is obtained in the limiting case when the process region size vanishes. In the subsonic case an $r^{-1/2}$ singularity appears, which implies that the energy flow to the crack edge is very insensitive to the size of the process region as long as this is small, compared to other significant dimensions of the body, for instance the crack length. In the intersonic case, however, a weaker singularity results, an $r^{-\gamma}$ singularity both for stresses and displacement gradients. This implies vanishing energy flow to the edge.

There is one curious exception: $\beta = \sqrt{2}k$. This gives a square root singularity. The general expression for the shear stress at $y = 0$ [10], Eqn. (9), reduces to exactly the same expression as for the subsonic case when $\gamma = 1/2$. The same applies to the expression for $\partial u/\partial x$ for $y = 0$.

For the intersonic case equivoluminal waves lag behind the edge. Their front, of course, is given by the angle θ_s found from the relation

$$
\sin \theta_s = k/\beta, \quad \pi/2 < \theta_s < \pi.
$$

(11)

At a distance r in front of a leading mode II edge, such that $r \gg d$ and $r \ll b$ one obtains

$$\tau_{r\theta} = Ar^{-\gamma} \left\{ \frac{\sin 2\theta}{2a_1} [(1 + a_1^2) \sin \phi_1/T_1 - U \cdot (a_2^2 - 1) \sin \phi_2/T_2] \right.$$

$$\left. - \frac{\cos 2\theta}{4a_1 a_2} [4a_1 a_2 \cos \phi_1/T_1 - U \cdot (a_2^2 - 1)^2 \sin \phi_2/T_2] \right\}, \tag{12}$$

where

$$A = \sin \pi\gamma \int_0^d \frac{\tau_0(-s)}{s^{1-\gamma}} \, ds$$

$$a_1^2 = 1 - \beta^2, \quad a_1 > 0$$

$$a_2^2 = \beta^2/k^2 - 1, \quad a_2 > 0$$

$$\phi_1 = \gamma \tan^{-1}(a_1 \tan \theta)$$

$$\phi_2 = \gamma \tan^{-1}[a_1 \tan \theta/(1 + a_2 \tan \theta)]$$

$$T_1 = (\cos^2\theta + a_1^2 \sin^2\theta)^{\gamma/2}$$

$$T_2 = [(\cos \theta + a_2 \sin \theta)^2 + a_1^2 \sin^2\theta]^{\gamma/2}$$

$$U = \begin{cases} 0 \text{ for } |\theta| < \theta_s \\ 1 \text{ for } |\theta| > \theta_s, \end{cases}$$

The terms representing equivoluminal waves are those containing the step function U. Again one notices the remarkable particularity of the case $\beta = \sqrt{2}k$, i.e., $a_2 = 1$: the equivoluminal waves disappear.

The significance of the square root singularity should perhaps not be too much emphasized. After all, it is a mathematical abstraction without clear physical significance. On the other hand, the strength of the singularity obtained after mathematical idealization tells something about the situation near the edge; a very low value of the singularity exponent signals very small stress gradients near the crack edge.

Investigation shows that the shear stress is not invariably largest in the straight forward direction. Also high tensile stresses occur. It is therefore doubtful whether propagation of a mode II crack without branching is possible at all in the intersonic region. However, directional stability can be maintained if the crack propagates in a weak layer.

3.2. Mode I

Usually modes I and II yield very analogous results, connected via simple substitutions. Conclusions made for one can usually be carried over to the other. The intersonic case is an exception. This seems to be due to the fact that mode II (as mode III) is associated with shear processes, and therefore the equivoluminal (shear) wave velocity, c_r, has a special significance. Equations (4) and (6) show that $dW/dS \to 0$ when $V \to c_r$ and that dW/dS changes sign at $V = c_r$ for a leading mode II edge.

For mode I a formal calculation [10] leads to an expression for dW/dS that is given by (6) after the substitutions

$$\gamma \to 1/2 + \gamma$$

$$f_1(\beta) \to -\left(\frac{1 - \beta^2}{\beta^2 - k^2}\right)^{1/2} f_1(\beta).$$

This implies that dW/dS does not vanish or change sign at $V = c_r$ for a mode I edge. dW/dS is negative in the intersonic region and, consequently, mode I propagation at constant or smoothly varying velocity in this region is not possible.

4. Supersonic velocities

The case of supersonic edge velocities is rather trivial. Crack growth must be provoked by a moving stress field. Wedging, where the mean crack edge velocity is determined by the

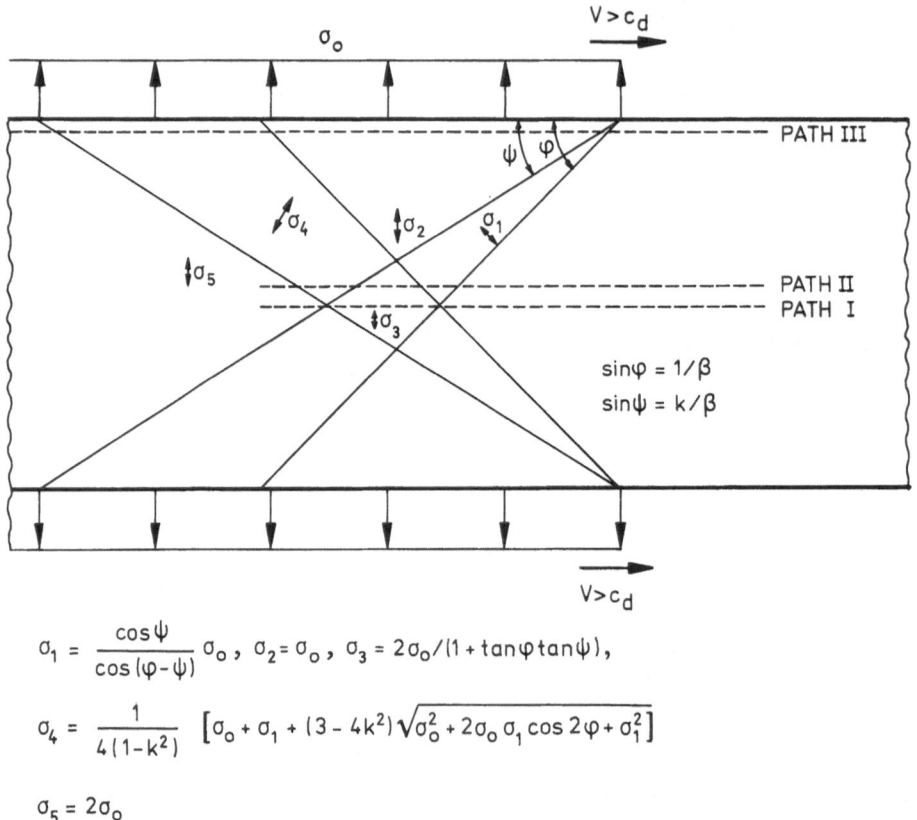

$$\sigma_1 = \frac{\cos\psi}{\cos(\varphi - \psi)}\,\sigma_0\,, \quad \sigma_2 = \sigma_0\,, \quad \sigma_3 = 2\sigma_0/(1 + \tan\varphi\tan\psi),$$

$$\sigma_4 = \frac{1}{4(1 - k^2)}\left[\sigma_0 + \sigma_1 + (3 - 4k^2)\sqrt{\sigma_0^2 + 2\sigma_0\sigma_1\cos 2\varphi + \sigma_1^2}\,\right]$$

$$\sigma_5 = 2\sigma_0$$

Fig. 6. Dynamic loading of a strip with supersonic load front velocity. The stresses in different regions are shown.

wedge velocity is one example. Another example, at which the crack surfaces are not subjected to direct loading, is an I-girder with moving carriages pressing the flanges apart. A mode I crack can thus be propagated along the web. This kind of loading is schematically shown in Fig. 6, where a long strip is subjected to two symmetrically moving loads, producing a normal stress σ_0 at the upper and lower boundaries of the strip.

Irrotational and equivoluminal waves are initiated from the loads. They traverse the strip and are later reflected at the boundaries. In the primary region of both irrotational and equivoluminal waves the normal stress at surfaces parallel to the strip direction is σ_0. The mass velocity, $\sigma_0/\varrho c_d$, where ϱ is the density, is perpendicular to the strip direction. In the primary region between the fronts of irrotational and equivoluminal waves the mass velocity is perpendicular to the front of irrotational waves. When the equivoluminal wave arrives another mass velocity is added. This mass velocity is parallel to the front of the equivoluminal

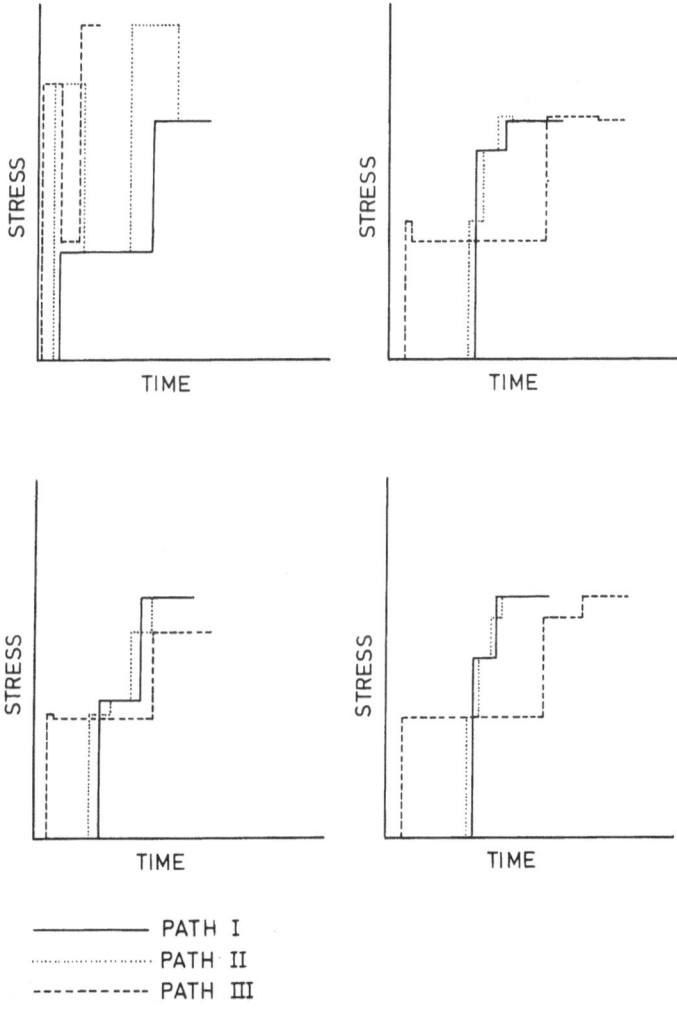

Fig. 7. Examples of stress histories for particles on paths I, II, and III; see Fig. 3. Upper figures: Poisson's ratio = 0.47. Load edge velocity is $1.02\,c_d$ (left figure) and $1.5\,c_d$ (right figure). Lower figures: Poisson's ratio = 0.3. Load edge velocity is $1.2\,c_d$ (left figure) and $1.5\,c_d$ (right figure).

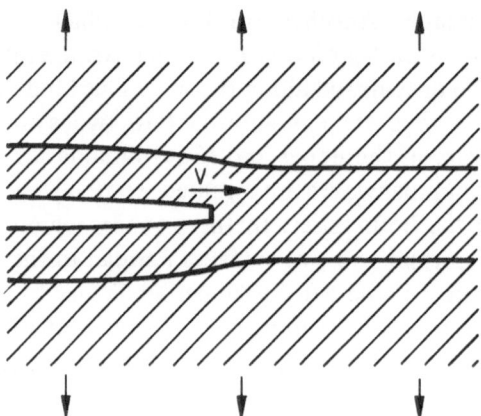

Fig. 8. Crack propagation in a layer, whose irrotational wave velocity is lower than the Rayleigh wave velocity in the medium outside.

wave. Together the two mass velocities combine to the known mass velocity in the region behind the front of the equivoluminal wave. In this way one easily finds the stresses in the different regions shown in Fig. 6.

When the possibility of fracture is considered, one finds that the situation is indeed very complex, even before reflections occur. Fracture can be initiated at almost any location, dependent on the Poission's ratio, the load velocity and the stress needed for initiation of fracture, with due regard, of course, also to the delay time for fracture. A study of Fig. 7, which gives a few examples of the stress history along different paths before reflections occur, confirms this view of complexity. It should also be noted that the direction of maximum principal stress is not always normal to the strip direction.

An example which does not involve moving loads is provided by an infinite plate containing a layer whose irrotational velocity is lower than the Rayleigh wave velocity in the outside medium; see Fig. 8. A mode I crack can then be propagated in the layer at subsonic velocity, although this is a sub-Rayleigh velocity with respect to the medium outside. This situation was studied previously by the author [13] and it was shown that the stresses at the crack edge are finite and that the crack edge is blunted. This is what would be expected: propagation of the crack edge cannot be initiated by waves travelling straight forwards from the crack edge. Rather it is a question of stress waves impinging from positions in the outside medium in front of the edge. Due to the assumption of continuity and homogeneity in the theoretical treatment [13], the edge moves with constant velocity and, since there is no stress singularity, the edge is blunted. In this way a finite, positive energy flow to the edge is possible. In the real case one would expect a great number of microcracks, individually opened by the somewhat backwards going stress waves and coalescing after a while. However, one would also expect that these microcracks are distributed over a fairly large region, perhaps across the whole layer if no precautions are taken; for instance providing the layer with grooves to guide the crack.

The conclusion to be drawn is that crack growth at fairly constant velocity is possible in the supersonic region. Anyway, it is not forbidden from the energy point of view. This conclusion is valid for all three modes.

5. Forced growth at forbidden velocities

What happens if one attempts to force a mode I crack to run at a velocity in the forbidden region, i.e., between the Rayleigh wave velocity and the velocity of irrotational waves? One can imagine one of the artifices discussed in the preceding section to be used. It is assumed that branching or volume cracking can be prevented, for instance by using grooves.

The requirement of a positive energy flow to a leading region seems to be inescapable. Thus the edge cannot be propagated at a constant or even (in the limiting case of extremely small scale yielding) smoothly varying velocity in a forbidden region. What remains is the possibility that the crack edge velocity alternates between an allowed velocity (or smoothly varying velocities) in the sub-Rayleigh region and an allowed velocity (or smoothly varying velocities) in the subsonic region. In this way the *average* velocity can end up in a velocity in the forbidden region.

When a wedge is driven at a sub-Rayleigh velocity the crack edge position will be a distance ahead of the edge; see Fig. 9. The distance depends upon the fracture criterion, for instance expressed as a relation between critical stress intensity factor and crack edge velocity. The smaller the critical stress intensity factor, the larger the distance between the crack edge and the wedge.

Now, what happens if the wedge velocity is increased beyond the Rayleigh wave velocity, but below the irrotational wave velocity? Again, the result ought to be velocities alternating between the sub-Rayleigh velocity and subsonic velocity. However, if the wedge is thin enough it comes into direct contact with the crack edge. Then energy is transferred directly from the wedge to the crack edge, i.e., edge propagation at any velocity is possible.

If one considers that the process region has a finite extension, then the mechanism of direct energy transfer can be regarded simply as wedging of this region itself, i.e., a sufficiently sharp wedge pierces the whole process region. (For a thicker wedge driven at supersonic velocity very high local compression will be caused, implying increasing propagation velocity of irrotational waves. Therefore the wedge velocity will not be supersonic locally.)

For a mode II edge the forbidden region is the subsonic super-Rayleigh region. An attempt to force the crack to move with a velocity in this region should result in edge velocities alternating between sub-Rayleigh velocities and velocities above the equivoluminal wave velocity.

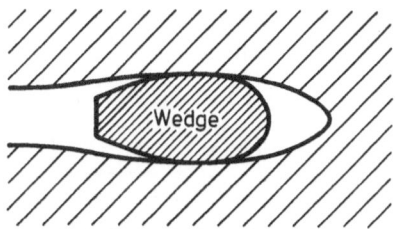

Fig. 9. Wedging at sub-Rayleigh velocities.

6. Intersonic edge propagation under mode II conditions

Starting with a formal study of constant sub-Rayleigh edge velocities, one notices remarkable differences between unidirectional (one leading and one trailing edge) and extensional (two leading edges) slip propagation [12]. In the unidirectional case the shear stress in front of the leading edge along the prospective plane of propagation decreases monotonically with increasing distance from the edge. In the extensional case a peak appears some distance ahead of the edge. It is, in fact, travelling with the velocity of equivoluminal waves [12]. The peak stress, as well as the width of the peak, increases with increasing velocity, and becomes quite substantial when the Rayleigh wave velocity is approached. Figure 10 shows one example. At a sufficiently high edge velocity slip might be initiated somewhere at the peak, i.e., well ahead of the main slipping region! This implies a sudden jump of the crack edge. If such jumps occur repeatedly the apparent mean edge propagation velocity might perhaps overshoot the Rayleigh wave velocity.

This opens interesting possibilities. If the velocity of a leading mode II edge can enter the intersonic region, it cannot be excluded that it might remain there. This situation seems to be probable if the crack length (i.e., the length of the slipping region) can be very long, as, for instance, at an earthquake event. Then the energy flow to the edge regions will eventually be very large, which will favor very high edge velocities.

There is also a mechanism by which unidirectional mode II slip might propagate at an intersonic velocity. The mechanism by which unidirectional slip originates seems to be the following one (for a plane case). Slip starts at some point in a region with high shear stress and propagates in both directions, i.e., as extensional slip. Both edges accelerate as the crack length increases. If one edge runs into a region with smaller ambient shear stress it may decelerate and come to a stop. But an edge of a slipping region must be moving all the time. Therefore a stop implies immediate reversal of the edge direction, i.e. the edge turns from a leading to a trailing one. The velocity of this trailing edge depends upon the displacement gradients in front of the edge. If these gradients are very small the edge velocity becomes very high. The magnitude of the gradients depends, among other things, upon the distribution of the ambient shear stresses, which previously caused deceleration of the leading edge. If the gradients are sufficiently small the trailing edge might well accelerate to a velocity in the intersonic region.

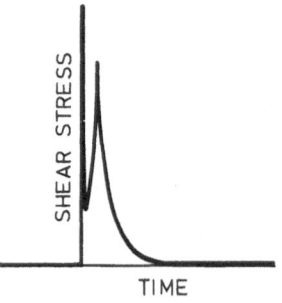

Fig. 10. Peak shear stress ahead of the edge at extensional mode II propagation at a velocity of about 94 percent of the Rayleigh wave velocity. The peak travels with the velocity of equivoluminal waves.

References

1. M.L. Williams, "Surface singularities resulting from various boundary conditions in angular corners of plates under bending", U.S. National Congress of Applied Mechanics, Illinois Institute of Technology, Chicago, Ill., June 1951.
2. M.L. Williams, *Journal of Applied Mechanics* 19 (1952) 526–528.
3. M.L. Williams, *Journal of Applied Mechanics* 24 (1957) 109–113.
4. G.I. Barenblatt, *Journal of Applied Mathematics and Mechanics PMM* 23 (1959) 622–634.
5. K.B. Broberg, *Geophysical Journal of the Royal Astronomical Society* 52 (1978).
6. J.D. Achenbach and Z.P. Bazant, *Journal of Applied Mechanics* 42 (1975) 183–189.
7. K.B. Broberg, *Journal of Applied Mechanics* 31 (1964) 546–547.
8. L.B. Freund, *Journal of Geophysical Research* 84 (1979) 2199–2209.
9. R. Burridge, G. Conn and L.B. Freund, *Journal of Geophysical Research* 85 (1979) 2209–2222.
10. K.B. Broberg, "Velocity peculiarities at slip propagation", Report from the Division of Engineering, Brown University, Providence, R.I. (May 1980).
11. K.B. Broberg, in *Fundamentals of Deformation and Fracture, Eshelby Memorial Symposium*, B.A. Bilby, K.J. Miller and J.R. Willis (eds.), Cambridge University Press, Cambridge (1985).
12. K.B. Broberg, *Journal of Technical Physics* 26 (1985) 275–284.
13. K.B. Broberg, in *Proceedings of an International Conference on Dynamic Crack Propagation*, G.C. Sih (ed.), Noordhoff International Publishing, Leyden (1972) 461–499.

Résumé. Selon les caractéristiques de propagation du bord d'une fissure, on peut distinguer divers registres distincts de vitesses de fissuration. Le cas le plus simple est celui du Mode III, où l'on ne rencontre que les vitesses subsoniques et supersoniques. Les Modes I et II conduisent à distinguer quatre vitesses différentes. Lorsqu'on analyse le champ du voisinage de l'extrémité de la fissure en ses bords moteurs, on trouve que certaines vitesses sont proscrites. La caractéristique la plus importante du champ est le transfert d'énergie depuis ou vers le bord. Il existe une distinction claire entre les Modes I et II: en Mode I, toute la gamme comprise entre les vitesses des ondes de Rayleigh et d'une onde irrotationnelle sont prescrites; en Mode II, seule est proscrite la gamme subsonique supérieure à l'onde de Rayleigh.

Lorsqu'on tente de provoquer une propagation des bords d'une fissure à une vitesse dans une gamme proscrite, le résultat semble être une propagation de bord à des vitesses oscillant entre deux gammes non proscrites, de part et d'autre de la vitesse visée.

En étudiant les contraintes en avant du bord d'une fissure de Mode II se développant selon les deux directions imposées, on constate que le bord peut spontanément s'accélérer suivant un ressaut, en passant d'une vitesse inférieure à une onde de Rayleigh à une une vitesse transonique.

International Journal of Fracture 39: 15–24 (1989)
© Kluwer Academic Publishers, Dordrecht

Measurement of dominant eigenvalues in cracked body problems

C.W. SMITH, J.S. EPSTEIN and M. REZVANI

Department of Engineering Science and Mechanics, Virginia Polytechnic Institute and State University, Blacksburg, Virginia 24061, USA

Received 20 September 1987; accepted in revised form 1 April 1988

Abstract. A refined optimal method consisting of the tandem application of frozen stress photoelasticity and high density moire interferometry for studying three dimensional effects in cracked bodies is briefly reviewed. It is then employed to measure the dominant eigenvalue at the right angle intersection of a straight front crack with a free surface under mode I loading. The variation of the eigenvalue through a transition zone near the free surface is also determined. The free surface result is found to be in agreement with analytical results.

1. Introduction

The eigenfunction series approach to cracked body problems was formalized by Williams [1] in 1957. This analysis placed emphasis upon the two dimensional nature of fracture analysis at that time. Subsequently, with the rapid development of digital computers many numerical solutions were proposed for three dimensional cracked body problems where the stress intensity factor (SIF) varied along the crack front. Since these solutions were approximate, and virtually no exact three dimensional (3-D) solutions were available for finite cracked bodies, the first author and his associates began an effort to develop two independent experimental methods of analysis for studying three dimensional effects in cracked bodies so that results from the two methods could be used as a check upon one another before being compared to results of 3-D cracked body analyses. These methods were frozen stress photoelasticity [2, 3] and high density moire interferometry [4]. Subsequent to their individual refinements, these methods have been integrated into a single method consisting of their tandem application to produce dual estimates of the stress intensity factor distributions in three dimensional cracked body problems. In this paper, after briefly reviewing the experimental methods and algorithms used to convert optical data into fracture parameters, we shall use these methods to extract estimates of the dominant eigenvalue from the region where a crack front intersects a free surface at right angles.

2. Optical methods

When optical methods are applied to cracked body problems, some equipment modifications may be anticipated in order to enhance near tip measurement. They will now be briefly described.

2.1. Frozen stress analysis

This method of analysis was introduced by Oppel [5] in 1936. It involves the use of a transparent plastic which exhibits, in simplest concept, diphase mechanical and optical properties. That is, at room temperature, its mechanical response is viscoelastic. However, above its "critical" temperature, its viscous coefficient vanishes, and its behavior becomes purely elastic, exhibiting a modulus of elasticity of about 0.2 percent of its room temperature

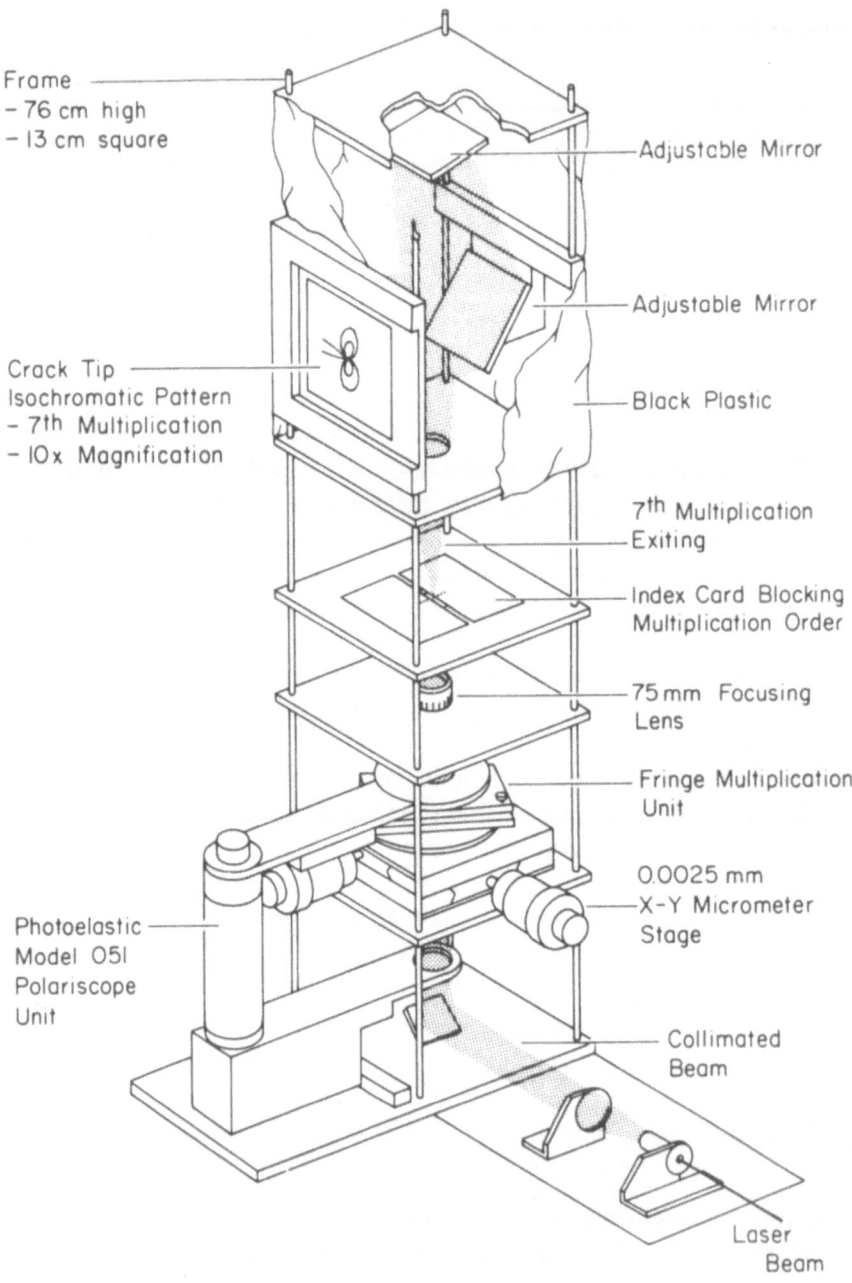

Fig. 1. Modified crossed circular polariscope.

value and a stress fringe sensitivity of twenty times its room temperature value. Thus, by loading the photoelastic models above critical temperature, cooling under load, and then removing the load, negligible elastic recovery occurs at room temperature and the stress fringes and deformation produced above critical temperature are retained. Moreover, the "frozen model" may be sliced without altering its condition.

In order to determine useful optical data from frozen stress analysis, one needs to suppress deformations near the crack tip in the photoelastic material in its rubbery state above critical temperature and to be able to produce the same crack shape and size produced in the prototype. In order to accomplish the first objective, applied loads are kept very small, and a polariscope modified to accommodate the tandem application of partial mirror fringe multiplication and Tardy compensation is employed. Such a polariscope is pictured in Fig. 1, which is self-explanatory. Normally, fifth multiples of fringe patterns are read to a tenth of a fringe thus providing adequate data within about 1 mm of the crack tip to two hundredths of a fringe order.

Natural crack shapes are obtained by introducing a starter crack at the desired location in the photoelastic model of the structure before stress freezing by striking a sharp blade held normal to the crack surface with a hammer. The starter crack will emanate from the blade tip and propagate dynamically a short distance into the model and then arrest. Further growth to the desired size is produced when loaded monotonically above critical temperature. The shape of the crack is controlled by the body geometry and loads. By comparing crack shapes grown in photoelastic models by this process to those grown under tension–tension fatigue loads in steel, excellent correlation has been obtained [6] even when some crack closure was present on the free surface of the latter. It appears that the cracked body geometry and loads control the crack shape in thick, reinforced bodies and that the stress ratio R (as long as it is positive) and plasticity or closure effects are of secondary importance.

Artificial cracks which were used in the experiments described in the sequel, are made by machining into the body a desired shape, maintaining a vee-notch tip with an included angle not exceeding 30 deg. With this angle, near tip stress fields are essentially the same as for branch cuts.

By removing thin slices of material which are oriented mutually orthogonal to the crack front and the crack plane locally, analysis of these slices will yield the distribution of the maximum shear stress in the *nz* plane of Fig. 2. Then, by computing this stress from the near tip mode I singular stress field equations, including the contribution of the regular stresses in the near tip zone as constants, one can arrive at an algorithm for extracting the SIF for each slice.

2.2. Moire interferometry

This method of analysis was introduced by Weller et al. [7] in 1948. As was the case with the frozen stress method, some modification of the usual approach is desirable in order to obtain accurate near tip data. In the present case, a "virtual" grating was constructed optically by reflecting part of an expanded laser beam from a mirror so as to intersect the unreflected part of the beam, forming walls of constructive and destructive interference which serve as the master grating (Fig. 3). The specimen grating, a reflective phase grating, is viewed through the virtual grating as it (the former) deforms in order to see the moire fringes proportional to the in-plane displacement normal to the grating. By photographing the moire fringe

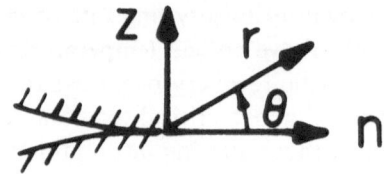

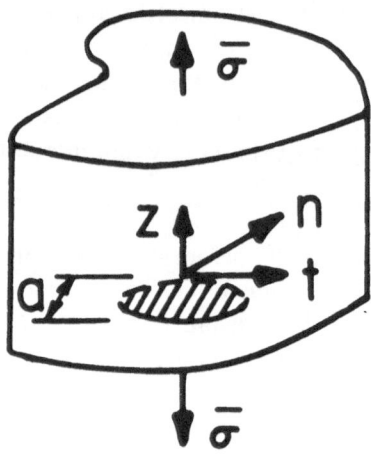

Fig. 2. Mode I crack tip notation.

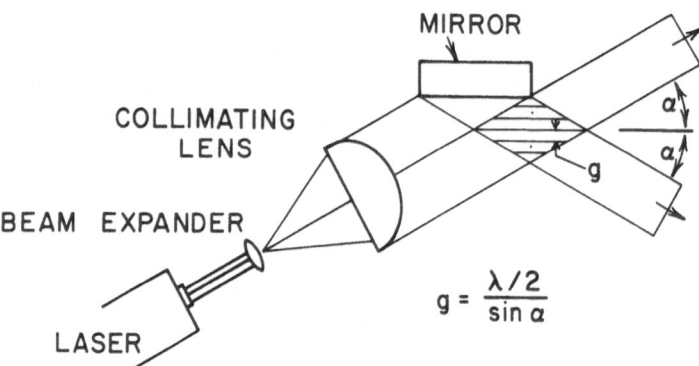

Fig. 3. Optical arrangement for virtual grating for Moire interferometry.

patterns produced on the surface of a frozen slice after it has been annealed to its stress free state, the inverse of the displacement fields produced in the plane of the slice by stress freezing may be measured.

3. Analytical considerations

Once optical data are obtained, algorithms must be constructed to convert the data into desired fracture parameters. One may begin by reviewing the two parameter approach which has been employed extensively [2, 3, 4] in LEFM with photoelastic data.

3.1. LEFM frozen stress algorithm – two parameter approach

By choosing a data zone sufficiently close to the crack tip that a Taylor Series Expansion of the non-singular stresses can be truncated to the leading terms, one may write for a straight front crack [2]

$$\sigma_{ij} = \frac{K_1}{(2\pi r)^{1/2}} f_{ij}(\theta) - \sigma_{ij}^0(\theta), \quad i, j = 1, 2 \tag{1}$$

where σ_{ij} are the stress components in the $x_1 x_2$ plane normal to the crack border, K_1 is the mode I stress intensity factor and r, θ are the polar coordinates with origin at the crack tip. Along $\theta = \pi/2$, one has, for the case where $(\sigma^0)^2$ is small relative to $8\tau_m^2$ (see [2]) (where σ^0 is proportional to the contribution of the nonsingular stress to the maximum shear stress, τ_{max} in the $x_1 x_2$ plane)

$$\tau_{max} = \frac{K_1}{(8\pi r)^{1/2}} + \frac{\sigma^0}{\sqrt{8}}. \tag{2}$$

Now defining an "apparent" SIF,

$$K_{AP} = \tau_{max}(8\pi r)^{1/2} \tag{3}$$

and normalizing with respect to $\bar{\sigma}(\pi a)^{1/2}$ where $\bar{\sigma}$ represents the remote stress and a the crack length, one has

$$\frac{K_{Ap}}{\bar{\sigma}(\pi a)^{1/2}} = \frac{K_1}{\bar{\sigma}(\pi a)^{1/2}} + \frac{\sigma^0}{\bar{\sigma}} \left(\frac{r}{a} \right)^{1/2}, \tag{4}$$

which suggests an elastic linear zone (ELZ) in a plot of $K_{Ap}/(\bar{\sigma}(\pi a)^{1/2})$ vs $(r/a)^{1/2}$ with a slope of σ^0/σ. Experience shows this zone to lie between $(r/a)^{1/2}$ values of approximately 0.2 to 0.4 (or above) in two dimensional problems. By extracting optical data from this zone and extrapolating across a near tip nonlinear zone, an accurate estimate of $K_1/\bar{\sigma}(\pi a)^{1/2}$ can be obtained. This is illustrated in Fig. 4 for two slices at different locations in a compact bending specimen. This approach may be directly extended to 3-D problems involving curved crack fronts by replacing i, j values of 1, 2 by n, z where the nz plane is normal to the crack border at each point along the flaw border and the n, z, t coordinate system is a local Cartesian system which follows the crack border (Fig. 2). In three dimensional problems the outer boundary of the data zone is usually restricted to $(r/a)^{1/2} \approx 0.4$ or less.

3.2. LEFM moire algorithm – two parameter approach

Corresponding to (1), one has for displacements for the two dimensional case [8]

$$u_i = CKG_i(\theta)r^{1/2} \quad \begin{cases} -\dfrac{\sigma^0}{E_s} r \cos \theta & i = 1 \\[2ex] +\dfrac{\sigma^0}{E_s} r \sin \theta & i = 2, \end{cases} \tag{5}$$

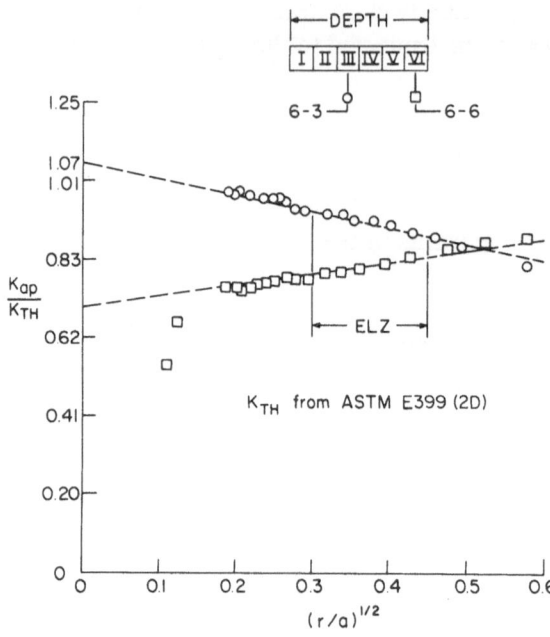

Fig. 4. Determination of SIF from photoelastic data from slices from a compact bending specimen.

where C is a constant containing the elastic constants, σ^0 is a nonsingular stress term and E_s is the shear modulus.

In the frozen stress algorithm, the effect of the nonsingular stress was included since it was independent of r. However, in (5), one may, for sufficiently small values of r, neglect the effect of the σ^0 terms. Thus, along $\theta = \pi/2$, for a three dimensional problem, one may write

$$u_z = C'Kr^{1/2}, \tag{6}$$

where a plot of $u_z/r^{1/2}$ vs. $r^{1/2}$ should yield a straight horizontal line whose ordinate is K. If one is not sufficiently close to the crack tip, however, the σ^0 terms do make some contribution in the data zone so one usually uses the form

$$u_z = C'K_{Ap}r^{1/2} \tag{7}$$

and extrapolates the linear part of u_z vs $r^{1/2}$ to the origin as before.

3.3. Algorithms accounting for free boundary effect

When a crack intersects a free surface at right angles, the dominant eigenvalue is altered. This phenomenon was noted by Sih [9] in 1971 and its effect in the neighborhood of the intersection was studied by Folias [10] in 1975. The value of the dominant singularity at the free surface where a quarter infinite crack intersects the free surface of a half space at right angles was obtained by Benthem [11] in 1980. A number of numerical solutions have been directed towards this problem which were recently reviewed by Burton et al. [12]. Using Benthem's three dimensional variables separable eigenfunction expansion of the σ_{ij} and u_i near the crack tip at the free surface for a quarter infinite crack intersecting a half space at right angles and the LEFM results as a guide, one can construct the following functional forms for the near tip σ_{ij} and u_i:

$$\sigma_{ij} = \frac{F_{ij}(\lambda_\sigma, \theta)}{r^{\lambda_\sigma}} - \sigma_{ij}^0 \quad (i, j = n, z), \tag{8}$$

where $\sigma_{ij}^0 = \sigma^0$ for $i, j = n$ and zero for all other values of i, j.

$$u_i = D_i(\theta)r^{\lambda_u}, \tag{9}$$

where λ_σ and λ_u represent the lowest dominant eigenvalues in the local stress and displacement field equations, respectively, and $|\lambda_\sigma| = 1 - \lambda_u$. Here similar functional forms are implied near the crack tip for σ_{ij} and u_i as in LEFM but λ is undetermined. From (9), one may, along $\theta = \pi/2$ write,

$$u_z = D_z r^{\lambda_u}$$

$$\log u_z = \log D_z + \lambda_u \log r \quad \text{(all logs are natural)} \tag{10}$$

and determine λ_u from a plot of $\log u_z$ vs $\log r$ in the linear range [13].

Independent determination of λ_σ may also be carried out but the procedure is not as straightforward due to the presence of the second term in (8) and the highly three dimensional nature of the near tip stress state at the free surface. If one evaluates $\tau_{\max}$ from (8) along $\theta = \pi/2$ as done in LEFM, one obtains:

$$\tau_{\max}^{nz} = \frac{\lambda_\sigma K_{\lambda_\sigma}}{\sqrt{2\pi r^{\lambda_\sigma}}} + \frac{\sigma^0}{2} \sin (\lambda_\sigma + 1) \frac{\pi}{2} = \lambda_\sigma \frac{(K_{\lambda_\sigma})_{Ap}}{\sqrt{2\pi r^{\lambda_\sigma}}}, \tag{11}$$

where K_{λ_σ} may be designated a stress eigenfactor. Defining $\tau_0 = (\sigma^0/2) \sin (\lambda_\sigma + 1)(\pi/2)$, one obtains

$$\log (\tau_{\max}^{nz} - \tau_0) = \log \left\{ \frac{\lambda_\sigma K_{\lambda_\sigma}}{\sqrt{2\pi}} \right\} - \lambda_\sigma \{\log r\}. \tag{12}$$

Thus, if one can determine τ_0, one can plot $\log (\tau_{\max} - \tau_0)$ vs. $\log r$ and obtain λ_σ as the slope of the linear range. Since LEFM is expected to prevail away from the free surface, one may conjecture that the value of τ_0 as determined from LEFM away from the free surface should suffice as an adequate correction for the non-singular field effect except near the free surface. Such a determination of λ_σ is shown in Fig. 6 for an interior slice where $\lambda_\sigma \approx 0.5$.

From (2) one sees that, in LEFM, $\tau_0 = \sigma^0/\sqrt{8}$. Thus τ_0 can be determined from the slope of the normalized K_{Ap} vs $r^{1/2}$ graph in the interior of the body (see (4)). Near the free surface, however, since λ_σ is unknown, one cannot determine τ_0 in the same way. Instead one relies upon the result obtained from moire analysis of the free surface, which suggests that $\tau_0 \approx 0$.

4. Experimental results

Since the frozen stress method allows analysis through the thickness normal to a free surface, it was decided to employ a thick enough test geometry to insure simulation of the half space employed in Benthem's analysis. For this purpose, compact bending specimens* were employed, which contained straight front artificial cracks as pictured in Fig. 5.

These models were made of a suitable stress freezing photoelastic material and were stress frozen under load after which thin slices were removed with slice surfaces mutually orthogonal to the crack front and the crack surface (i.e., parallel to the $x_1 x_2$ plane). After analyzing these slices photoelastically, a 1200 line/mm grating was applied to one surface of each slice (including the free surface of the beam) and the slices were annealed and then viewed through a "virtual" grating to obtain moire displacement fields.

For the interior slices, τ_0 values were determined from LEFM but for the surface slice, guided by LEFM and moire studies, τ_0 was set equal to zero. Data were fitted to (10) and (12) to obtain λ_μ and λ_σ, respectively.

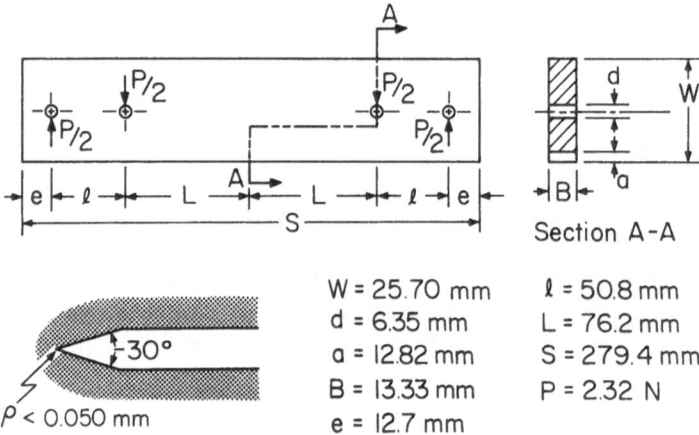

W = 25.70 mm ℓ = 50.8 mm
d = 6.35 mm   L = 76.2 mm
a = 12.82 mm   S = 279.4 mm
B = 13.33 mm   P = 2.32 N
e = 12.7 mm

ρ < 0.050 mm

Fig. 5. Compact bending specimen geometry.

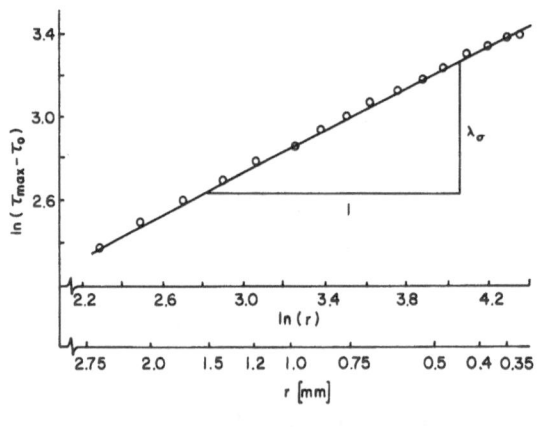

DISTANCE FROM CRACK
SURFACE NORMAL TO CRACK TIP

Fig. 6. Determination of λ_σ from photoelastic data from compact bending specimen.

---

* Four point bend specimens conforming to ASTM E399.

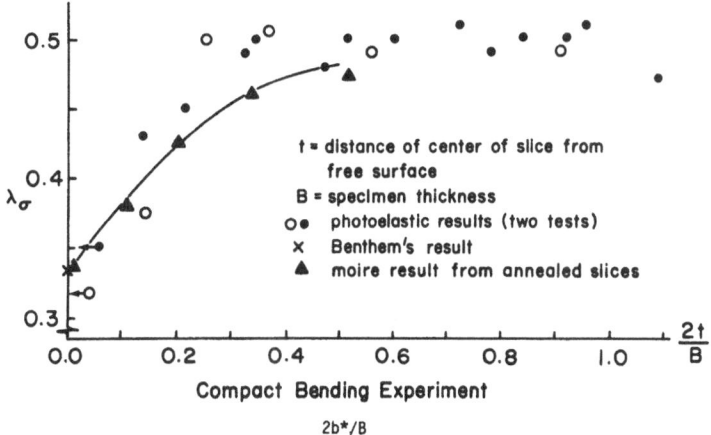

Fig. 7. Distribution of λ_σ for compact bending specimen.

Figure 7 presents a composite of all results, both photoelastic and moire, which were obtained from the compact bending tests. The results from the photoelastic surface slice using $\tau_0 = 0$ are shown with an arrow and compare favorably with Benthem's result at the free surface. The moire result agrees closely with Benthem's result at the free surface.

5. Discussion of results and summary

Although Fig. 7 clearly shows some scatter in the photoelastic results and some difference (of the order of the scatter) between the photoelastic and moire results, both sets of data clearly show a transition region near the free surface where λ_σ varies from the free surface value of 0.33 (also obtained by Benthem) and the classical value of one half in the interior. This transition zone is not predicted by Benthem's analysis but is suggested by Folias to be present. Studies of the type presented here have also been conducted on surface flaws and have yielded similar results [14] without considering the influence of τ_0 and [15] in which its influence is included. The fact that the singularity order appears to be transitional for the polar coordinate system employed for data analysis may imply that, within the transition zone, the separation of variables method may not be valid.

In summary, an optical method of analysis consisting of the tandem application of frozen stress photoelasticity and high density moire interferometry for measuring near tip three dimensional effects in cracked bodies has been reviewed. It was then demonstrated how it could be applied to quantify the loss in the LEFM inverse square root singularity in the neighborhood of the right angle crack border – free surface intersection for compact bending specimens. Results agreed with Benthem's result at the free surface but suggest a transition zone inside the free surface.

Benthem's analysis predicted the effect to be greatest for nearly incompressible materials and the material used here exhibited a value of Poisson's ratio of 0.48. The practical significance of these results lies in the use of surface measurements near crack tips in such materials (rubbers, gasket materials, rocket motor grain, etc.). Use of such data in LEFM equations can lead to substantial error in SIF prediction.

Acknowledgements

The authors are pleased to note the influence on their work of Dr M.L. Williams, Dr J.P. Benthem, Professor E.S. Folias, Professor J.L. Swedlow, the contributions of G. Nicoletto, and W.R. Lloyd and the support of the National Science Foundation and the US Air Force Astronautics laboratory for parts of these studies.

References

1. M.L. Williams, *Journal of Applied Mechanics* 24 (1957) 109–114.
2. C.W. Smith, in *Experimental Techniques in Fracture Mechanics 2*, Society for Experimental Stress Analysis, A.S. Kobayashi (ed.) (1975) 3–58.
3. C.W. Smith, in *Mechanics of Fracture*, Vol. 7, G.S. Sih (ed.) Martinus-Nijhoff (1981) 163–187.
4. C.W. Smith and A.S. Kobayashi, in *Handbook on Experimental Mechanics*, A.S. Kobayashi (ed.) Society for Experimental Mechanics, Prentice Hall, Inc. (1987) 891–956.
5. G. Oppel, *Forsch. Geb. Ingenieurw.* 7 (1936) 240–248.
6. C.W. Smith, *Journal of Experimental Mechanics* 20, No. 4 (1980) 126–133.
7. R. Weller and B.M. Shepard, *Proceedings of Society for Experimental Stress Analysis* VI, No. 1 (1948) 35–38.
8. J. Eftis, N. Subramonian, and H. Liebowitz, *Journal of Engineering Fracture Mechanics* 19, No. 1 (1977) 189–210.
9. G.C. Sih, *International Journal of Fracture Mechanics* 7, No. 1 (1971) 39–62.
10. E.S. Folias, *Journal of Applied Mechanics* 42, Series E, No. 3 (1975) 663–672.
11. J.P. Benthem, *International Journal of Solids and Structures* 16 (1980) 119–130.
12. W.S. Burton, G.V. Sinclair, J.A. Solecki, and J.L. Swedlow, *International Journal of Fracture* 25 (1984) 3–32.
13. C.W. Smith and J.S. Epstein, in *Proceedings of Vth International Congress on Experimental Mechanics*, June 1984, 102–110.
14. C. Ruiz and J.S. Epstein, *International Journal of Fracture* 28 (1985) 231–238.
15. C.W. Smith and M. Rezvani, *Proceedings of a Joint International Conference of the British Society for Strain Measurement and the Society for Experimental Mechanics*, August 1987, in press.

Résumé. On expose brièvement une méthode optique consistant à appliquer en parallèle la photo-élasticité sous contraintes figées et l'interférométrie sur bandes de Moiré à hautes densités, pour l'étude des effets tridimensionnels dans les solides fissurés. On utilise ensuite cette méthode à la mesure de l'eigenvalue dominante à l'intersection suivant un angle droit d'un front de fissure droit avec une surface libre, sous une sollicitation de Mode I. On détermine également la variation de l'eigenvalue dans une zone de transition au voisinage de la surface libre. Le résultat relatif à cette dernière est en bonne concordance avec les résultats analytiques.

International Journal of Fracture 39: 25–34 (1989)
© Kluwer Academic Publishers, Dordrecht

On the stress singularities at the intersection of a cylindrical inclusion with the free surface of a plate

E.S. FOLIAS

Department of Mathematics, University of Utah, Salt Lake City, Utah 84112, USA

Received 27 September 1987; accepted in revised form 1 April 1988

Abstract. Utilizing the form of a general 3D solution, the author investigates analytically the stress field in the neighborhood of the intersection of a cylindrical inclusion and a free surface. The inclusion is assumed to be of a homogeneous and isotropic material and is to be embedded in an isotropic plate of an arbitrary thickness. The stress field is induced by a uniform tension applied on the plate at points far remote from the inclusion (see Fig. 1).

The displacement and stress fields are derived explicitly and a stress singularity is shown to exist for the case when the inclusion is stiffer than the plate material. Moreover, the stress singularity is shown to be a function of the respective ratios of the shear moduli and Poisson's.

The special case of $G_2 \to 0$, $G_2 = G_1$ and $G_2 \to \infty$ are also investigated.

1. Introduction

Quite often in engineering practice, structures are composed of two elastic materials with different properties which are bonded together over some surface. Such type of problem has been investigated from a 2D point of view by many researchers and the results can be found in the literature. For example, in 1927 Knein [1] considered the plane strain problem of an orthogonal elastic wedge bonded to a rigid base. In 1955, Rongved [2] investigated the problem of two bonded elastic half-spaces subjected to a concentrated force in the interior. Subsequently, in 1959, Williams [3] studied the stress field around a fault or a crack in dissimilar media. The work was then generalized in 1965 by Rice and Sih [4] to include arbitrary angles.

It was not until 1968 that Bogy [5] considered the general problem of two bonded quarter-planes of dissimilar isotropic, elastic materials subjected to arbitrary boundary tractions. The problem was solved by an application of the Mellin transform in conjunction with the Airy stress function. In 1971, the same author [6] extended his work to also include dissimilar wedges of arbitrary angles. Shortly thereafter, Hein and Erdogan [7], using the same method of solution, independently reproduced the results by Bogy. Finally, in 1975 Westmann [8] studied the case of a wedge of an arbitrary angle which was bonded along a finite length to a half-space. His analysis showed the presence of two singularities close to each other. Thus, elimination of the first singular term does not lead to a bounded stress field since the second singularity is still present.

Based on 3D considerations,* in 1979 Luk and Keer [9] investigated the stress field in an elastic half space containing a partially embedded axially-loaded, rigid cylindrical rod. The

---

* Due to the symmetry of the applied load, the problem is mathematically 2D.

problem was formulated in terms of Hankel integral transforms and was finally cast into a system of coupled singular integral equations the solution of which was sought numerically. The authors were able, however, to extract in the limit from the integral equations the characteristic equation governing the singular behavior at the intersection of the free surface and that of the rigid inclusion. Their result was in agreement with that obtained by Williams [10] for a right-angle corner with fixed-free boundary conditions.

In 1980 Haritos and Keer [11] investigated the stress field in a half-space containing an embedded rigid block under conditions of plane strain. The problem was formulated by cleverly superimposing the solutions to the problem of horizontal and vertical line inclusions beneath an elastic half-space. By isolating the pertinent terms, the authors were able to extract directly from the integral equations the order of the stress singularity at both corners. Both results are in agreement with the Williams solution. Moreover, the authors point out the importance of the second singularity to the results of the load transfer problems.

Finally, in 1986 Folias [12], utilizing the form of a general 3D solution for the equilibrium of linear elastic layers which he developed in 1975 [13], derived explicitly the 3D displacement and stress fields at the intersection of a hole and a free surface. The analysis revealed that the stresses at the corner at proportional to $\varrho^{\alpha-2}$ where ϱ represents the local radius from the corner and $\alpha = 3.73959 \pm i\, 1.11902$. It is interesting to note that the root is precisely the same as that obtained by Williams in his classic paper [10] for a 90 deg material corner with free–free stress boundaries (plane strain). An extension of the analysis to other angles of intersection revealed the same analogy between 3D and 2D. Thus the Williams solution has further applicability than was originally thought.

The same general 3D solution can now be used to solve the corresponding problem of a cylindrical inclusion. Moreover, the method is also applicable to non-symmetric applied loads while retaining a rather simple mathematical character.

2. Formulation of the problem

Consider the equilibrium of a homogeneous, isotropic, linear elastic plate which occupies the space $|x| < \infty, |y| < \infty$ and $|z| \leqslant h$ and contains a cylindrical inclusion of radius a whose generators are perpendicular to the bounding planes $z = \pm h$. It is assumed that the cylindrical inclusion is made of an isotropic and homogeneous material of different elastic properties than those of the plate. The plate is subjected to a uniform tensile load σ_0 in the direction of the y-axis and parallel to the bounding planes (see Fig. 1).

In the absence of body forces, the coupled differential equations governing the displacement functions $u_i^{(m)}$ are

$$\frac{1}{1 - 2v_m} \frac{\partial e^{(m)}}{\partial x_i} + \nabla^2 u_i^{(m)} = 0; \quad i = 1, 2, 3 \quad m = 1, 2, \tag{1}$$

where ∇^2 is the Laplacian operator, v_m is Poisson's ratio, $u_i^{(1)}$ and $u_i^{(2)}$ represent the displacement functions in media 1 (plate) and 2 (inclusion) respectively, and

$$e^{(m)} \equiv \frac{\partial u_i^{(m)}}{\partial x_i} \quad i = 1, 2, 3; \quad m = 1, 2. \tag{2}$$

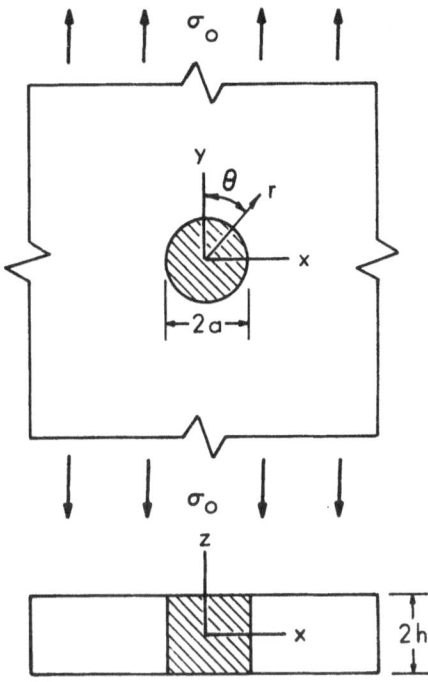

Fig. 1. Infinite plate of arbitrary thickness with cylindrical inclusion.

The stress-displacement relations are given by Hooke's law as

$$\sigma_{ij}^{(m)} = \lambda_m e_{kk}^{(m)} \delta_{ij} + 2G e_{ij}^{(m)},\tag{3}$$

where λ_m and G_m are the Lamé constants describing media 1 and 2.

3. Method of solution

The main objective of this analysis is to derive an asymptotic solution valid in the immediate vicinity of the corner points where the interface meets the free surface of the plate. For this purpose, we assume the complementary displacement field to be of the form [12, 13]:

$$u^{(m)} = \frac{1}{1 - 2v_m} \frac{\partial}{\partial x} \left\{ 2(1 - v_m) f_2^{(m)} + h \frac{\partial f_1^{(m)}}{\partial z} + z \frac{\partial f_2^{(m)}}{\cdot \partial z} \right\}\tag{4}$$

$$v^{(m)} = \frac{1}{1 - 2v_m} \frac{\partial}{\partial y} \left\{ 2(1 - v_m) f_2^{(m)} + h \frac{\partial f_1^{(m)}}{\partial z} + z \frac{\partial f_2^{(m)}}{\partial z} \right\}\tag{5}$$

$$w^{(m)} = \frac{1}{1 - 2v_m} \frac{\partial}{\partial z} \left\{ -2(1 - v_m) f_2^{(m)} + h \frac{\partial f_1^{(m)}}{\partial z} + z \frac{\partial f_2^{(m)}}{\partial z} \right\},\tag{6}$$

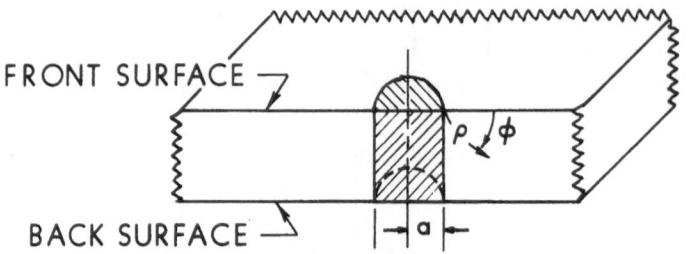

Fig. 2. Definition of local coordinates at the corner.

where the functions $f_1^{(m)}$ and $f_2^{(m)}$ are three dimensional harmonic functions. If we furthermore assume that

$$f_j^{(m)} = r^{-1/2} H_j^{(m)}(r - a, h - z)e^{i20}; \quad j = 1, 2 \tag{7}$$

then the functions $H_j^{(m)}$ must satisfy the following equation:

$$\frac{\partial^2 H_j^{(m)}}{\partial(r - a)^2} + \frac{\partial^2 H_j^{(m)}}{\partial(h - z)^2} - \frac{15}{4(a + r - a)^2} H_j^{(m)} = 0. \tag{8}$$

It is found convenient at this stage to introduce the local coordinate system (see Fig. 2)

$$r - a = \varrho \cos \phi$$

$$h - z = \varrho \sin \phi$$

in view of which, (8) may now be written as

$$\frac{\partial^2 H_j^{(m)}}{\partial \varrho^2} + \frac{1}{\varrho} \frac{\partial H_j^{(m)}}{\partial \varrho} + \frac{1}{\varrho^2} \frac{\partial^2 H_j^{(m)}}{\partial \phi^2} - \frac{15 H_j^{(m)}}{4a^2[1 + (\varrho/a) \cos \phi]^2} = 0. \tag{9}$$

Under the assumption that the radius of the inclusion is sufficiently large, so that the condition $\varrho \ll a$ is meaningful, we seek the solution to (9) in the form

$$H_j^{(m)} = \sum_{n=0}^{\infty} \varrho^{\alpha+n} F_{jn}^{(m)}(\phi), \tag{10}$$

with α a constant. Without going into the mathematical details, we construct the following series expansion in ascending powers of ϱ:

$$H_j^{(m)} = \varrho^{\alpha}\{A_j^{(m)} \cos (\alpha\phi) + B_j^{(m)} \sin (\alpha\phi)\}$$

$$+ \varrho^{\alpha+1}\{C_j^{(m)} \cos (\alpha + 1)\phi + D_j^{(m)} \sin (\alpha + 1)\phi\} + O(\varrho^{\alpha+2}), \tag{11}$$

where the constants α, $A_j^{(m)}$, $B_j^{(m)}$, $C_j^{(m)}$ and $D_j^{(m)}$ are to be determined from the boundary conditions. Specifically,

at $\phi = 0$: $\sigma_{zz}^{(1)} = \tau_{xz}^{(1)} = \tau_{yz}^{(1)} = 0$ (12)

at $\phi = \pi$: $\sigma_{zz}^{(2)} = \tau_{xz}^{(2)} = \tau_{yz}^{(2)} = 0$ (13)

at $\phi = \dfrac{\pi}{2}$: $u_j^{(1)} = u_j^{(2)}$; $j = 1, 2, 3$ (14)

$$\sigma_{rr}^{(1)} = \sigma_{rr}^{(2)}, \tau_{r\theta}^{(1)} = \tau_{r\theta}^{(2)}, \tau_{rz}^{(1)} = \tau_{rz}^{(2)}.$$ (15)

Substituting (11) into (12) and (13) one finds that all terms up to the order $O(\varrho^{\alpha-2})$ are satisfied if one assumes the following combinations vanish:

$$B_1^{(1)} = 0$$ (16)

$$-h(\alpha + 1)A_1^{(1)} - B_2^{(1)} = 0$$ (17)

$$A_1^{(2)} \sin(\alpha\pi) - B_1^{(2)} \cos(\alpha\pi) = 0$$ (18)

$$[A_2^{(2)} - h(\alpha + 1)B_1^{(2)}] \tan(\alpha\pi) - B_2^{(2)} + h(\alpha + 1)A_1^{(2)} = 0.$$ (19)

Similarly, the displacement and stress boundary conditions at the cylindrical surface are satisfied if we assume the following combinations vanish

$$\frac{1}{1 - 2v_1}\left\{(-\alpha - 1 + 2v_1)A_2^{(1)} \tan\left(\frac{\alpha\pi}{2}\right) - (-\alpha - 1 + 2v_1)B_2^{(1)}\right.$$

$$\left. + h(\alpha + 1)\left[A_1^{(1)} + B_1^{(1)} \tan\left(\frac{\alpha\pi}{2}\right)\right]\right\}$$ (20)

$$- \frac{1}{1 - 2v_2}\left\{(-\alpha - 1 + 2v_2)\left[A_2^{(2)} \tan\left(\frac{\alpha\pi}{2}\right) - B_2^{(2)}\right]\right.$$

$$\left. + h(\alpha + 1)\left[A_1^{(2)} + B_1^{(2)} \tan\left(\frac{\alpha\pi}{2}\right)\right]\right\} = 0$$

$$\frac{1}{1 - 2v_1}\left\{(\alpha - 2 + 2v_1)A_2^{(1)} + (\alpha - 2 + 2v_1)B_2^{(1)} \tan\left(\frac{\alpha\pi}{2}\right)\right.$$

$$\left. + h(\alpha + 1)\left[A_1^{(1)} \tan\left(\frac{\alpha\pi}{2}\right) - B_1^{(1)}\right]\right\}$$

$$- \frac{1}{1 - 2v_2} \left\{ (\alpha - 2 + 2v_2) \left[A_2^{(2)} + B_2^{(2)} \tan \left(\frac{\alpha\pi}{2} \right) \right] \right.$$

$$\left. + h(\alpha + 1) \left[A_1^{(2)} \tan \left(\frac{\alpha\pi}{2} \right) - B_1^{(2)} \right] \right\} = 0 \tag{21}$$

$$\alpha A_2^{(1)} - (1 - \alpha) \tan \left(\frac{\alpha\pi}{2} \right) B_2^{(1)} - \beta \left\{ \alpha A_2^{(2)} + \alpha \tan \left(\frac{\alpha\pi}{2} \right) B_2^{(2)} \right.$$

$$\left. - \tan \left(\frac{\alpha\pi}{2} \right) \left[A_2^{(2)} \sin (\alpha\pi) - B_2^{(2)} \cos (\alpha\pi) \right] \right\} = 0 \tag{22}$$

$$- (\alpha - 1) A_2^{(1)} \tan \left(\frac{\alpha\pi}{2} \right) + (\alpha - 2) B_2^{(1)} - \beta \left\{ \left[-(\alpha - 1) \tan \left(\frac{\alpha\pi}{2} \right) \right. \right.$$

$$\left. \left. + \sin (\alpha\pi) \right] A_2^{(2)} + [(\alpha - 1) - \cos (\alpha\pi)] B_2^{(2)} \right\} = 0, \tag{23}$$

where for simplicity we have defined

$$\beta \equiv \frac{1 - 2v_1}{1 - 2v_2} \frac{G_2}{G_1}. \tag{24}$$

The characteristic value α may now be determined by setting the determinant of the algebraic system (20)–(24) equal to zero. Once the roots have been determined, the complete displacement and stress fields can be constructed in ascending powers of ϱ.

4. Discussion of the results

Without going into the mathematical details, the characteristic values α can easily be determined with the aid of a computer. Although the equation has an infinite number of complex roots, only the one with a $1 < \min \operatorname{Re} \alpha < 2$ is relevant. In general, the characteristic values of α depend on the material properties of both the plate as well as the inclusion.

The analysis clearly shows that, in the neighborhood of the interface and the free surface, the stress field is proportional to $\varrho^{\alpha-2}$ and that for certain material properties it is singular. Moreover, the first root is found to be precisely the same as that of the corresponding 2D case [6]. Figures 3, 4 and 5 depict typical results for various material properties. Finally, in the limit as the shear modulus (i) $G_2 \to 0$ and (ii) $G_2 \to \infty$ one recovers the results corresponding to a (i) hole (i.e., $\alpha = 3.73959 \pm i1.11902$) and (ii) a perfectly rigid inclusion (i.e., $\alpha = 1.7112 + i0.$), respectively. It is interesting to note that in both limit cases the exponent α is the same as that obtained by Williams [10] for a 90 degree material angle with (i) free–free and (ii) fixed–free boundaries in plane strain. Finally, as $G_2 \to G_1$ and $v_2 \to v_1$, the solution of a continuous plate is recovered, a result that clearly meets our expectations.

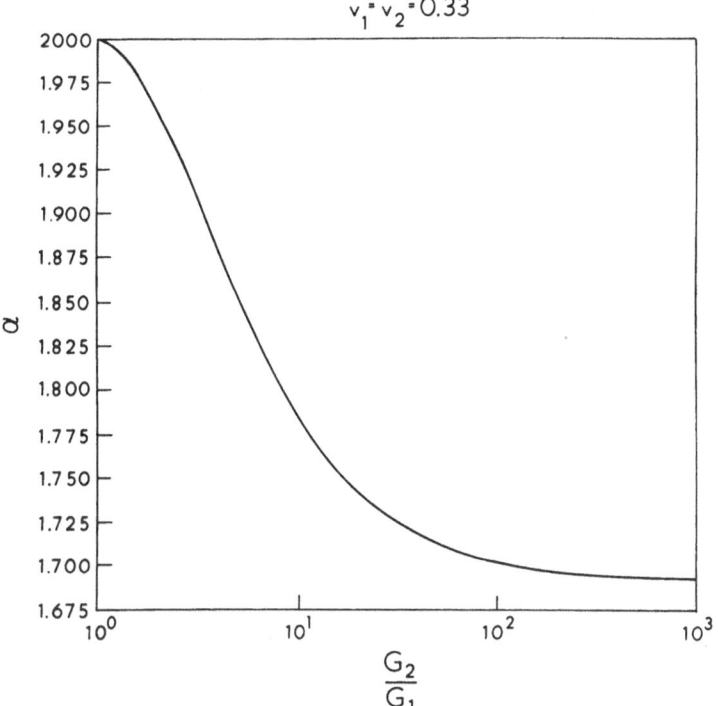

Fig. 3. Strength of the singularity vs $G_2 \backslash G_1$ for $v_1 = v_2 = 0.33$.

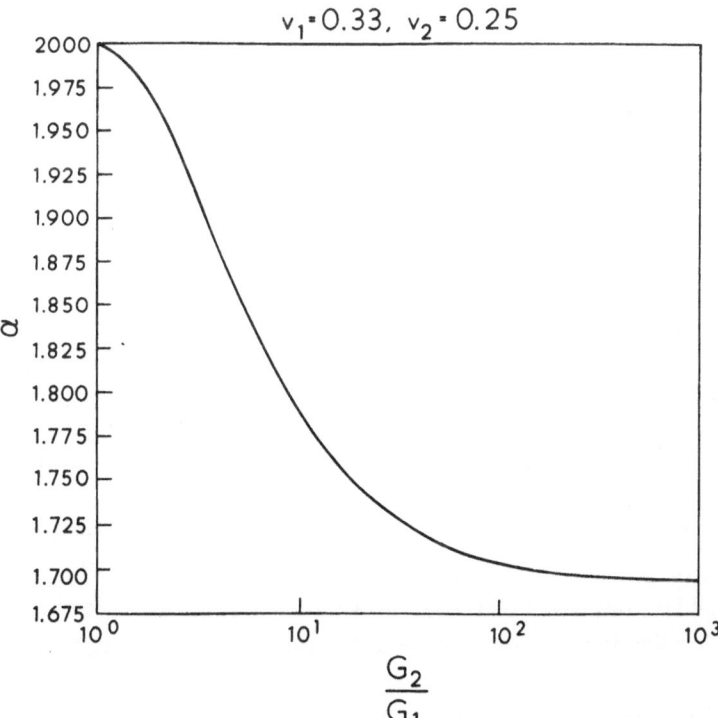

Fig. 4. Strength of the singularity vs $G_2 \backslash G_1$ for $v_1 = 0.33$, $v_2 = 0.25$.

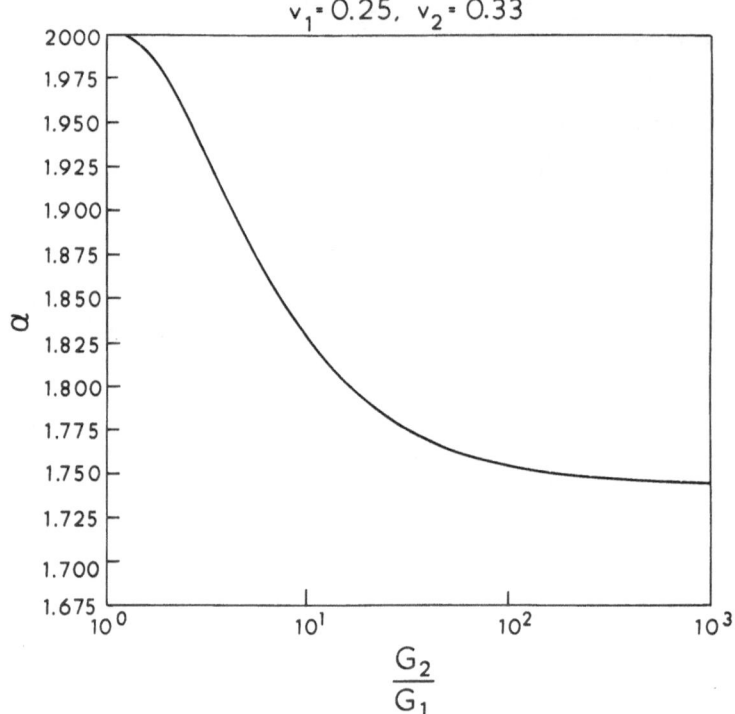

Fig. 5. Strength of the singularity vs $G_2 \backslash G_1$ for $v_1 = 0.25$, $v_2 = 0.33$.

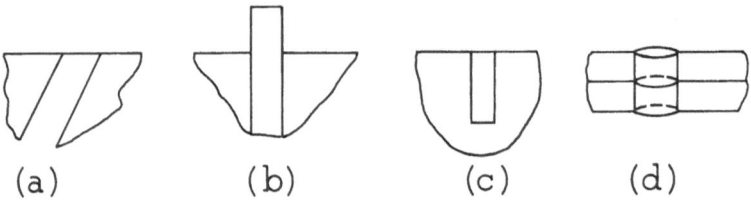

Fig. 6. Geometrical applications.

An extension of this analysis to other angles of intersection with the free surface reveals the same results as those predicted by Bogy [6] for the case of plane strain. A few cases of practical interest which come to mind are shown in Fig. 6(a)–(c). Similarly, it can be shown that the same results apply at the intersection of an interface and the free surface of a hole in a laminated plate consisting of homogeneous and isotropic laminates (see Fig. 6d). Finally, by way of a conjecture, one may now deduce that the same analogy exists for the corresponding cases consisting of anisotropic materials.

It should also be noted that the analysis confirms the presence of another (a little bit weaker) singularity which was first pointed out by Westmann [8]. While it is true that the singular stress field is dominated by the largest singularity, the presence of two singular terms has important implications to the problem of adhesion as well as the problem of load diffusion from one material to another [11].

The results of this analysis are also of importance to the field of composite materials. In general, composite structures are designed in such a way as to carry the load along the

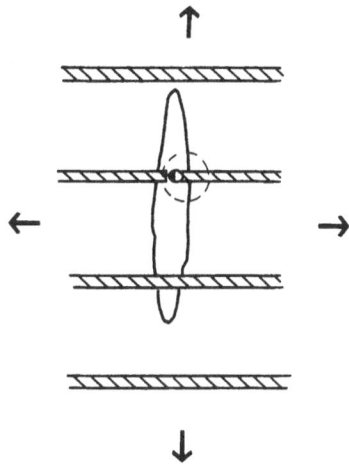

Fig. 7. Possible composite failure mode.

direction of their fibers. Quite often, however, a small portion of that load will be applied, at least locally, in a direction perpendicular to the fibers (e.g., a pressurized vessel). As a practical matter, let us consider a region where the matrix and a few fibers have cracked in a manner depicted in Fig. 7. Conditions in the neighborhood of the circled fiber are very similar* to those assumed in the present paper. There are, however, two important differences. First, the dense distribution of fibers induces stress interactions which lead to higher stress levels. Second, the typical diameter of a fiber is approximately 0.003 inches, a magnitude that may be thought of as of the same order as ϱ. In that case, (9) is no longer separable and a different form of the solution must be sought.**

In closing, it is noteworthy to point out the importance of the general 3D solution [13] for it reveals the inherent form of the solution at such neighborhoods and permits a simple analytical approach, à la Williams [10], for the determination of the stress singularities.

Acknowledgement

This work was supported in part by the Air Force Office of Scientific Research Grant No. AFOSR-87-0204. The author wishes to thank Lt. Col. G. Haritos for this support.

References

1. M. Knein, "Zur Theorie der Druckversuchs," Abhandlungen der Aerodynamische Inst. u.d. Technische Hochschule, Aachen, Germany, Vol. 7 (1927) 43–62.
2. L. Rongved, "Force Interior to One of Two Joined Semi-Infinite Solids," Second Midwestern Conference on Solid Mechanics (September 1955) 1–13.
3. M.L. Williams, *Bulletin of the Seismological Society of America* 49 (1959) 199.
4. J.R. Rice and G.C. Sih, *Journal of Applied Mechanics, Transactions ASME*, Series E32 (1965) 418.
5. D.B. Bogy, *Journal of Applied Mechanics, Transactions ASME* 35 (1968) 460–466.

* Here the author assumed that the same analogy between 3D and 2D exists for anisotropic materials, too.
** This matter is presently under investigation.

6. D.B. Bogy, *Journal of Applied Mechanics* 38 (1971) 377–386.
7. V.L. Hein and F. Erdogan, *International Journal of Fracture* 7 (1971) 317–330.
8. R.A. Westmann, *International Journal of Engineering Science* 13 (1975) 369.
9. V.K. Luk and L.M. Keer, *International Journal of Solids and Structures* 15 (1979) 805–827.
10. M.L. Williams, *Journal of Applied Mechanics, Transactions ASME* 19 (1952) 526.
11. G.K. Haritos and L.M. Keer, *International Journal of Solids and Structures* 16 (1979) 19–40.
12. E.S. Folias, *International Journal of Fracture* 35 (1987) 187–194.
13. E.S. Folias, *Journal of Applied Mechanics* 42 (1975) 663–674.

Résumé. En utilisant la forme d'une solution générale à trois dimensions, l'auteur analyse le champ de contraintes au voisinage de l'intersection d'une inclusion cylindrique et d'une surface libre. On suppose que l'inclusion est constituée d'une matériau homogène et isotrope, et qu'elle est insérée dans une plaque isotrope d'épaisseur arbitraire. Le champ de contraintes et dû à une tension uniforme appliquée en des points suffisamment éloignés de l'inclusion.

On exprime de manière explicite les champs de déplacements et de contraintes, et on montre qu'il existe une singularité de la contrainte dans le cas où l'inclusion est dans un matériau plus rigide que celui de la plaque. En outre, on montre que la singularité de la contrainte est fonction des rapport respectifs des modules de cisaillement et de Poisson.

On étudie également les cas spéciaux $G_2 \to 0$, $G_2 = G_1$ et $G_1 \to \infty$.

International Journal of Fracture 39: 35–43 (1989)
© Kluwer Academic Publishers, Dordrecht

Some viscoelastic wave equations

DANG DINH ANG* and ALAIN PHAM NGOC DINH

Department of Mathematics, University of Orleans, France

Received 22 September 1987; accepted 1 April 1988

Abstract. The authors study the problem of existence, uniqueness and asymptotic behavior for $t \to \infty$ of (weak or strong) solutions of equations in the form

$$u_{tt} - \lambda \Delta u_t - \sum_{i=1}^{N} \partial/\partial x_i \sigma_i(u_{x_i}) + f(u, u_t) = 0$$

$$\lambda \geqslant 0, \ (x, t) \in \Omega \times (0, T)$$

$$\Omega = \text{a domain in } \mathbb{R}^n,$$

with various boundary and initial conditions on $u(x, t)$. The case $\lambda > 0$ corresponds to a nonlinear Voigt model (for σ_i nonlinear). The case $\lambda = 0$, $N = 1$ and $f(u, u_t) = |u_t|^\alpha \operatorname{sgn} u_t$, $0 < \alpha < 1$ with nonhomogeneous boundary conditions corresponds to the motion of a linearly elastic rod in a nonlinearly viscous medium. The method followed is the Galerkin method.

1. Introduction

Consider the problem $(P_\lambda)(\lambda \geqslant 0)$ consisting of the equation:

$$u_{tt} - \lambda \Delta u_t - \sum_{i=1}^{N} \partial/\partial x_i \sigma_i(u_{x_i}) + f(u, u_t) = 0 \tag{1}$$

$$(x, t) \in \Omega \times (0, T)$$

$$u = 0 \text{ on } \partial\Omega \tag{2}$$

$$u(x, 0) = \tilde{u}_0(x), \ u_t(x, 0) = \tilde{u}_1(x), \tag{3}$$

where $u_t = \partial u/\partial t$, $u_{x_i} = \partial u/\partial x_i$.

In (1), Ω is a bounded domain in $\mathbb{R}^n$ with a sufficiently smooth boundary $\partial\Omega$, σ_i ($i = 1, \ldots N$) are continuous functions satisfying certain monotonic and other conditions to be specified later. Equations of the type (1) with $f = 0$, were given the first systematic treatment by Greenberg, MacCamy and Mizel [4] in the case of space dimension $N = 1$. They were proposed by the authors (loc. cit.) as field equations governing the longitudinal motions of a homogeneous bar of uniform cross section and constant density in its rest

---

* On leave from Hochiminh City University, Hochiminh City, Viet Nam.

configuration. For instance if x is the position of a cross section in the rest configuration and $u(x, t)$ and $\tau(x, t)$ are the displacement of the section and the stress on the section at time t, then the equation of motion (if the density is one) is

$$u_{tt} = \partial \tau / \partial x. \tag{4}$$

If the stress depends nonlinearly on u_x, $\tau = \sigma(u_x)$ then (4) is the nonlinear wave equation:

$$u_{tt} = \sigma'(u_x)u_{xx}. \tag{5}$$

In §2 (1) is studied with $\lambda > 0$ and $f(u_t) = |u_t|^{\alpha}$ sgn u_t, $0 < \alpha < 1$ [1]. For $\tilde{u}_0$ in $H^1_0(\Omega)$, $\tilde{u}_1$ in $L^2(\Omega)$ and σ_i in $C(\mathbb{R}, \mathbb{R})$ non decreasing, $\sigma_i(0) = 0$ and inducing mappings of $L^2(\Omega)$ into itself, taking bounded sets into bounded sets, the problem admits a global weak solution. If in addition the σ_i's are assumed locally Lipschitzian, then the solution is unique. For $N = 1$, $\tilde{u}_0$ in $H^1_0(\Omega) \cap H^2(\Omega)$, u_1 in $L^2(\Omega)$, σ_i in $C^1(\mathbb{R}, \mathbb{R})$ with $\sigma_i' > 0$ and locally Hölder continuous, there exists a unique strong solution $u(t)$ of the initial and boundary value problem i.e., $t \to u(t)$ is continuous on $t \geq 0$ to $H^1_0(\Omega) \cap H^2(\Omega)$ and twice continuously differentiable on $t > 0$ to $L^2(\Omega)$.

In §3 consider the problem with $\lambda > 0$, the function f being a function of u: $f(u)$. With $\tilde{u}_0$ in $H^1_0(\Omega)$ and $\tilde{u}_1$ in $H^1_0(\Omega)$ and $\tilde{u}_1$ in $L^2(\Omega)$, under a certain local Lipschitzian condition on f, a local existence and uniqueness theorem is proved [2], using the method of successive linearizations.

If we strengthen the above hypotheses and assume that $1 \leq N \leq 3$, $\tilde{u}_0 \in H^1_0(\Omega) \cap H^2(\Omega)$ and $\tilde{u}_1 \in L^2(\Omega)$ while f is to satisfy no other condition than $f' \geq -c$ for $c > 0$ "small", then, the unique solution $u(t)$ exists for all $t \geq 0$, with the property that $\|u_t\|$ and $\|\Delta u(t)\|$ decay exponentially to 0 as $t \to \infty$, generalizing a result of Webb [10]. Here and elsewhere $\| . \|$ stands for the $L^2(\Omega)$ norm.

In the case $N = 1$, $\lambda = 0$, $\sigma(x) = x$ and $f(u_t) = |u_t|^{\alpha}$ sgn u_t, $0 < \alpha < 1$, we study (1) with a nonhomogeneous condition namely

$$u_x(0, t) = g(t), u(1, t) = 0. \tag{2'}$$

Then the global existence and uniqueness of the initial and boundary value problem (1), (2′) and (3) with $\tilde{u}_0 \in H^1_0(\Omega)$ and $\tilde{u}_1 L^2(\Omega)$ can be proved by using a Volterra nonlinear integral equation and the monotonicity generated by the nonlinear term [3].

Let $L_2 = L_2(\Omega)$, $H^1_0 = H^1_0(\Omega)$, $H^2 = H^2(\Omega)$. Here $H^1_0(\Omega)$ and $H^2(\Omega)$ denote the usual Sobolev spaces on Ω. Let $\langle . \rangle$ denote either the L_2-inner product of the pairing of a continuous linear functional with an element of a function space. Let $\| . \|_x$ be a norm on a Banach space X, let X^* be its dual. We denote by $L^p(0, T; X)$, $1 \leq p \leq \infty$, the space of functions f on $(0, T)$ to X' such that

$$\|f\|_{L^p(0,T,X)} = \left(\int_0^T \|f(t)\|_X^p \, dt \right)^{1/p} \quad \text{for} \quad 1 \leq p \leq \infty$$

$$\|f\|_{L^\infty(0,T,X)} = \text{ess sup } \|f(t)\|_X \quad \text{for} \quad p = \infty.$$

2. Problem 1

$$u_{tt} - \Delta u_t - \sum_{i=1}^{N} \partial/\partial x_i \sigma_i(u_{x_i}) + |u_t|^\alpha \operatorname{sgn} u_t = 0.$$

Consider a special basis of $H_0^1(\Omega)$: $w_1, w_2, \ldots, w_n, \ldots$ formed by the eigenfunctions of Laplacian Δ on $H_0^1(\Omega)$. Let $(w_1, w_2, \ldots, w_n)$ be the linear space generated by $w_1, w_2, \ldots, w_n$ Let

$$u^{(n)}(t) = \sum_{k=1}^{n} c_{k,n}(t)w_k(x) \tag{6}$$

be a solution of the following system

$$\langle u_{tt}^{(n)}, w_k \rangle + \sum_{i=1}^{N} \langle \sigma_i(u_{x_i}^{(n)}(t), (w_k)x_i \rangle - \langle \Delta u_t^{(n)}(t), w_k \rangle + \langle f(u_t^{(n)}(t)), w_k \rangle = 0 \tag{7}$$

$$1 \leqslant k \leqslant n$$

$$u^{(n)}(0) = u_{0n}, u_t^{(n)}(0) = u_{1n}, \tag{8}$$

where $u_{0n} \to \tilde{u}_0$ in H_0^1 strongly, $u_{1n} \to \tilde{u}_1$ in L_2 strongly.
Clearly a solution $u^{(n)}(t)$ exists on a sufficiently small interval $(0, T_n)$.
By using the fact that $\langle f(v), v \rangle \geqslant 0$ and $\int_\Omega h_i(u_{x_i}^{(n)}(t))(x)\,dx \geqslant 0$ where $h_i(z) = \int_0^z \sigma_i(s)\,ds$ we can prove the following a priori estimate:

$$\|u_t^{(n)}(t)\|^2 + \int_0^t \|\nabla u_t^{(n)}(s)\|^2\,ds \leqslant c \tag{9}$$

$$0 \leqslant t \leqslant T_n$$

where c is independent of n.
    If we define the nonlinear operator $A: H_0^1 \to H^{-1}(\Omega) = H^{-1}$ (the dual of H_0^1)

$$Au: - \sum_{i=1}^{N} (\sigma_i(u_{x_i}))x_i \tag{10}$$

then, it follows from (9) and the hypotheses on σ_i that

$$\|Au^{(n)}(t)\|_* \leqslant M_T \quad 0 \leqslant t \leqslant T, \tag{11}$$

where $\|.\|_*$ is the dual norm on H^{-1} and M_T a constant independent of n, but depending on T. We also need an estimate on the $u_{tt}^{(n)}$. From the approximated problem:

$$u_{tt}^{(n)} - \Delta u_t^{(n)}(t) + Au^{(n)}(t) + f(u_t^{(n)}(t)) = 0. \tag{12}$$

We can deduce

$$\int_0^T \|u_{tt}^{(n)}(t)\|_{H^{-1}}^2 \, dt \leqslant M_T \quad 0 \leqslant t \leqslant T. \tag{13}$$

On the other hand we have

$$\|f(u_t^{(n)}(t))\| \leqslant c \quad 0 \leqslant t \leqslant T \tag{14}$$

since Ω is bounded and $0 < \alpha < 1$.

From the above a priori estimates and using a Lemma on compactness [5], we can extract from the sequence $\{u^{(n)}\}$ a subsequence, still denoted by $\{u^{(n)}\}$, such that:

$$u^{(n)} \to u \text{ strongly in } L_2(Q) \text{ with } Q = \Omega \times (0, T) \tag{15}$$

$$u_t^{(n)} \to u_t \text{ strongly in } L_2(Q) \tag{16}$$

$$Au^{(n)} \to \xi \text{ in } L^\infty(0, T; H^{-1}) \text{ weak }_* \tag{17}$$

$$f(u_t^{(n)}) \to X \text{ in } L^\infty(0, T; L_2) \text{ weak }_* \tag{18}$$

Letting $n \to \infty$ in (7) we find that u satisfies the equation:

$$d/dt\langle u_t(t), v\rangle + \langle \nabla u_t(t), \nabla v\rangle + \langle \xi(t), v\rangle + \langle X(t), v\rangle = 0$$

$$\text{a.e. } t \text{ in } (0, T) \text{ for all } v \text{ in } H_0^1 \tag{19}$$

and the initial conditions (3).

So, we shall have proved the existence of a weak or distributional solution of (1)–(3) once we have shown that:

$$\xi = Au \text{ and } X = f(u). \tag{20}$$

The proof of (20) relies on the following:

$$\overline{\lim_{n \to \infty}} \int_0^t \langle Au^{(n)}(s), u^{(n)}(s)\rangle \, ds \leqslant \int_0^t \langle \xi(s), u(s)\rangle \, ds \quad t \text{ in } (0, T) \tag{21}$$

and

$$\lim_{n \to \infty} \int_0^T \langle f(u_t^{(n)})(s), u_t^{(u)}(s)\rangle \, ds = \int_0^T \langle X(s), u_t(s)\rangle \, ds. \tag{22}$$

Equation (21) is derived from (15), (16) and from the fact that the operator A is a monotone operator from H_0^1 to H^{-1}.

In the same way the function $u \to f(u)$ generates a monotone operator on L_2 to L_2 which involves with (16) and (18) the equality (22). Using finally the hemicontinuity of A and f, then we get (20).

Let us consider the problem of global existence of strong solution of (1)–(3). To this end, we shall have to strengthen conditions on the initial data and on σ_i's. The role of the space dimension is important and we shall limit ourselves to $N = 1$. With the latter restriction, it will be sufficient, from our point of view, to place the following requirements on the initial data and on the coefficients of the field equation:

$$\tilde{u}_0 \text{ in } H_0^1 \cap H^2, \, \tilde{u}_1 \text{ in } L_2 \tag{23}$$

$$\sigma_i \text{ in } C^1(\mathbb{R}, \mathbb{R}), \, \sigma_i' > 0, \, \sigma_i(0) = 0 \tag{24}$$

and σ_i' locally Hölder continuous.

Then, we have the following.

THEOREM 1. *Let $N = 1$ and let (23) and (24) hold. Then, there exists a unique solution $u(t)$ of (1)–(3) with the following properties:*

    $t \rightarrow u(t)$ is continuous on $t \geqslant 0$ to $H_0^1 \cap H^2$

                continuously differentiable on $t > 0$ to $H_0^1 \cap H^2$.

                continuously differentiable on $t \geqslant 0$ to L_2.

    and twice continuously differentiable on $t > 0$ to L_2.

The idea of the proof is as follows. We take the weak solution $w(t)$ of (1)–(3) (which will presently be shown to exist) and then, using the analytic theory of semi-group and the uniqueness of the solution of (1)–(3), we prove that $w(t)$ is in fact the strong solution of the theorem by considering the differential equation:

$$u_t = \Delta u + \tilde{u}_1 - \Delta \tilde{u}_0 + G(w) + F(w) \tag{25}$$

with the initial condition

$$u(0) = \tilde{u}_0 \tag{26}$$

and where

$$G(w(t)) = \int_0^t Aw(s)\mathrm{d}s, \, F(w(t)) = -\int_0^t f(w_t(s))\mathrm{d}s.$$

3. Problem 2

$$u_{tt} - \Delta u_t - \Delta u + f(u) = 0$$

● First we shall make the following assumption:

$f: H_0^1(\Omega) \rightarrow L^2(\Omega)$ satisfies $\tag{27}$

for each bounded set B or $H_0^1(\Omega)$, there exists $k_B > 0$ such that

$$\|f(y) - f(z)\| \leqslant k_B \|\nabla y - \nabla z\| \quad \forall y, z \in B$$

Then we have the following:

THEOREM 2. *Suppose f satisfies (27) and let $\tilde{u}_0 \in H_0^1(\Omega)$, $\tilde{u}_1 \in L^2(\Omega)$, then there exists a $u \in C(0, T; H_0^1)$ and such that $u_t \in C(0, T; L_2) \cap L^2(0, T; H_0^1)$. Furthermore $u(t)$ is the limit of the sequence $\{u^{(n)}(t)\}$ of solutions of the following linear and boundary value problems:*

$$u_{tt}^{(n)} - \Delta u_t^{(n)} - \Delta u^{(n)} = -f(u^{(n-1)}) \quad n \geqslant 1, u_0 \equiv 0$$

$$u^{(n)} = 0 \text{ on } \partial\Omega \tag{28}$$

$$u^{(n)}(0) = \tilde{u}_0, u_t^{(n)}(0) = \tilde{u}_1$$

The sequence $\{u^{(n)}(t)\}$ converges uniformly to $u(t)$ in $C(0, T; H_0^1)$ for $n \to \infty$. On the other hand the sequence $\{u_t^{(n)}(t)\}$ converges uniformly to $u_t(t)$ in $C(0, T; L_2) \cap L^2(0, T; H_0^1)$.

The idea of the proof can be shown as follows. By induction we show that the sequences $\{u^{(n)}\}$ and $\{u_t^{(n)}\}$ exist and belong to a bounded set of $C(0, T; H_0^1)$ and $C(0, T; L_2) \cap L^2(0, T; H_0^1)$ respectively for a suitable T. Then we demonstrate that the sequence $\{u^{(n)}\}$ constructed in (28) is a Cauchy sequence in $C(0, T; H_0^1)$. Using the Lipschitzian property of f and passing to the limit as $n \to \infty$ in the variational equation associated with the initial and value problem (28), we find that the limit u of the sequence $\{u^{(n)}\}$ satisfies the equation:

$$\mathrm{d}/\mathrm{d}t \langle u_t(t), v \rangle + \langle \nabla u_t(t), v \rangle + \langle \nabla u(t), v \rangle = -\langle f(u(t), v \rangle$$

   a.e. on $(0, T)$ and $\forall v \in H_0^1$

$$u(0) = \tilde{u}_0 \tag{29}$$

$$u_t(0) = \tilde{u}_1.$$

● Second we shall consider the problem of global existence and asymptotic behavior for $t \to \infty$. To this end, we shall limit ourselves, in what follows, to the case $1 \leqslant N \leqslant 3$, and furthermore, we shall restrict the hypotheses on f and on the regularity of the initial data. Thus we shall consider the following conditions on f:

$$f \in C^1(\mathbb{R}, \mathbb{R}), f(0) = 0 \tag{30}$$

$$(f(u) + \varepsilon u) \cdot u \geqslant 0 \quad \text{for all} \quad |u| \geqslant a, \tag{31}$$

with $0 < \varepsilon < 1$ satisfying $\varepsilon \alpha^2 < 1$ where $\alpha > 0$ is such that

$$\|u\| \leqslant \alpha \|\nabla u\| \text{ and } \|\nabla u\| \leqslant \alpha \|\Delta u\|, \forall u \in H^2 \cap H_0^1$$

$$f' \geqslant -c, c > 0. \tag{32}$$

Let

$$\tilde{u}_0 \in H_0^1 \cap H^2 \quad \text{and} \quad \tilde{u}_1 \in L_2. \tag{33}$$

For each $T > 0$ there exists under the hypotheses (30)–(33) a unique solution $u(t)$ of the initial and boundary value problem (1)–(3). The main result for the problem of asymptotic behavior for $t \to \infty$ can be expressed as follows:

THEOREM 3. *Let (30) and (33) hold, c satisfying the following conditions*:

$$0 < c < 1/2 \tag{34}$$

$c \, \alpha^2 < 1 \quad (\alpha \text{ being as in (31))},$

then the solution $u(t)$, which exists for all $t \geqslant 0$, decays exponentially to 0 as $t \to \infty$ in the following sense: there exists an $M > 0$ and $\gamma > 0$ such that:

$$\|u_t(t)\|^2 + \|\Delta u(t)\|^2 \leqslant M e^{-\gamma t} \text{ for all } t \geqslant 0.$$

For the proof of Theorem 3 we write $f(u) = g(u) - cu$, then $g'(u) \geqslant 0$ and hence f satisfies (31) and the solution $u(t)$ exists for all $t \geqslant 0$. Then, we show that

$$\|u_t(t)\|^2 + \|\Delta u(t)\|^2 \leqslant M, \quad \text{for all } t \geqslant 0, \tag{35}$$

where M is a constant.

Finally taking the inner product of (1) first with $u_t(t) e^{\gamma t}$ and then with $-\beta \Delta u(t) e^{\gamma t}$ and integrating with repect to the time variable from 0 to t we find, taking (35) into account, a suitable choice for β and γ (c satisfying to (34)), which implies that there exists an $M > 0$ such that:

$$\|u_t(t)\|^2 + \|\Delta u(t)\|^2 \leqslant M e^{-\gamma t} \quad \text{for all} \quad t \geqslant 0.$$

4. Problem 3

$$u_{tt} - \Delta u + |u_t|^\alpha \operatorname{sgn} u_t = 0$$

Problem 3 (posed to the first named author D.D.A. by R.A. Schapery) is associated with nonhomogeneous mixed conditions (2′) and initial conditions (3). Here we shall make the following assumptions:

$$\tilde{u}_0 \in H^1(\Omega),\ \tilde{u}_1 \in L^2(\Omega),\ \Omega\ =\ (0,\ 1)$$

$$g(t),\ g'(t) \in L^2(0,\ T);\ g(0)\ \text{exists}. \tag{36}$$

The method used here is the Galerkin method associated with a Volterra nonlinear integral inequation namely:

$$\sigma_n(t) \leqslant D_1(t)\ +\ D_2(t) \int_0^t \sigma_n^\alpha(\Theta)\,\mathrm{d}\Theta \tag{37}$$

where

$$\sigma_n(t)\ =\ \|u^{(n)}(t)\|_v^2\ +\ \|u_t^{(n)}(t)\|^2\ +\ \int_0^t |u_t^{(n)}(0,\ \Theta)|^2\mathrm{d}\Theta,$$

$$V\ =\ \{v \in H^1(\Omega) | v(1)\ =\ 0\},$$

$u^{(n)}(t)$ being the approximate solution on a "basis" $(v_1,\ \ldots,\ v_n)$ of V and $D_1(t)$ and $D_2(t)$ two positive continuous functions.

Equation (37) leads to the required a priori estimates. To pass to the limit we shall require the following Lemma [3]:

LEMMA 1. *Let u be the solution of the following problem*:

$$u_{tt}\ -\ \Delta u\ +\ X\ =\ 0$$

$$u_x(0,\ t)\ =\ g(t),\ u(1,\ t)\ =\ 0$$

$$u(0)\ =\ \tilde{u}_0,\ u_t(0)\ =\ \tilde{u}_1$$

$$u \in L^\infty(0,\ T;\ V)\ and\ u_t,\ X \in L^\infty(0,\ T;\ L_2),$$

then we have

$$1/2a(\tilde{u}_0,\ \tilde{u}_0)\ +\ 1/2\|\tilde{u}_1\|^2\ -\ \int_0^s \langle X,\ u_t\rangle\mathrm{d}\Theta\ -\ \int_0^s g(\Theta)u_t(0,\ \Theta)\,\mathrm{d}\Theta$$

$$\leqslant\ 1/2a(u(s),\ u(s))\ +\ 1/2\|u_t(s)\|^2 \quad \text{a.e. } s \in [0,\ T]$$

with $a(u,\ v)\ =\ \int_0^1 (\partial u/\partial x \cdot \partial v/\partial x)\mathrm{d}x.$

Finally we have the following result:

THEOREM 4. *For each $T > 0$, the initial and boundary value problem (1) (2') (3) under the assumptions (36) has a unique solution $u \in L^\infty(0,\ T;\ V)$ such that $u_t \in L^\infty(0,\ T;\ L_2)$.*

Note that for the equation of the form

$$u_{tt} - \Delta u = \varepsilon f(t, u, u_t), (x, t) \in (0, 1) \times (0, T) \tag{38}$$

associated to the conditions (2), (3) a theorem on local existence is proved [9] and an asymptotic expansion of order 2 in $\varepsilon(\varepsilon > 0)$ is obtained, for ε sufficiently small and $f \in C^1([0, \infty] \times \mathbb{R}^2)$. The linear recursive schemes developed in §3 enable us to use a perturbation technique based on the ideas of best uniform approximation by polynomials, which considerably extends the classical Tau Method of Lanczos. This Tau Method has been developed by Ortiz [6] and computational procedures for the numerical treatment for partial differential equations with polynomial coefficients have been discussed by Ortiz and Samara [8]. Ortiz and Pham Ngoc Dinh [7] have discussed the numerical solution of a semi-linear hyperbolic problem of the following type:

$$u_{tt} - \Delta u = f(t, u) \tag{39}$$

and sufficient conditions for the quadratic convergence are given in this paper.

Acknowledgements

The first named author (D.D.A.) wishes to acknowledge the support given him by the University of Orleans, France, in the course of this investigation.

References

1. Dang Dinh Ang and A. Pham Ngoc Dinh, *SIAM Journal on Mathematical Analysis* 19 (1988) 337–347.
2. Dang Dinh Ang and A. Pham Ngoc Dinh, On the strongly damped wave equation: $u_{tt} - \Delta u - \Delta u_t + f(u) = 0$, 19 (1988).
3. Dang Ding Anh and A. Pham Ngoc Dinh, *Nonlinear Analysis* T.M.A. 12 (1988) 581–592.
4. J.M. Greenberg, R.C. MacCamy and V. Mizel, *Journal of Mathematics and Mechanics* 17 (1968) 707–728.
5. J.L. Lions, Quelques méthodes de résolution des problèmes aux limites non linéaires, Dunod, Gauthier-Villars (1969).
6. E.L. Ortiz, *SIAM Journal on Numerical Analysis* 6 (1969) 480–492.
7. E.L. Ortiz and A. Pham Ngoc Dinh, *SIAM Journal on Mathematical Analysis* 18 (1987) 452–464.
8. E.L. Ortiz and H. Samara, *Computers and Mathematics with Applications* 10 (1984) 5–13.
9. A. Pham Ngoc Dinh and Nguyen Thanh Long, *Demonstratio Mathematica* 19 (1986) 45–63.
10. G.F. Webb, *Canadian Journal of Mathematics* 32 (1980) 631–643.

Résumé. En suivant la méthode de Gaberkin, les auteurs ont étudié le problème de l'existence, de l'unicité et du comportement asymptotique lorsque $t \to \infty$ des solutions des équations d'état des ondes visco-élastiques, pour diverses conditions initiales et aux limites de $U(x, t)$. On analyse les cas auxquels correspondent des valeurs positives (modèle non linéaire de Voigt) ou nulle du paramètre, ce dernier cas étant représentatif du mouvement d'un barreau élastique linéaire dans un milieu visqueux non linéaire, monyennant l'adoption de diverses conditions aux limites.

International Journal of Fracture 39: 45–62 (1989)
© Kluwer Academic Publishers, Dordrecht

Application of fracture mechanics and half-cycle theory to the prediction of fatigue life of aerospace structural components

WILLIAM L. KO

NASA Ames Research Center Dryden Flight Research Facility, Edwards, CA 93523, USA

Received 20 September 1987; accepted 1 April 1988

Abstract. The service life of aircraft structural components undergoing random stress cycling was analyzed by the application of fracture mechanics. The initial crack sizes at the critical stress points for the fatigue crack growth analysis were established through proof load tests. The fatigue crack growth rates for random stress cycles were calculated using the half-cycle method. A new equation was developed for calculating the number of remaining flights for the structural components. The number of remaining flights predicted by the new equation is much lower than that predicted by the conventional equation.

This report describes the application of fracture mechanics and the half-cycle method to calculate the number of remaining flights for aircraft structural components.

Symbols

A	crack location parameter
a	crack length of edge crack, one-half the crack length of through-thickness crack, or depth of surface crack
a_l	crack size (length or depth) after the lth flight
a_c^o	limit crack length (or depth) associated with the operational peak load
a_c^P	initial fictitious crack length (or depth) established by the proof load tests
C	material constant in Walker crack growth rate equation
c	half-length of surface crack
E	complete elliptic function of the second kind
F_l	number of remaining flights after the lth flight
$\bar{F}_l$	number of remaining flights after the lth flight predicted from the newly developed equation of remaining flights
f	operational peak stress factor ($f < 1$)
i	integer associated with half-cycles, or critical stress points
j	integer associated with half-cycles
K_1	mode I stress intensity factor
K_{1c}	mode I critical stress intensity factor
K_{max}	mode I stress intensity factor associated with σ_{max}, $AM_K\sigma_{max}\sqrt{\pi a/Q}$
k	modulus of elliptic function
l	integer associated with flights
M_K	flaw magnification factor
m	Walker exponent associated with K_{max}
N	number of stress cycles available for operations
N_l	number of stress cycles used during the lth flight
n	Walker exponent associated with stress ratio R
Q	surface flaw shape and plasticity factor
R	stress ratio, $\sigma_{min}/\sigma_{max}$
t	thickness of plate
V_A	front hook vertical load
V_{BL}	left rear hook vertical load

V_{BR} right rear hook vertical load

Δa_l amount of crack growth during the lth flight

δa_i crack growth increment resulting from one cycle of constant amplitude stress cycling under loading magnitude of ΔK_i and R_i

ΔK mode I stress intensity amplitude, $AM_K(\sigma_{max} - \sigma_{min})\sqrt{\pi a/Q}$

σ_∞ remote uniaxial tensile stress

σ_∞^O uniaxial tensile stress associated with the peak operational load level

σ_∞^P proof tensile stress induced by the proof loads

σ_i tensile stress at critical stress point i

σ_i^* peak value of σ_i induced by the proof loads

σ_{max} maximum stress of the stress cycle

σ_{min} minimum stress of the stress cycle

σ_U tensile strength

σ_Y yield stress

τ_U ultimate shear strength

ϕ angular coordinate for semielliptical surface crack

1. Introduction

Aircraft structural components commonly contain flaws, defects, or anomalies of various shapes; these either are inherent in the basic material or are introduced during the manufacturing and assembly processes. A large percentage of service cracks found in aircraft structures are initiated from crack nucleation sites such as tool marks, manufacturing defects, and surface microinclusions [1]. Under the combined influences of environment and service loading, these flaws may grow to reach catastrophic sizes, resulting in serious reduction of service life or complete loss of the aircraft. Thus, to a great extent, the integrity of the aircraft structure is dependent upon the safe and controlled growth of cracks as well as the achievement of residual strength in their presence.

The operational life (service life) of aircraft structural components is affected by the magnitude and cumulative effects of external loads coupled with any detrimental environmental action. The presence of moisture, chemicals, suspended contaminants, and naturally occurring elements such as rain, dust, and seacoast atmosphere can cause deterioration in structural strength due to premature cracking and acceleration of subcritical crack growth [2, 3].

As aircraft structures begin to age (that is, as flight hours accumulate), existing subcritical cracks or new cracks can grow in some high-stress points of the structural components. The usual approach is to inspect the structures periodically at certain intervals. However, even after inspection there may be some undetected cracks in a structure. To ensure that the structure still has integrity for future flights, proof load tests are usually conducted on the ground. The purpose of proof load tests is to load a structure to certain proof load conditions (slightly lower than the design limit load conditions) to test its integrity. If there should exist undetected cracks in the structural component that are larger than the critical crack sizes, that structural component will surely fail during the proof load tests and will be replaced. This process can reduce the chance of catastrophic accidents during flight. If all structural components survive the proof load tests, then fracture mechanics can be applied to estimate fatigue life (number of remaining flights) for each critical structural component by using the initial crack size established for each structural component during the proof load tests and then using the stress cycles (obtained through strain gauge measurements) for each structural component during the first flight after the proof load tests.

2. Theory

Fracture mechanics

The top part of Fig. 1 shows the most common types of cracks: through-thickness crack, surface crack, and edge crack. According to fracture mechanics, the stress intensity factor K_I for the mode I deformation (tension mode) associated with any type of crack can be expressed as

$$K_I = A M_K \sigma_\infty \sqrt{\frac{\pi a}{Q}}, \tag{1}$$

where A is the crack location parameter ($A = 1$ for the through-thickness crack, $A = 1.12$ for both the surface and the edge cracks (see Fig. 1)); M_K is the flaw magnification factor (for a very shallow crack $M_K = 1$; as the depth of the crack reaches the back surface of the plate, $M_K = 1.6$ (see Fig. 1)); σ_∞ is the remote uniaxial tensile stress; a is one-half the crack length

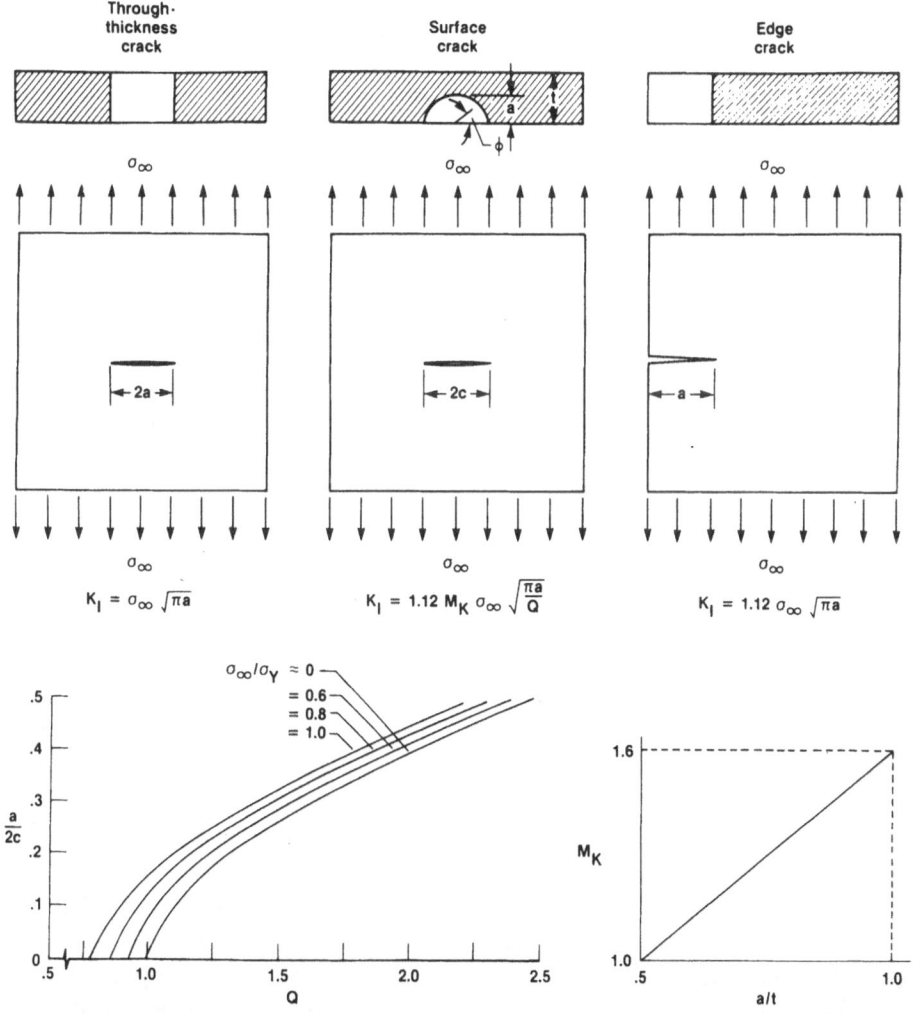

Fig. 1. Three types of cracks and the plots of surface flaw shape factor and flaw magnification factor.

for the through-thickness crack, or the length of the edge crack, or the depth of the surface crack (see Fig. 1); and Q is the surface flaw shape and plasticity factor given by

$$Q = [E(k)]^2 - 0.212(\sigma_\infty/\sigma_Y)^2, \tag{2}$$

where σ_Y is the yield stress and $E(k)$ the complete elliptic function of the second kind, defined as

$$E(k) = \int_0^{\pi/2} \sqrt{1 - k^2 \sin^2\phi} \; d\phi, \tag{3}$$

where ϕ is the angular coordinate for a semielliptical surface crack, defined by Fig. 1, and the modulus k of the elliptic function is defined by

$$k = \sqrt{1 - (a/c)^2}, \tag{4}$$

where c is the half-length of the surface crack (see top center of Fig. 1). The bottom part of Fig. 1 shows the plots of $a/2c$ as functions of Q for different values of σ_∞/σ_Y and the plot of M_k as a function of a/t, where t is the plate thickness.

Proof load tests

The purpose of proof load tests is to load the entire aircraft structure (or its components) to certain proof load levels to test structural integrity and to establish initial fictitious crack sizes associated with critical structural components for fatigue life analysis. The proof load levels are usually slightly lower than the design limit load conditions associated with different maneuvers. If there exists in a certain structural component a previously undetected crack that is larger than the critical crack size associated with the proof load, that component will certainly fail during the proof load tests and will be replaced. Thus, a catastrophic accident during flight can be avoided. If the entire structure survives the proof load tests, then the critical stress point of the structural component has been subjected to a proof tensile stress σ_∞^P induced by the proof loads. If K_{Ic} denotes the critical stress intensity factor (or material fracture toughness) of the structural component material, the maximum crack length a_c^P the structural component can carry under the proof loads without failure (or rapid crack extension) may be calculated from (1) by setting $K_I = K_{Ic}$, and $\sigma_\infty = \sigma_\infty^P$. In reality, there may not be any cracks that developed during proof tests; however, it is assumed that a fictitious crack of length a_c^P has been created at the critical stress point of the structural component during the proof load tests. During actual operations, the structural component will be subjected to much lower stress levels than the proof stress level σ_∞^P. If σ_∞^O is defined as the peak operational stress level (highest peak of the stress cycles), then according to (1), the structural component can carry a fictitious crack of size a_c^O, which is much larger than a_c^P. The value a_c^O thus determined is considered to be the limit crack size toward which the initial crack a_c^P is allowed to grow after repeated operations. The crack size difference, $a_c^O - a_c^P$, is then the crack size increase permitted for the structural component in repeated operations. The left-hand plot in Fig. 2 shows crack length, a, as a function of normalized stress σ_∞/σ_U, where σ_U is the tensile strength of the material. It is seen that the lower the

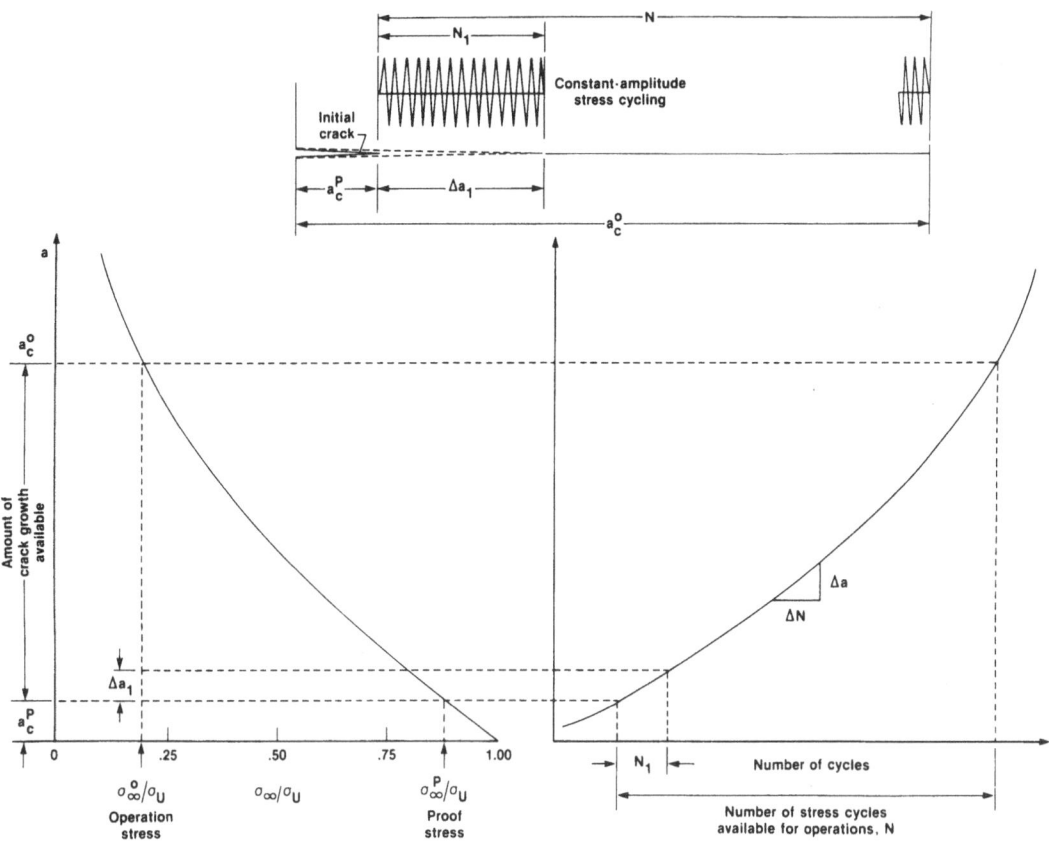

Fig. 2. Crack length as functions of applied stress and number of cycles.

operational stress level, the larger the limit crack size available for the structural component. The right-hand plot in Fig. 2 shows crack length, *a*, as a function of number of constant stress cycles *N*.

Remaining flights

If the structural component is cycled under constant stress amplitude (in idealized case for the purpose of discussion) for N_1 cycles during the first flight with the associated crack growth of Δa_1, then the number of remaining flights F_1 before the limit crack size is reached may be estimated from the following conventional equation (see Fig. 2):

$$F_1 = \frac{a_c^O - a_c^P}{\Delta a_1} = \frac{N}{N_1}. \tag{5}$$

Equation (5) is based on the assumption that the amount of crack growth per flight for all subsequent flights will be equal to Δa_1 of the first flight. In reality, the amount of crack growth per flight will steadily increase with the number of flights accumulated, and the actual number of remaining flights will be less than the value F_1 predicted by (5) if the number of remaining flights is large (that is, $F_1 \gg 1$). As discussed in the following section, for a relatively low range of F_1, (5) may give a reasonably accurate prediction of the number of remaining flights. The amount of crack growth Δa_1 in (5) may be calculated from the following Walker equation for fatigue crack growth rate under constant amplitude stress cyclings:

$$\frac{da}{dN} = C(K_{max})^m(1 - R)^n = C(\Delta K)^m(1 - R)^{n-m}, \tag{6}$$

where C, m, and n are material constants and K_{max}, ΔK, and R are, respectively, maximum stress intensity factor, stress intensity amplitude, and stress ratio, given by

$$K_{max} = AM_K\sigma_{max}\sqrt{\frac{\pi a}{Q}} \tag{7}$$

$$\Delta K = AM_K(\sigma_{max} - \sigma_{min})\sqrt{\frac{\pi a}{Q}} \tag{8}$$

$$R = \sigma_{min}/\sigma_{max}, \tag{9}$$

where σ_{max} and σ_{min} are, respectively, the maximum and the minimum stresses of the constant amplitude stress cycles.

In reality, the stress cycles encountered during operations at the critical stress points of the structural component are not constant amplitude stress cycles. To apply (6) to (9) to variable amplitude stress cycles, different methods must be developed. In this report a half-cycle method [7–10] is used in the calculation of the fatigue crack growth rate for variable amplitude stress cycles.

Half cycle theory

The top part of Fig. 3 shows an example of random stress cycles (variable amplitude loading history). The stress history curve is the combination of a series of both increasing and decreasing load half-cycles of different loading magnitude (ΔK, R), as shown in the lower part of Fig. 4. The half-cycle theory (or half-wave theory) [7–10] states that the damage (or crack growth) caused by each half-cycle of either increasing or decreasing load is assumed to equal one-half the damage caused by a complete cycle of the same loading magnitude (ΔK, R). This means that the damage caused by the complete cycle could be equally divided between the two phases of increasing and decreasing loads. The loading sequence thus can be resolved into half-cycle groups of increasing and decreasing loads (see lower part of Fig. 3). Each half-cycle (either increasing or decreasing load) can then be considered as a half-cycle of the constant amplitude cyclings under the same loading magnitude (ΔK, R) and can be computed separately in time sequence to estimate the corresponding damage. The half-cycle theory thus permits accurate evaluation of the load spectrum from a recorded load time history. If a_l ($l = 1, 2, 3, \dots$) is the final crack length after N_l ($l = 1, 2, 3, \dots$) random stress cyclings in the *l*th flight, then according to the half-cycle theory, a_l may be calculated from

$$a_l = a_{l-1} + \sum_{i=1}^{2N_l} \frac{\delta a_i}{2} = a_{l-1} + \Delta a_l \tag{10}$$

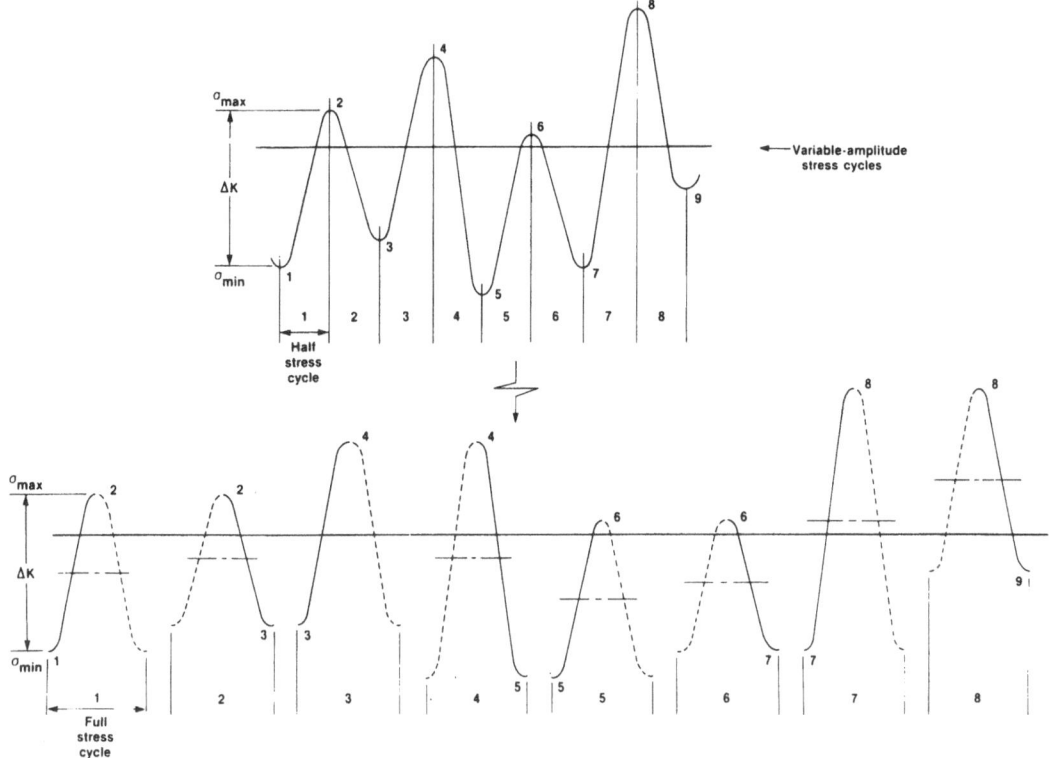

Fig. 3. Resolution of random stress cycles into half stress cycles of different stress ranges.

with

$$\Delta a_l = \sum_{i=1}^{2N_l} \frac{\delta a_i}{2} = a_l - a_{l-1} \quad \text{and} \quad a_{1-1} = a_c^P, \tag{11}$$

where $\delta a_i/2$ is the crack growth increment induced by the ith half-cycle under the loading magnitude ΔK_i and R_i; $\delta a_i/2$ is assumed to equal the crack growth increment induced by a half-cycle of the constant amplitude stress cycle fatigue test under the same loading magnitude (ΔK_i, R_i). Thus, by using (6) to (9), $\delta a_i/2$ may be calculated from

$$\frac{\delta a_i}{2} = \frac{1}{2}\left[\frac{da}{dN}\right]_i = \frac{C}{2}[(K_{max})_i]^m(1 - R_i)^n = \frac{C}{2}(\Delta K_i)^m((1 - R_i)^{n-m}. \tag{12}$$

The crack length, a, associated with ΔK_i (see (8)) will be the summation of the initial crack length and all the crack growth increments created by all the previous half-cycles:

$$a = a_c^P + \sum_{j=1}^{i-1} \frac{\delta a_j}{2} \quad (i \geqslant 2). \tag{13}$$

Similar to the case of constant amplitude stress cycling (see (5)), the number of remaining flights F_l ($l = 1, 2, 3, \ldots$) after the lth flight of random stress cycling may be calculated

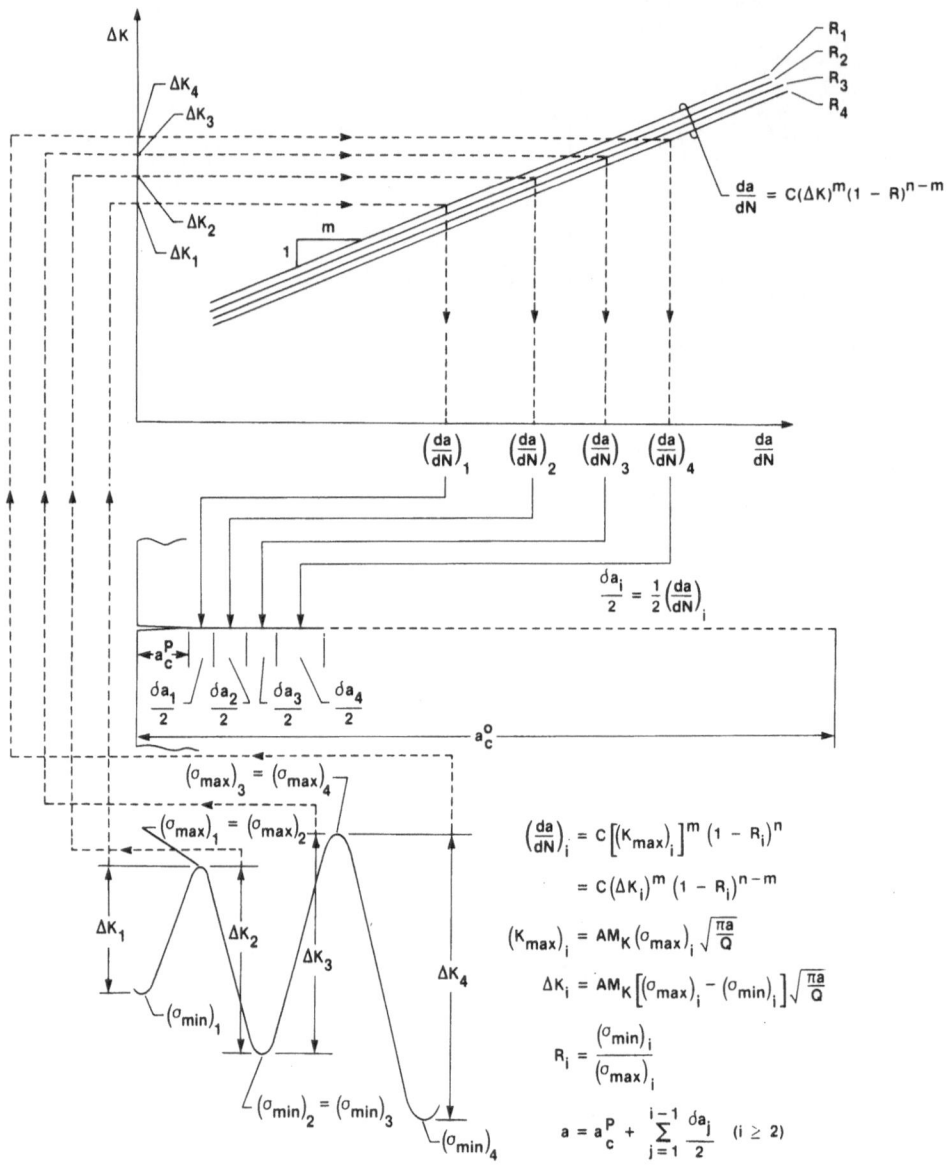

Fig. 4. Graphical evaluation of crack increments for random stress cycles using the half-cycle method.

from

$$F_l = \frac{a_c^o - a_{l-1}}{\Delta a_l} = \frac{a_c^o - a_{l-1}}{a_l - a_{l-1}} = \frac{a_c^o - a_{l-1}}{\displaystyle\sum_{i=1}^{2N_l} \frac{\delta a_i}{2}}. \tag{14}$$

Figure 4 graphically illustrates how to evaluate the crack growth increment $\delta a_i/2$ associated with the ith half-cycle of the random stress cycling by using the plots of ΔK_i as functions of $\mathrm{d}a/\mathrm{d}N$ for different values of R.

As mentioned previously, (5) and (14) both may give relatively accurate values for F_l when the number of remaining flights is relatively low. When predicting large F_l, (5) and (14) must be modified because Δa_l increases with the flights accumulated.

New equation for remaining flights

For the case of constant amplitude stress cycling, the amount of crack growth Δa_l ($l = 1, 2, 3, \ldots$) during the lth flight (N_l cyclings) may be obtained by integrating (6):

$$\Delta a_l = C\left(AM_K\sigma_{\max}\sqrt{\frac{\pi}{Q}}\right)^m (1 - R)^n N_l(a_{l-1})^{m/2} \quad (l = 1, 2, 3, \ldots), \tag{15}$$

where (7) was used.

For simplicity, if $\sigma_{\max}$, R, and N_l remain the same for all the flights, then the following crack growth rates can be established by using (15), assuming $\Delta a_l/a_{l-1} \ll 1$:

$$\frac{\Delta a_2}{\Delta a_1} = \left(\frac{a_1}{a_c^P}\right)^{m/2} = \left(\frac{a_c^P + \Delta a_1}{a_c^P}\right)^{m/2} = 1 + \frac{m}{2}\frac{\Delta a_1}{a_c^P} + \cdots \tag{16}$$

$$\frac{\Delta a_3}{\Delta a_1} = \left(\frac{a_2}{a_c^P}\right)^{m/2} = \left(\frac{a_c^P + \Delta a_1 + \Delta a_2}{a_c^P}\right)^{m/2} = 1 + 2\left(\frac{m}{2}\frac{\Delta a_1}{a_c^P}\right) + \cdots \tag{17}$$

$$\frac{\Delta a_4}{\Delta a_1} = \left(\frac{a_3}{a_c^P}\right)^{m/2} = \left(\frac{a_c^P + \Delta a_1 + \Delta a_2 + \Delta a_3}{a_c^P}\right)^{m/2} = 1 + 3\left(\frac{m}{2}\frac{\Delta a_1}{a_c^P}\right) + \cdots \tag{18}$$

$$\frac{\Delta a_l}{\Delta a_1} = \left(\frac{a_{l-1}}{a_c^P}\right)^{m/2} = \left(\frac{a_c^P + \Delta a_1 + \Delta a_2 + \Delta a_3 + \cdots + \Delta a_{l-1}}{a_c^P}\right)^{m/2}$$

$$= 1 + (l - 1)\left(\frac{m}{2}\frac{\Delta a_1}{a_c^P}\right) + \cdots \tag{19}$$

If the available crack size, $a_c^O - a_c^P$, can allow $\bar{F}_1$ number of remaining flights, then

$$\frac{a_c^O - a_c^P}{\Delta a_1} = \overbrace{\frac{\Delta a_1 + \Delta a_2 + \Delta a_3 + \cdots + \Delta a_l + \cdots + \Delta a_{F_1}}{\Delta a_1}}^{F_1 \text{ terms}}, \tag{20}$$

where the left-hand side is F_1, the number of remaining flights predicted by assuming that $\Delta a_1 = \Delta a_2 = \Delta a_3 = \cdots = \Delta a_l = \cdots = \Delta a_{F_1}$ (see (14) for $l = 1$). Substituting (14) and (16) to (19) into Eqn. (20),

$$F_1 = \overbrace{(1 + 1 + 1 + \cdots + 1)}^{F_1 \text{ terms}} + \frac{m}{2}\frac{\Delta a_1}{a_c^P}\overbrace{[1 + 2 + 3 + \cdots + (\bar{F}_1 - 1)]}^{(F_1 - 1) \text{ terms}} \tag{21}$$

or

$$F_1 = \bar{F}_1 + \frac{m}{2} \frac{\Delta a_1}{a_c^P} \frac{(\bar{F}_1 - 1)[(\bar{F}_1 - 1) + 1]}{2}. \tag{22}$$

Equation (22) may be rearranged into the form

$$\frac{m}{4} \frac{\Delta a_1}{a_c^P} \bar{F}_1^2 + \left(1 - \frac{m\Delta a_1}{4a_c^P}\right) \bar{F}_1 - F_1 = 0. \tag{23}$$

Solving for $\bar{F}_1$ and neglecting higher order terms, there results

$$\bar{F}_1 \approx \frac{2a_c^P}{m\Delta a_1} \left(\sqrt{1 + \frac{m\Delta a_1}{a_c^P} F_1} - 1 \right), \tag{24}$$

which gives the relationship between $\bar{F}_1$ and F_1.

If the prediction of the number of remaining flights is based on the crack growth Δa_l that occurred during the lth flight, (24) takes the form

$$\bar{F}_l \approx \frac{2a_{l-1}}{m\Delta a_l} \left(\sqrt{1 + \frac{m\Delta a_l}{a_{l-1}} F_l} - 1 \right). \tag{25}$$

Equations (24) and (25) both apply to the case of constant amplitude stress cycling. However, they may be used for the case of variable amplitude stress cycling without introducing significant error. Figure 5 shows the plot of (24) (that is, $\bar{F}_1$ as a function of F_1) for $m = 3.6$

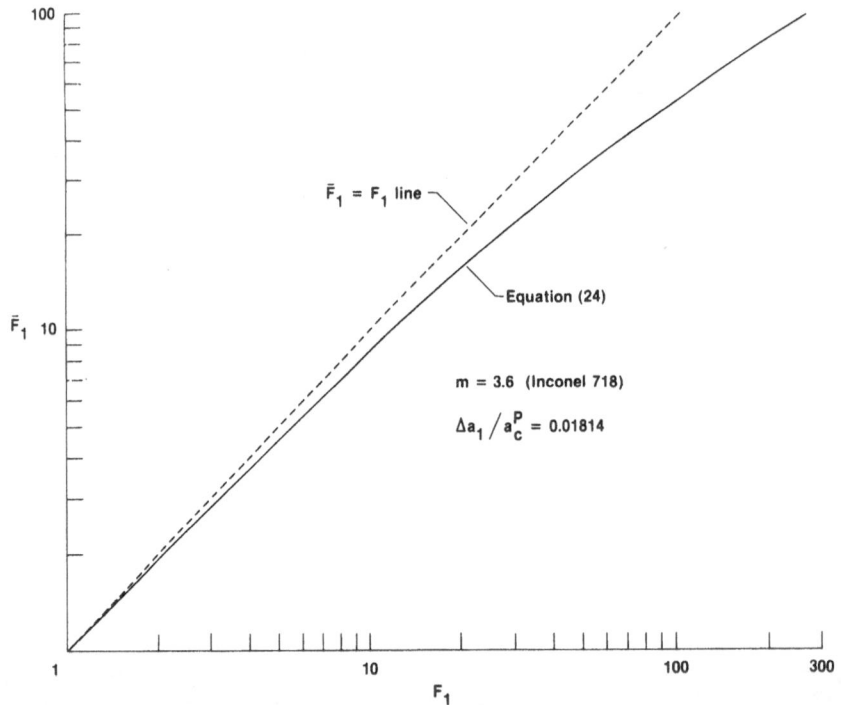

Fig. 5. Number of remaining flights predicted from Ko equation of remaining flights.

(Inconel 718 alloy) and $\Delta a_1/a_c^P = 0.01814$ for the example described in the following sections. Some typical values of $\bar{F}_1$ and F_1 are compared in the following tabulation:

F_1:	1	10	50	100	150	200	500	1000
$\bar{F}_1$:	1	9	33	53	70	84	147	218

The ratio $\bar{F}_1/F_1$ is 0.53 for $F_1 = 100$ and decreases to 0.218 for $F_1 = 1000$. Thus, (5) certainly overpredicts the number of remaining flights, and safety factors ranging from 2 to 4 must be used depending on the range of the value of F_1.

3. Example problem

For this report, the example problem chosen for the fatigue analysis using the half-cycle method is the severe fatigue problem of the three hooks of the NASA B-52-008 carrier aircraft pylon used to carry the space shuttle solid rocket booster drop test vehicle (SRB/DTV) shown in Fig. 6. The 49,000 lb SRB/DTV was attached to the pylon with one front hook and two rear hooks (see Fig. 6). The SRB/DTV was carried up to a high altitude and released to test the performance of the solid rocket booster main parachute. The shapes of the front hook and the two rear hooks are shown in Figs. 7 and 8, respectively. The front hook is made of Inconel 718 alloy and the two rear hooks of AMAX MP35N alloy (AMAX Speciality Corporation). Table I shows material properties of the two alloys. Because of the great weight of the SRB/DTV, the three hooks had serious fatigue life problems. Fracture mechanics and the half-cycle method can be applied to predict the service life (number of remaining flights) of the three hooks that carried the SRB/DTV. Reference [10] presents the detailed fatigue analysis of the NASA B-52 aircraft pylon major components.

Critical stress points

Before conducting fatigue crack growth analysis, the locations of the critical stress points for the three hooks had to be determined. This was done by performing NASTRAN finite element stress analysis of the three hooks [11]. The critical stress point of each hook is

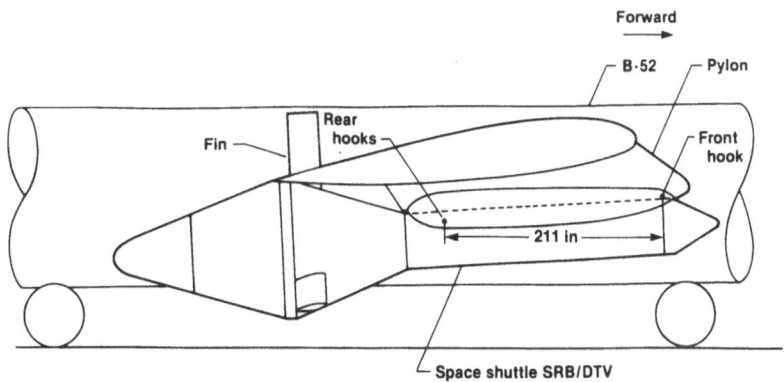

Fig. 6. Geometry of space shuttle solid rocket booster drop test vehicle (SRB/DTV) attached to B-52 pylon (view looking inboard from right side).

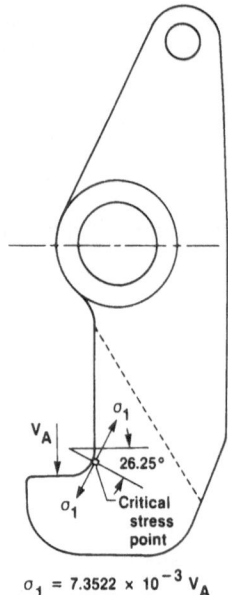

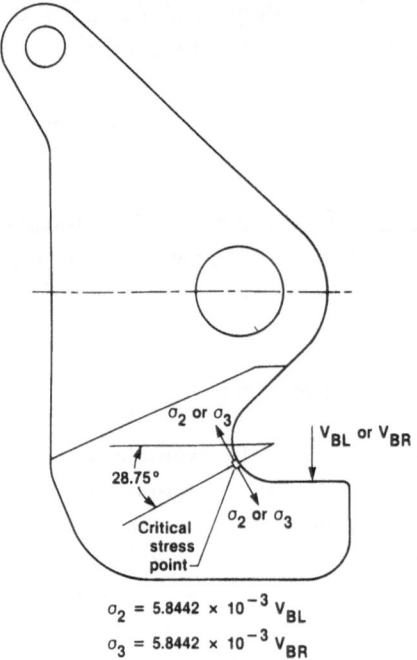

Fig. 7. Front hook and the location of critical stress point σ_1.

Fig. 8. Rear hook (left or right) and the location of critical stress point (σ_2 or σ_3).

Table I. Material properties of front and rear hooks

Stress point	Part name	Material	σ_U (ksi)	σ_Y (ksi)	$\tau_U{}^a$ (ksi)	C $\left(\dfrac{\text{in}}{\text{cycle}} (\text{ksi}\sqrt{\text{in}})^{-m} \right)$	m	n	K_{1_c} (ksi $\sqrt{\text{in}}$)
σ_1	Front hook	Inconel 718 alloy	175	145	135	9.220×10^{-12}	3.6	2.16	125
σ_2	New left rear hook	AMAX MP35N alloy	250	235	141	2.944×10^{-11}	3.24	1.69	124
σ_3	New right rear hook	AMAX MP35N alloy	250	235	141	2.944×10^{-11}	3.24	1.69	124

^{a} τ_U is the ultimate sheer strength.

located at the inner circular boundary of the hook. Figures 7 and 8 show the exact locations of the critical stress points of the three hooks. Through the NASTRAN stress analysis, the relationships between the stress at the critical stress point and the hook loads were established as

$$\sigma_1 = 7.3522 \times 10^{-3} V_A \quad \text{for front hook,} \tag{26}$$

$$\sigma_2 = 5.8442 \times 10^{-3} V_{BL} \quad \text{for left rear hook} \tag{27}$$

$$\sigma_3 = 5.8442 \times 10^{-3} V_{BR} \quad \text{for right rear hook,} \tag{28}$$

where σ_1, σ_2, and σ_3 are, respectively, the stresses (in kips, or 10^3 lb, per square inch (ksi)) at the critical stress points of the front hook, left rear hook, and right rear hook: and V_A, V_{BL}, and V_{BR} are the corresponding hook vertical loads in pounds. During proof load tests and during flight, V_A, V_{BL}, and V_{BR} were measured by means of strain gages installed near the critical stress points of the hooks [10 and 11]. Equations (26) to (28) were used to generate stress cycles for the fatigue crack growth analysis using the strain-gage-measured values of V_A, V_{BL}, and V_{BR}.

Initial and operational crack sizes

During the proof load tests, the three hooks were loaded to their respective peak proof loads to establish the initial crack size a_c^P for each critical stress point. The peak proof stresses σ_i^* ($i = 1, 2, 3$) at the critical stress points induced by the peak proof hook loads may be calculated from (26) to (28). The proof crack size a_c^P at the critical point of each hook established by the proof load tests may be calculated from (1) by setting $K_1 = K_{1_c}$:

$$a_c^P = \frac{Q}{\pi} \left(\frac{K_{1_c}}{A M_K \sigma_i^*} \right)^2, \tag{29}$$

where $A = 1.12$ for the surface crack, $M_K = 1$ (which was obtained from the lower right plot of Fig. 1 for $a/t \ll 1$ because the depth of the crack is very small compared with the depth of the hook), and the value of Q, the surface flaw shape and plasticity factor, will be determined as follows.

The surface crack is assumed to be semielliptical in shape with an aspect ratio of $a/2c = 1/4$. (This value is based on the observation of surface cracks of the fractured old rear hooks.) Taking $a/2c = 1/4$ and $\sigma_\infty/\sigma_Y = \sigma_i^*/\sigma_Y = 1$ (because the growth of plastic zones around the critical stress points was neglected, the values of σ_i^* calculated for the three hooks slightly exceeded the corresponding yield stresses σ_Y), the curve for $\sigma_\infty/\sigma_Y = 1$ in the lower left plots of Fig. 1 gives $Q = 1.25$.

If the peak value of the stress cycles during operation (or flight) is $f\sigma_i^*$, where f is the operational peak stress factor ($f < 1$), then the operational limit crack size a_c^O may be calculated from

$$a_c^O = \frac{Q}{\pi} \left(\frac{K_{1_c}}{A M_K f \sigma_i^*} \right)^2 = \frac{a_c^P}{f^2}. \tag{30}$$

Part of Table II shows the peak proof hook loads, the peak proof stresses σ_i^*, and the proof crack sizes a_c^P at the critical stress points. Note that for all three hooks, σ_i^* exceeded the failure stresses σ_U of the hook materials (see Table I), yet the three hooks did not fail during the proof load tests. The reason is that (25) to (27), established by the NASTRAN analysis, are for a purely elastic case without consideration of plastic deformations. In reality the plastic zone can develop around the critical stress point, and therefore the hook can actually carry a greater load than the brittle failure load. In the present fatigue crack growth analysis, only the elastic case is considered.

Table II. Crack sizes and remaining flights

Hook	Proof hook load (lb)	Proof stress at critical stress point, σ_i (ksi)	Proof crack size a_c^p (in)	Crack growth Δa_1 (in)	Operational peak stress factor f	Operational crack size $a_c^o = \dfrac{a_c^p}{f^2}$ (in)	Remaining flights F_1	Remaining flights $\bar{F}_1$
Front hook	$V_A = 36{,}520$	268.502	0.0990	0.0118295	0.5450	0.333	128	63
Left rear hook	$V_{BL} = 44{,}110$	257.786	0.0734	0.0005887	0.5946	0.2076	227	91
Right rear hook	$V_{BR} = 44{,}230$	258.487	0.0730	0.000705	0.5986	0.2037	169	75

Load spectra

To perform the fatigue crack growth analysis, the load spectra (stress cycles) for the three critical points of the hooks must be obtained first. Using the strain-gage-measured values of the three hook loads, V_A, V_{BL}, and V_{BR}, during the first test flight, the three stresses σ_i ($i = 1, 2, 3$) may be calculated by using (26) to (28). Figures 9 to 11 [10] show portions of the loading histories (load spectra) calculated for the critical stress points of the three hooks during a takeoff run. Those load spectra were obtained by filtering the original load spectra down to 5 Hz to eliminate the small-amplitude high-frequency stress cycles that are

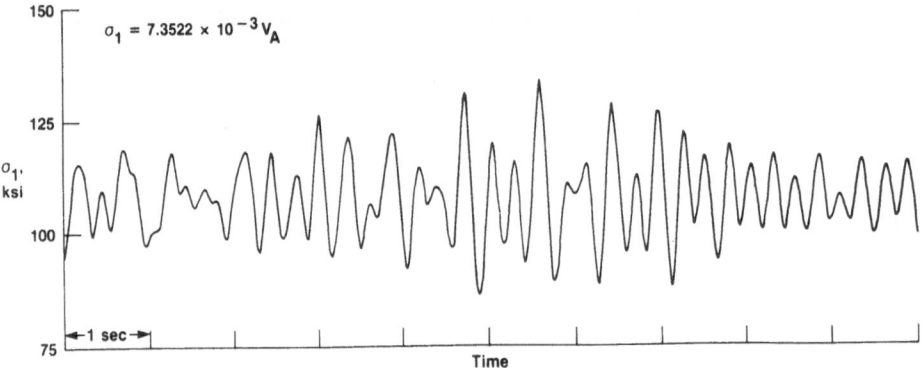

Fig. 9. Stress cycles for critical stress point σ_1 at front hook.

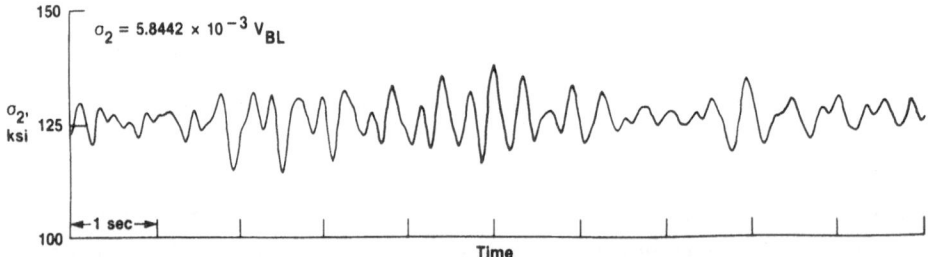

Fig. 10. Stress cycles for critical stress point σ_2 at left rear hook.

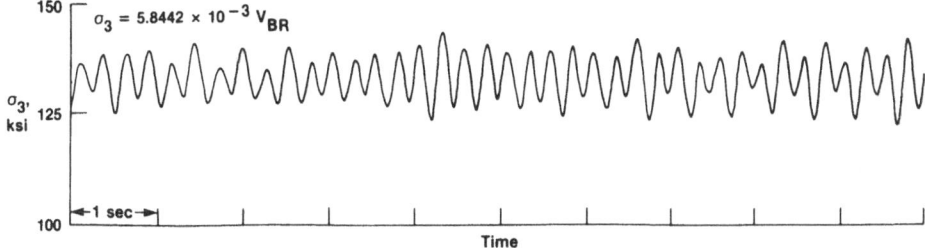

Fig. 11. Stress cycles for critical stress point σ_3 at right rear hook.

considered unimportant in the present fatigue life analysis. Notice that the load spectra for all three critical stress points exhibit a high degree of random cycling.

Calculations of crack growth

To apply the half-cycle method, the load spectra (see Figs. 9 to 11) were first resolved into a series of half-cycles of increasing and decreasing loads of different loading magnitude (ΔK, R) (see Fig. 3). The crack growth increments $\delta a_i/2$ per half-cycle were calculated from (12) in time sequence and summed (using (10)) to give the total amount of crack growth per flight for each critical stress point. Finally, (14) and (25) were used to calculate the number of remaining flights associated with each critical stress point. It must be emphasized that in using the half-cycle method, every half-cycle of different stress amplitude in the load spectrum is calculated, and thus the half-cycle method can give an accurate evaluation of the fatigue crack growth as compared with, say, the exceedence count method (for which some of the stress peaks lying below the mean line could be missed).

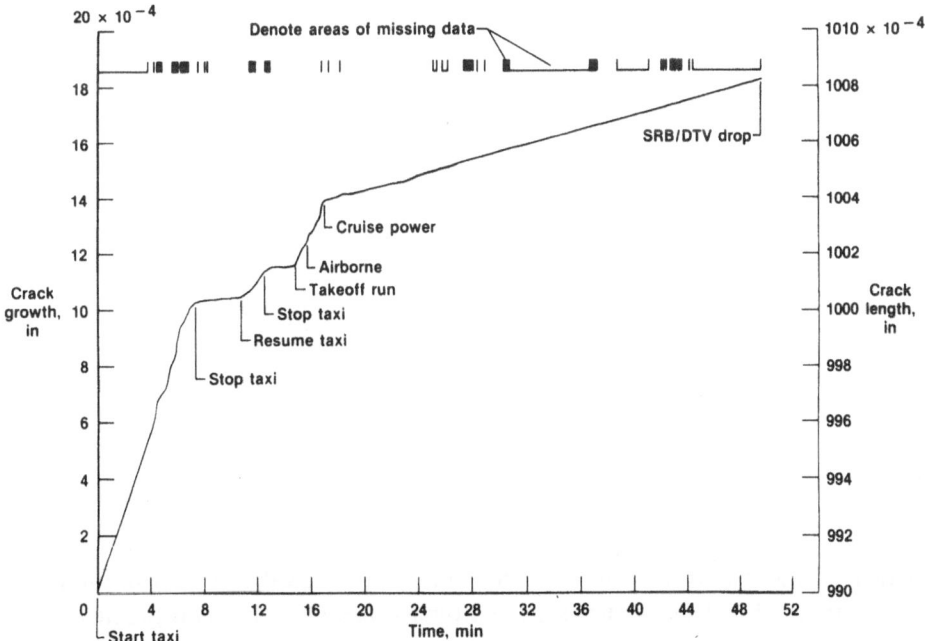

Fig. 12. Crack growth curve for stress point 1 (σ_1), flight 1.

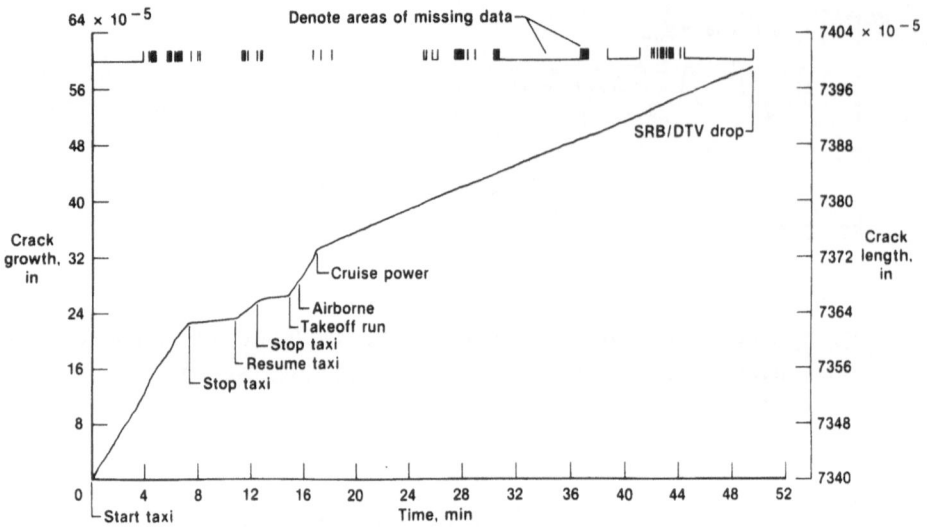

Fig. 13. Crack growth curve for stress point 2 (σ_2), flight 1.

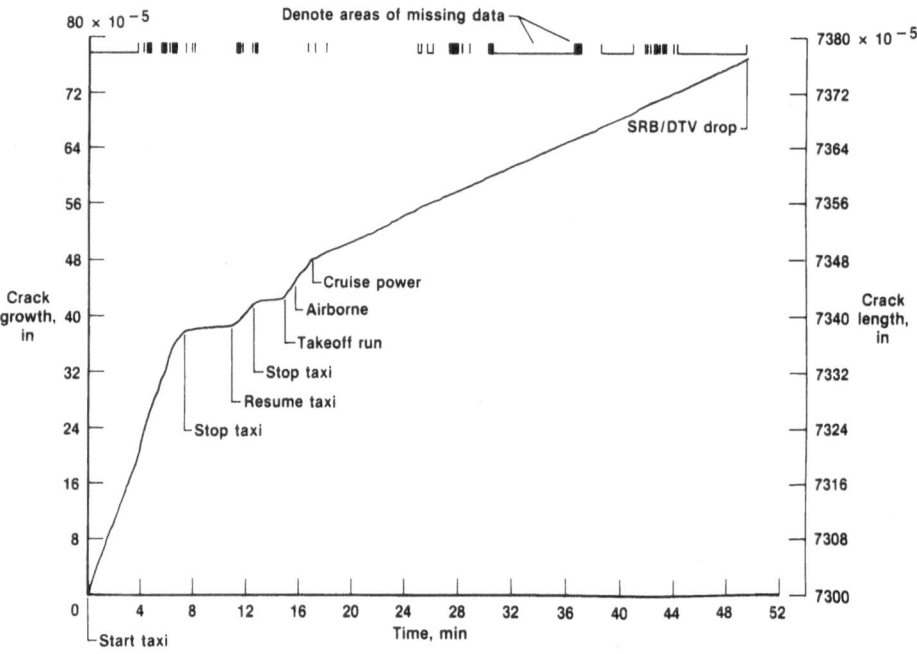

Fig. 14. Crack growth curve for stress point 3 (σ_3), flight 1.

Results

Figures 12 to 14, taken from [10], show the fatigue crack growth curves calculated for the three critical stress points for the first test flight; the maneuver transition points are indicated. Note that for the three hooks, the fatigue crack growth rate is greatest during the initial stage of taxiing and the takeoff run and becomes very low during cruising because of relatively low amplitude stress cyclings. Table II lists the amount of crack growth Δa_1,

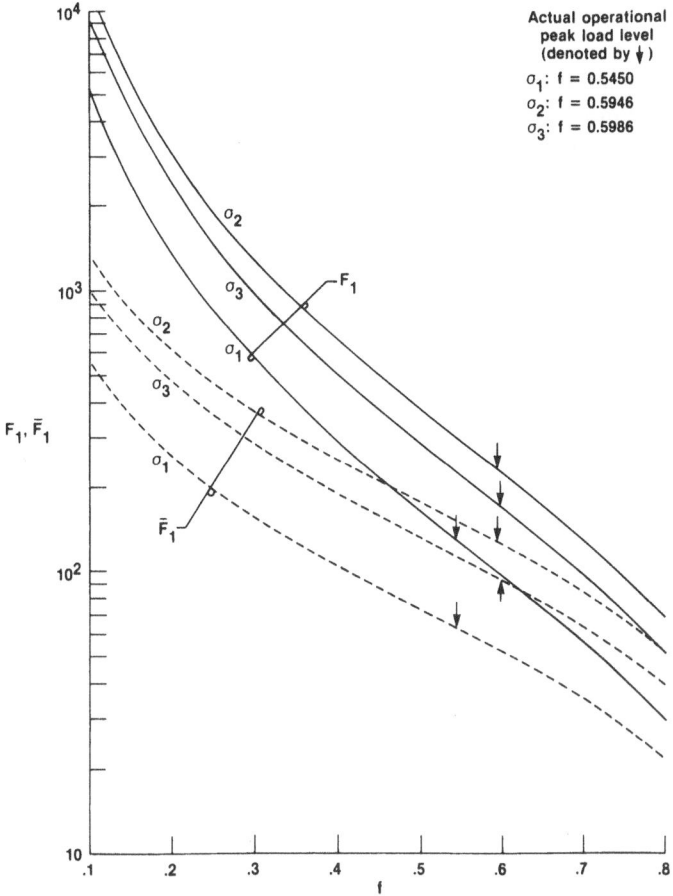

Fig. 15. Number of remaining flights as functions of operational peak stress factor.

operational peak stress factor f, operational crack size a_c^o, and remaining flights F_1 and $\bar{F}_1$ calculated respectively from (14) to (24). Figure 15 shows the plots of F_1 and $\bar{F}_1$ as functions of f for the three hooks. Note that the over-prediction of remaining flights based on F_1 (14) becomes more pronounced as the number of remaining flights increases. The arrows in Fig. 15 indicate the actual operational peak load levels. At these load levels, safety factors must be in the range of 2 to 2.5 if F_1 is used instead of $\bar{F}_1$.

4. Conclusion

Fracture mechanics and the half-cycle method were applied to the service life analysis of aircraft structural components. The initial crack sizes at the critical stress points of the structural components were determined by using proof load tests. The random stress cycle fatigue crack growth rates were calculated using the half cycle method. A new equation was developed for calculating the number of remaining flights for the structural components. The newly developed equation predicted the number of remaining flights more accurately (a much lower number) than did the conventional equation (which is based on the assumption that the amount of crack growth per flight remains constant).

Acknowledgements

The author gratefully acknowledges contributions by William W. Totton and Jules M. Ficke of Synernet Corporation in setting up computer programs for flight data reductions and crack growth computations.

References

1. D.R. Donaldson and W.F. Anderson, in *Proceedings of the Crack Propagation Symposium*, Vol. 2, Cranfield, (September 1961).
2. J.P. Butler, The Material Selection and Structural Development Process for Aircraft Structural Integrity Under Fatigue Conditions. AFFDL-TR-70-144, Conference Proceedings, Air Force Conference, Miami, Florida, December 15–16, 1969.
3. W. Astley and J. Scott, Engineering Practice to Avoid Stress Corrosion Cracking. (The Environment Encountered in Aircraft Service.) AGARD Conference Proceedings No. 53.
4. R.G. Forman, V.E. Kearney and R.M. Engle, *Journal of Basic Engineering, Transactions ASME*, D89 (1967) 459.
5. E.K. Walker, The Effect of Stress Ratio During Crack Propagation and Fatigue for 2024-T3 and 7075-T6 Aluminum, Effects of Environments and Complex Load History on Fatigue Life. ASTM STP 462 (1970) 1.
6. J. Schijve, *Engineering Fracture Mechanics*, 11 (1979) 167–221.
7. W.L. Starkey and S.M. Marco, *Transactions ASME* (1957) 1329–1336.
8. C. Incarbone, in *Fatigue of Aircraft Structures*, Proceedings of the Symposium held in Paris, 16th–18th May 1961, W. Barrois and E.L. Ripley (eds), Pergamon Press, New York (1963) 209–217.
9. S.H. Smith, in *Acoustic Fatigue in Aerospace Structures*, Syracuse University Press (1965) 331.
10. William L. Ko, Alan L. Carter, William W. Totton and Jules M. Ficke, Application of Fracture Mechanics and the Half-Cycle Method to the Prediction of Fatigue Life of B-52 Aircraft Pylon Components. NASA TM-88277 (1988).
11. William L. Ko and Lawrence S. Schuster, Stress Analyses of B-52 Pylon Hooks. NASA TM-84924 (1985).

Résumé. On a analysé en application de la mécanique de rupture la vie en service de composants de structures aéronautiques soumis à une sollicitation cyclique aléatoire. A l'aide d'essais à la charge d'épreuve, on a établi les dimensions initiales d'une fissure aux points critiques de sollicitation, pour analyser sa croissance par fatigue.

Les vitesses de croissance d'une fissure de fatigue dans le cas des charges aléatoires ont été calculées par la méthode des demi-cycles. On a mis au point une nouvelle expression pour calculer le nombre d'heures de vol résiduelles pour des éléments de structure. S'il résulte de cette nouvelle expression, ce nombre est beaucoup plus faible que celui prédit par les équations de vie résiduelle conventionnelles.

International Journal of Fracture 39: 63–77 (1989)
© Kluwer Academic Publishers, Dordrecht

Fatigue crack growth by crack tip cyclic plastic deformation: the unzipping model

H.W. LIU

Department of Mechanical and Aerospace Engineering, Syracuse University, Syracuse, NY 13244, USA

Received 1 October 1987; accepted in revised form 1 April 1988

Abstract. A fatigue crack may propagate by the mechanism of crack tip cyclic plastic deformation, by the mechanism of fracture of brittle particles and embrittled grain boundaries, or, often, by a combination of both.

Neumann and Vehoff have made in situ observations of alternate shear decohesions on two intersecting conjugate slip bands at a crack tip as the basic mechanism of fatigue crack growth. It is a mechanism by plastic deformation.

A *micro-mechanism* based finite element model is made to simulate the unzipping process of the crack tip shear decohesion mechanism. The calculated crack growth rates by the finite element model agree very well with the measured rates in the intermediate ΔK region of a number of materials.

1. Introduction

When a cracked metallic solid is stressed, plastic deformation will take place in the crack tip region. As the region near a crack tip is deformed plastically, the crack tip opens up and the crack grows. A crack may grow by the process of crack tip plastic deformation alone.

Crack tip opening displacement (CTOD) has been analyzed by a number of researchers [1–8]. The results of the finite element calculations [5, 6, 7, 8] have shown that

$$\text{CTOD} = d_n \frac{J}{\sigma_0}, \tag{1}$$

where J is the J-integral, σ_0 is the flow stress or yield stress of the solid, and d_n is a constant for a given material. d_n is strongly dependent on the strain hardening exponent. It is lower for a material that strain hardens strongly and vice versa. d_n is weakly dependent on the ratio σ_0/E. E is the Young's modulus. The value of d_n is close to 0.5.

The geometric relation between fatigue crack growth rate and crack tip opening displacement was pointed out by McClintock [9]:

$$\frac{da}{dN} = \tfrac{1}{2}\text{CTOD}. \tag{2}$$

However, it will be shown that only part of the total opening displacement at a crack tip contributes to fatigue crack growth and the rest of the CTOD only blunts the crack tip without any contribution to crack growth.

In this paper, several of the proposed physical processes of fatigue crack growth will be reviewed, the alternate shear slip fatigue crack growth mechanism (the unzipping mechanism) will be modelled using the finite element method, an expression for fatigue crack growth will be derived based on the alternate shear mechanism, and the calculated fatigue crack growth rate will be compared with the measured rate. It is a *micro-mechanism based quantitative study*.

2. Fatigue crack growth by crack tip shear decohesion

Orowan [10] proposed the alternate shear ruptures on two sets of intersecting slip planes as the mechanism for ductile fractures. The basic mechanism of alternate shear ruptures was adapted to fatigue crack tip blunting, striation formation, and fatigue crack growth by McEvily [11], Laird [12], Pelloux [13], Tomkins [14], Neumann [15, 16], and Vehoff and Neumann [17].

Laird [12] described a shear slip process that causes crack tip blunting and striation formation. Figures 1(a), (b) and (c) show the crack tip profile during the loading half of a fatigue cycle, and Figs. 1(d), (e) and (f) show the profile during the unloading half of the cycle. Figure 1(a) shows the crack tip configuration at zero load. Figure 1(b) shows the double-notch configuration at the crack tip. A notch is formed by the forward shear slip in one shear band during the loading half of a fatigue cycle and the reverse shear slip in a neighboring shear band during the unloading half of the fatigue cycle. The forward and the reverse shear slips also form the fatigue striation.

A crack tip is blunted by the plastic deformation within the entire crack tip plastic zone. According to Laird, the entire crack tip blunting contributes to crack growth. However, it will be shown that only the shear slip at the very tip of a crack will cause a crack to grow; the shear slip in the region behind a crack tip will cause blunting but not crack growth.

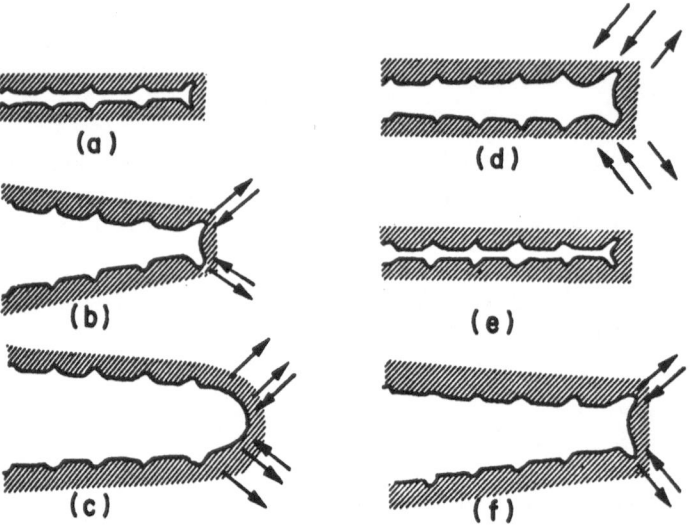

Fig. 1. The plastic blunting process of fatigue crack propagation in the stage II mode: (a) zero load, (b) small tensile load, (c) maximum tensile load, (d) small compressive load, (e) maximum compressive load, and (f) small tensile load. The double arrowheads in (c) and (d) signify the greater width of slip bands at the crack in these stages of the process. The stress axis is vertical. [12] (Reprinted by the permission of the American Society for Testing and Materials from copyright material.)

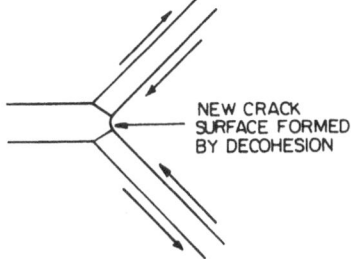

Fig. 2. The double shear-band model proposed by Tomkins. [14]

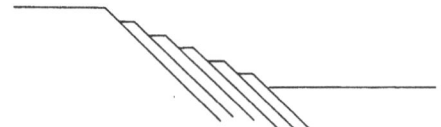

Fig. 3. Shear slip bands in a single crystal as shown schematically. [18]

The double-notch configuration can be caused by the double-shear band model as shown in Fig. 2. Based on the model, Tomkins [14] calculated the crack tip opening displacement and fatigue crack growth rate. According to the model, plastic deformation, i.e., shear slip, takes place only along these two shear bands; elsewhere, the deformation is elastic. In other words, the entire plastic deformation at a crack tip region takes place along these two shear bands, and the plastic deformation in the entire crack tip region contributes to crack growth. However, the constrained crack tip plastic deformation in the plane-strain state is "diffused," and only a small part of the crack tip plastic deformation contributes to crack growth. The double-shear band model overestimates the fatigue crack growth rate.

Plastic deformation in a crystalline solid is not continuous, and the shear slip in crystalline solids takes place on specific planes (slip planes) and in specific directions (slip directions). The slip bands in a single crystal do not have to coincide with the plane of the maximum shear stress. Shear slip in a single crystal is often concentrated in parallel slip bands separated by "elastic"slabs with very little or no sign of plastic deformation, as illustrated in Fig. 3 [18].

This discrete slip process at a crack tip was observed in situ by Neumann [15, 16] and Vehoff and Neumann [17] as the mechanism of fatigue crack growth in copper and Fe-2.4% Si single crystals. Figure 4 shows the slip bands at the crack tip in a Fe-2.4% Si single crystal when the applied stress is increased during a fatigue cycle. The applied tensile stress was in the $\langle 1\,1\,0 \rangle$ direction, the root of the notch was parallel to $\langle 1\,1\,1 \rangle$, and the crack propagated along $\langle 1\,1\,2 \rangle$. In Figs. 4(a), (b) and (c) the slip band A was active. The right hand side of the V-notch crack moved to the lower right, along the slip band A. Subsequently, during the same loading cycle, slip band *B* was active as shown in Figs. 4(d), (e) and (f) and the left hand side of the V-notch moved to the lower left. The apex of the V-notch remained "sharp".

Figure 5 shows the profile of a copper specimen during a sequence of two fatigue cycles. The newly formed crack surfaces (i.e., the V-notch) and the crack flank due to earlier growth are clearly shown. The region ahead of the crack tip was "elastic" and showed very little, if any, sign of active slip.

66   *H.W. Liu*

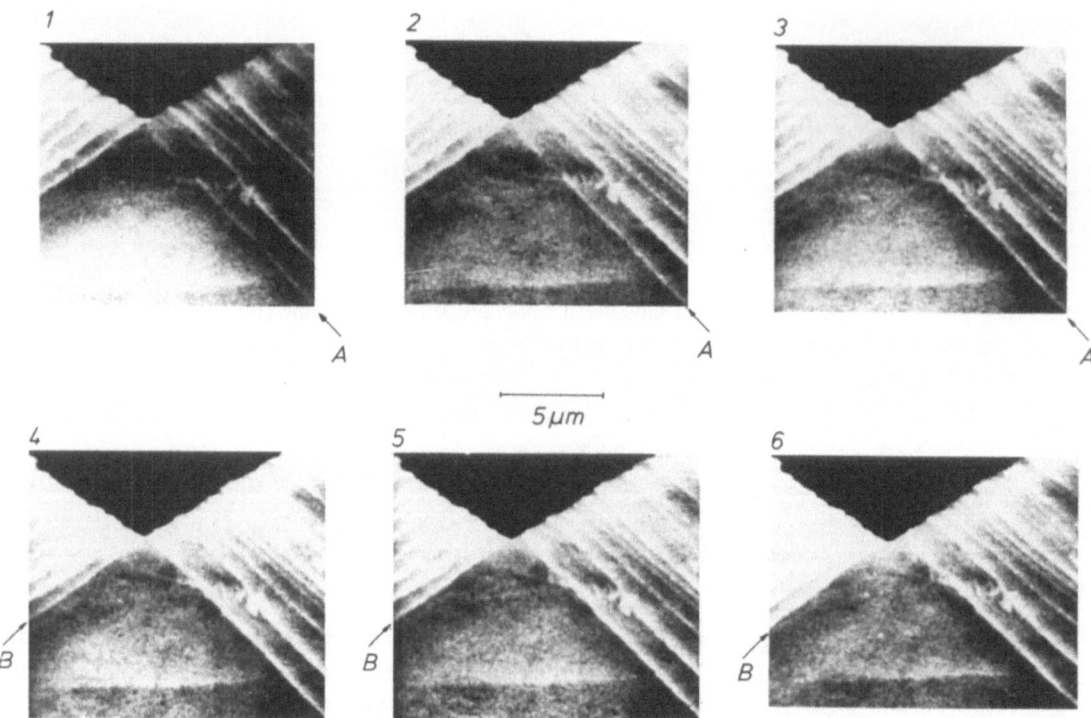

Fig. 4. Fatigue crack growth due to the alternate activation of two slip bands in a Fe-2.4% Si single crystal. In 4(a), (b), and (c), the slip band A is active and the right-hand side of the V-shaped crack tip moves to the lower right. In 4(d), (e), and (f), the slip band B is active and the left-hand side moves to the lower left. [17] (By the courtesy of Vehoff and Neumann with the permission of Pergamon Press.)

The pictures in Figs. 4 and 5 by Vehoff and Neumann [17] clearly show that during a fatigue cycle, a number of shear decohesions took place alternately on two sets of intersecting conjugate slip planes. During each increment of shear decohesion, the crack tip moved ahead by a small amount, δa. Fatigue crack growth per cycle is the sum of all of the small δa's during the cycle.

Each of the slip bands emanated from the crack tip and spanned across the entire ligament of a specimen. The small specimens were in the state of general-yielding, and the slip bands were in the "non-constrained" plastic deformation state. The slip band at the very tip of a notch experiences the highest force on its dislocations, and the shear decohesion process was localized in this band. This also contributes to the formation of the non-constrained slip line fields in Figs. 4 and 5.

The slip in a single crystal is highly anisotropic, and it follows a specific slip system. The stress–strain relation of a polycrystal averaged over a large number of randomly oriented grains can be considered isotropic. The shear deformation in an isotropic polycrystal takes place along the planes of maximum shear stresses. The shear deformation in a polycrystal is also "localized" to narrow bands along the planes of maximum shear stress. We'll call them shear bands in contrast to slip bands in single crystals. A shear band is the result of extensive shear slip in many slip bands in the numerous grains along the shear band in a polycrystal. The material between the neighboring shear bands is deformed plastically but at a much lower level.

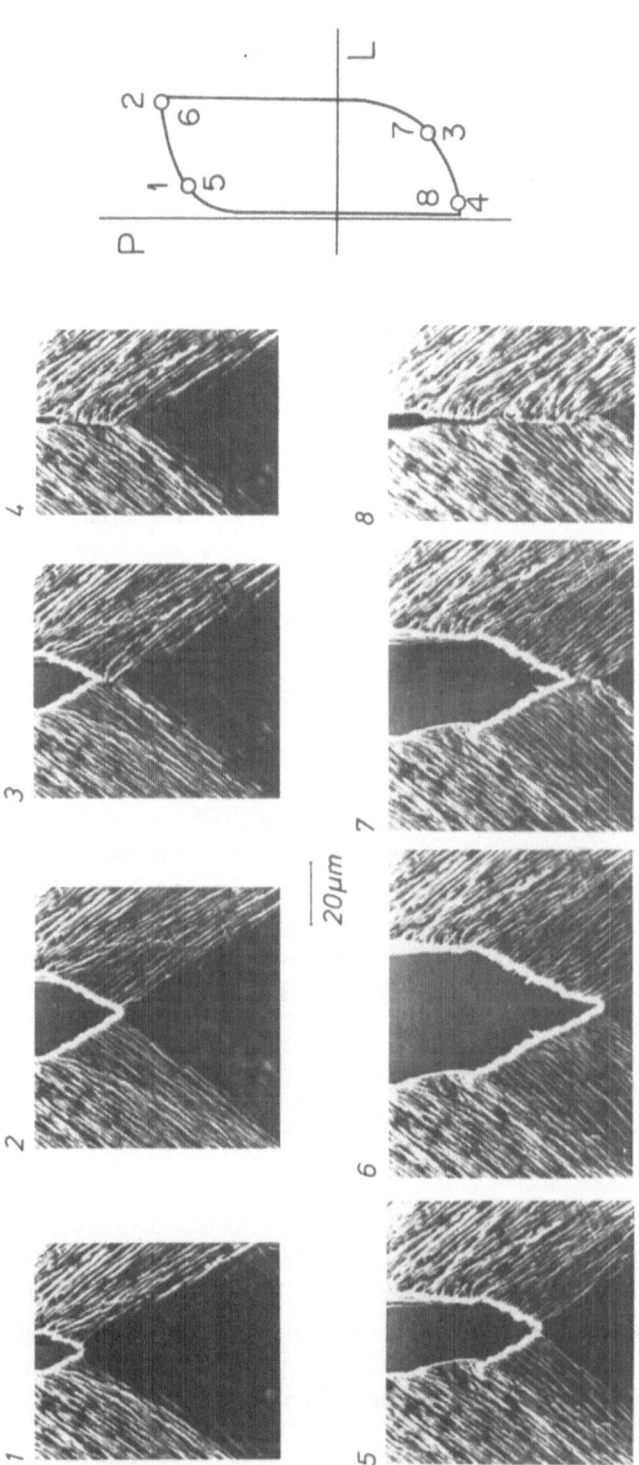

Fig. 5. (a) A sequence of two load cycles in a single crystal copper as observed on the specimen's side face during straining inside a SEM. Crack propagation from the top to the bottm. (b) Schematic force elongation curve with the subpicture numbers of Fig. 5(a) showing the approximate positions of the subpictures in the load cycle. (By the courtesy of Vehoff and Neumann and with the permission of the Pergamon Press.)

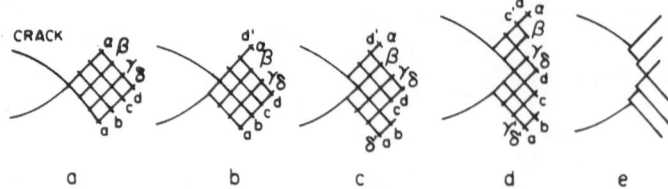

Fig. 6. Schematic figures of plane-strain shear-lines and the unzipping process of shear decohesion, crack tip opening, and crack growth during the rise of the fatigue load. [21]

The plastic deformation in a plane-strain crack tip plastic zone in small scale yielding is "constrained". A constrained plane-strain crack tip shear line field is shown schematically in Fig. 6. Figure 6(a) shows the intersecting shear bands ahead of a crack tip before shear decohesion takes place. As K increases, the shear bands are activated, the crack tip shear decohesion takes place, the crack tip opens up, and the crack grows. The crack grows by a shear deformation mechanism. Figure 6(b) shows the crack tip configuration after shear decohesion takes place on shear band "α", during the first increment δK_1. The region above the shear band moves to the upper right as shown by the tangential displacement from d to d'.

During the second increment δK_2, the shear decohesion process is shifted to shear band "b". During the increment δK_2, shear decohesion may continue to take place on shear band "α". However, after the crack tip moves ahead of shear band "α", the shear decohesion on "α" will blunt the crack tip but will not contribute to crack growth. During the increment δK_2, only the shear decohesion on shear band "b" at the very tip of the crack contributes to crack growth. Figure 6(c) shows the crack tip configuration after shear decohesions on both shear bands "α" and "b". Figure 6(d) shows the crack tip location after the shear decohesions on the conjugate shear bands "β" and "c".

During the crack tip decohesion process, the "slab" between two neighboring shear bands moves away from the crack tip like the tooth of a zipper during the process of unzipping.

In Fig. 6(d), the continued shear decohesions on shear bands "α" and "a" are not shown once they move behind the crack tip. Figure 6(e) shows the crack tip configuration caused by the shear decohesions on all of the shear bands including the decohesions after the shear bands have moved behind the crack tip.

3. The modelling of crack tip opening displacement and the unzipping model of fatigue crack growth

The crack tip opening displacement (CTOD) has been investigated extensively during the recent past [1–8]. Figure 7(a) shows the undeformed sharp crack tip. A^+ and A^- are the same point at the crack tip above and below the crack line. Figure 7(b) shows a blunted crack tip as depicted by a classic continuum model. The crack tip is blunted but the point at the crack tip remains intact, i.e., A^+ and A^- remain a single point. The physical process of shear decohesion is not explicitly taken into account. The calculated CTOD is only implicitly related to crack growth by the shear decohesion mechanism.

The CTOD of the strip yielding model was first studied by Dugdale [1] and subsequently by Goodyear and Fields [2], Bilby, Cottrell, and Swinden [3], Rice [4] and many others. The

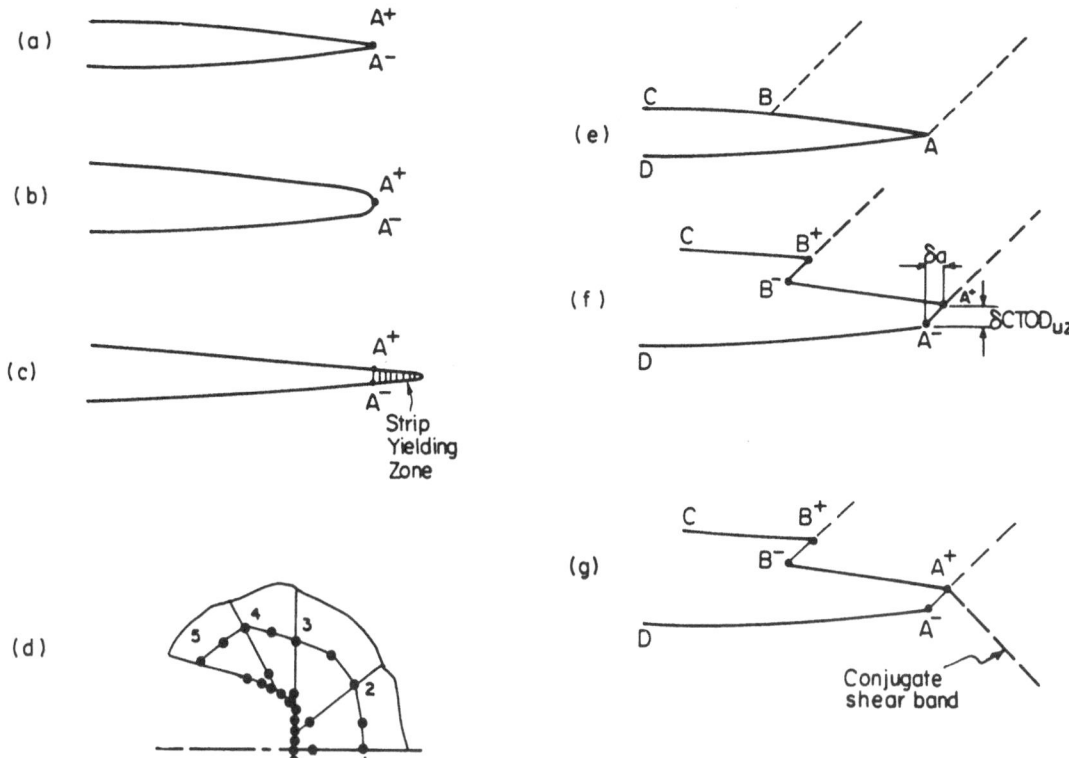

Fig. 7. The modelling of crack tip deformation. (a) A sharp crack tip before deformation, (b) a classic continuum model, the crack tip is intact, (c) Dugdale's strip yielding model with crack tip opening and blunting but not crack growth, (d) a FEM model with crack tip branching and crack tip blunting caused by the deformation of the crack tip elements, (e) two shear bands: one at the crack tip and one is behind the crack tip, (f) shear band deformation during a rising load, (g) the unzipping model showing that only the decohesion of the shear band at the crack tip contributes to crack growth.

Dugdale model is shown in Fig. 7(c). All of the crack tip plastic deformation is restricted to a narrow strip as shown by the shaded area in Fig. 7(c). According to the model, all of the crack tip plastic deformation contributes to the crack tip opening displacement. The crack tip is branched into two distinct points A^+ and A^-. However, the crack tip does not move ahead, and the physical process of crack growth by shear decohesion is not explicitly taken into account. The calculated CTOD gives the general blunting of the crack tip region, but not the crack growth. Again, the crack growth by the shear decohesion mechanism is only implicitly related to the calculated CTOD.

More recently, a special crack tip triangular element has been used to calculate crack tip opening displacement. The triangular element is formed by the collapsing of one side of a four-sided element. In the undeformed state, the nodal points along the collapsed side are condensed into a single point. All of the condensed nodes of the crack tip triangular elements meet at the crack tip. During the deformation, the condensed nodes are allowed to branch out. Crack tip blunting and crack tip opening displacement can be modelled by the branching of the condensed crack tip nodes as illustrated in Fig. 7(d). The crack tip opening displacement thus calculated includes the opening displacements of all of the collapsed crack tip elements. The single point at the crack tip is branched into many points along the blunted crack front. The branching of the crack tip point in a crack tip element is caused by the intense plastic deformation within the element. It is not related directly to the shear slip along an entire shear band as observed by Neumann [15, 16] and Vehoff and Neumann [17].

Figures 7(e), (f) and (g) show the geometric relations between crack tip blunting, opening, and advancing as envisioned by the unzipping model. Two shear bands, A and B, are shown in Fig. 7(e). Shear decohesion may take place on both of these two shear bands as shown in Fig. 7(f). When the shear decohesion takes place on the conjugate shear band at the crack tip (see Fig. 7(g)), only the shear decohesion at shear band A (i.e., the shear band at the very tip of the crack) contributes to crack growth. The shear decohesion at shear band B behind the crack tip will blunt the crack tip but does not cause the crack to grow. The shear decohesion at shear band A at the crack tip is related to the increments δa and δCTOD_{uz} as illustrated in Fig. 7(f).

Shear slip, i.e., the cause of plastic deformation in a crystal, takes place on slip planes. The displacements across a slip plane are discontinuous. The discontinuous shear slip at a crack tip causes shear decohesion and crack growth. A realistic model should take the discontinuous displacements across the decohesion plane into consideration.

Plane-strain crack tip plastic zone is not limited to a narrow strip. Rather it is diffused. Plastic deformation and shear decohesion take place within the entire diffused plastic zone. However, crack growth is caused by the decohesion of the shear band at the very tip of a crack. The shear decohesion elsewhere does not contribute to crack growth. Therefore, it is expected that the crack growth due to the unzipping process will be much less than that given by (2) with the CTOD obtained by the strip yield model or by the classic continuum models. The CTOD calculated by the Dugdale model or the classic continuum model gives the general contour of the crack tip, but these two models are not explicitly related to the actual physical process of fatigue crack growth.

A finite element model is made to simulate the unzipping fatigue crack growth process realistically and accurately. It incorporates the discrete shear decohesion process at a crack tip and it distinguishes the decohesion which contributes to crack growth from that of crack blunting.

4. Finite element modelling of the unzipping fatigue crack growth mechanism

As observed by Vehoff and Neumann [18], crack growth during a fatigue cycle is the sum of many small crack increments, δa's, caused by many consecutive alternate shear decohesions on two sets of intersecting conjugate shear bands. We want to calculate the small shear decohesion increment and its corresponding increments δCTOD_{uz} and δa as the applied stress intensity factor is increased by a small amount, δK. The fatigue crack growth rate per cycle is the sum of all of the δa's when the applied stress intensity factor increases from K_{min} to K_{max}.

The crack tip opening displacement as calculated by the Dugdale model and the classic continuum models is related to the applied K. The crack tip opening displacement caused by shear decohesion as depicted by the unzipping model is much smaller than that given by these classic models, but the functional relationship between these variables should remain the same, i.e.,

$$\text{CTOD}_{uz} = C(1 - v^2) \frac{K^2}{E\sigma_Y} \quad \text{(plane strain)}, \tag{3}$$

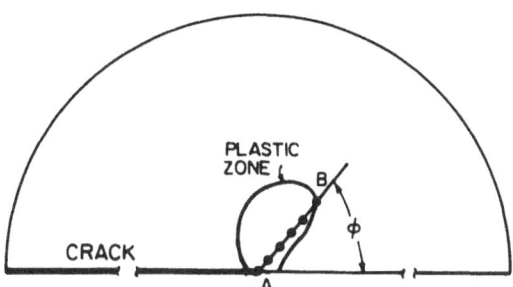

Fig. 8. Schematic figure of a crack tip plastic zone imbedded in a semi-circular crack tip region. The discontinuous shear displacement across the shear band *AB* causes crack growth.

where CTOD_{uz} is the crack tip opening displacement caused by the unzipping process, i.e., caused by the shear decohesions of the shear bands when the shear bands arrive at the crack tip successively. v, E and σ_Y are Poisson's ratio, Young's modulus, and tensile yield strength, respectively. C is a constant of proportionality. For the Dugdale's strip yielding model, C is equal to one.

Take the differential on both sides,

$$\delta\text{CTOD}_{uz} = 2C(1 - v^2)\frac{K\delta K}{E\sigma_Y}. \tag{4}$$

For the unzipping model, C is unknown, but it can be evaluated from the calculated value of δCTOD_{uz} and the imposed δK value.

The piecewise power hardening stress–strain relations are used for the finite element calculations

$$\sigma = E\varepsilon \qquad \text{for} \quad \sigma \leqslant \sigma_Y \tag{5a}$$

$$\frac{\sigma}{\sigma_Y} = \left(\frac{\varepsilon}{\varepsilon_Y}\right)^n \quad \text{for} \quad \sigma \geqslant \sigma_Y, \tag{5b}$$

where ε_Y and n are yield strain and strain hardening exponent respectively. Calculations are made for three different strain-hardening exponents: 0.1, 0.2 and 0.307. The values of the tensile yield strength and Young's modulus are $53.8\,\text{MN/m}^2$ and $68.9 \times 10^3\,\text{MN/m}^2$ respectively.

Finite element calculations are made for a semi-circular crack tip region, Fig. 8, with the linear elastic crack tip displacement field [19 and 20] imposed on the semi-circular boundary. The crack length is 2.54 cm, and the smallest element at the crack tip is 5.334×10^{-3} cm.

First, the cracked solid is located incrementally to a K-value of $1.10\,\text{MPa}\sqrt{\text{m}}$ ($1000\,\text{psi}\sqrt{\text{in}}$). The stresses and strains in the crack tip region are calculated and the *small* crack tip plastic zone, r_p, is delineated. A shear band, *AB*, is chosen as shown in Fig. 8. The shear band extends from the crack tip (*A*) to the boundary of the plastic zone (*B*). The shear band forms an angle ϕ with the crack line. Two different angles of ϕ are studied: 45° and 60°.

After the pre-determined K-level is reached, a small increment δK is applied. During the application of δK, the nodal points along the shear band are branched, and the branched nodal points are allowed to slip "freely" along the shear band but are not allowed to move away from the band. During the application of δK, the tangential nodal forces on these branched nodal points are kept constant. The normal stress along the shear band is allowed to increase in accordance with the imposed condition of zero relative displacements in the direction normal to the shear band. The normal stresses across the shear band are kept continuous. These prescribed boundary conditions along the shear band are made to be satisfied by repeated interations [21–23].

In order to have accurate results, the shear band AB should span 4, 5, or more elements. In the meantime, the condition of small scale yielding should be maintained. Therefore, a semi-circular region with an imposed linear elastic crack tip displacement field on the outer boundary is chosen.

In the finite element modelling, the shear band is treated as an exterior boundary. In order to facilitate the iteration, the finite element program allows the boundary conditions at the shear band and at the semi-circular boundary to be imposed independently. During the incremental loading up to the initial K-value of 1.10 MPa$\sqrt{m}$, the displacements at the nodal points along the band AB remain continuous. Only during the application of the subsequent increment δK, are the nodal points along AB allowed to branch.

While δK is applied, all of the elements are allowed to deform elastic-plastically according to the prescribed constitutive relations. The shear decohesion and the increment δCTOD_{uz} caused by the increment δK are thus calculated.

Normally the crack tip opening displacement calculation is very sensitive to the details of the crack tip elements, but the deformation away from the crack tip is much less sensitive to the details of the crack tip elements. It should be pointed out that the unzipping shear decohesion is the cumulative effect of the relative shear displacements over the entire shear band, AB. Therefore, the unzipping crack opening displacement and crack growth calculations are relatively insensitive to the details of the crack tip elements. Two different finite element programs and two different iteration procedures give the same results.

With a wide range of δK-values from 1.10×10^3 to 141×10^3 Pa$\sqrt{m}$ (1 to 128 psi$\sqrt{in}$), the first three significant figures of the evaluated constant C in (4) do not vary.

The effects of the inclined angle of the decohesion plane, ϕ, and the strain hardening exponent, n, on the ratio, $\delta\text{CTOD}_{uz}/\delta K$, are shown in Table 1. These values together with the known values of K, σ_Y, E, and v are used to evaluate the constant C in (4). The model gives a higher value of CTOD_{uz} for a higher strain hardening exponent and the larger inclined angle.

The calculation is made for the shear band above the crack line. The same calculated results are equally applicable for the conjugate shear band below the crack plane. In the case of small scale yielding, the total CTOD_{uz} is the sum of all of the increments δCTOD_{uz} as K increases from $K_{\min}$ to $K_{\max}$. The sum can easily be obtained by integration, i.e., (3) with the constant C given by the values in Table 1.

Table 1. The effects of strain-hardening exponents and inclined angles of decohesion planes on crack tip opening and advancing

ϕ	n	$\delta\text{CTOD}_{uz}/\delta K$ 10^6 [m/(MN/m$^{3/2}$)]	C in Eqn. (4)	A in Eqn (6)
	0.1	0.1946	0.121	0.0175
60°	0.2	0.2182	0.136	0.0185
	0.307	0.2388	0.148	0.0215
45°	0.307	0.1121	0.07	0.0175

Fatigue crack growth rate is related to the crack tip opening displacement caused by the shear decohesion. The geometric relationship between shear-decohesion crack-tip-opening-displacement and fatigue crack increment is shown in Fig. 7(f).

Equations (3) and (6) are for a monotonically increasing load. For cyclic fatigue load, K is replaced by ΔK, and σ_Y is replaced by $2\sigma_{Y(C)}$. $\sigma_{Y(C)}$ is the cyclic yield stress. Thus we have

$$\frac{da}{dN} = \tfrac{1}{2} \text{CTOD}_{uz} \cot \phi$$

$$= A(1 - v^2) \frac{\Delta K^2}{E \sigma_{Y(C)}}. \tag{6}$$

The values of A are tabulated in Table 1. "A" increases slightly over 20 percent as the strain hardening exponent increases from 0.1 to 0.307. The value of "A" for the 60 deg inclined angle is about 20 percent more than that of the 45 deg angle. Unless one is interested in fine details, the effects of the variations of both strain hardening exponent, n, and the inclined angle, ϕ, are negligible. A value of 0.02 is fairly representative for the results of the analyses.

$$\frac{da}{dN} = 0.02(1 - v^2) \frac{\Delta K^2}{E \sigma_{Y(C)}} \cong 0.02 \frac{\Delta J}{\sigma_{Y(C)}}. \tag{7}$$

The calculations are made for a semi-circular region. Strictly speaking, because of the asymmetry of a single shear band, an entire circular crack tip region should be used. However, it is judged that the difference will be minor.

5. Comparision with experimental results

Bates and Clark [24] related fatigue striation spacing with ΔK for a number of materials. Taking striation spacing as da/dN, we have

$$\frac{da}{dN} \text{ (in./cycle)} = 6 \left(\frac{\Delta K}{E}\right)^2. \tag{8}$$

From the data in the literature, Hahn et al. [25] have correlated da/dN with ΔK for a number of steels. They found that

$$\frac{da}{dN} \text{ (in./cycle)} = 8 \left(\frac{\Delta K}{E}\right)^2. \tag{9}$$

Barsom [26] measured fatigue crack growth in a number of steels.

$$\frac{da}{dN} \text{ (in./cycle)} = 0.66 \times 10^{-8} (\Delta K)^{2.25}. \tag{10}$$

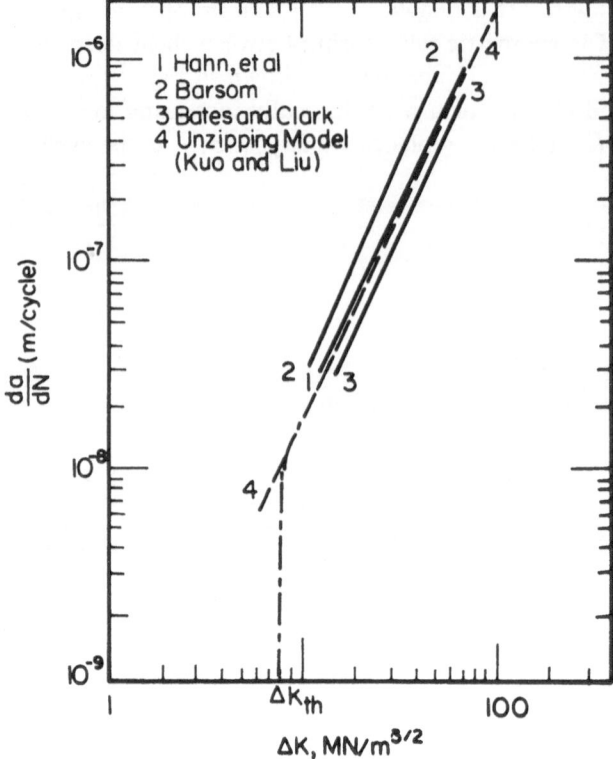

Fig. 9. The comparison of the calculated unzipping fatigue crack growth rate with the measured growth rate.

According to Bates and Clark and Hahn et al. da/dN can be written in the form

$$\frac{\mathrm{d}a}{\mathrm{d}N} = \gamma \left(\frac{\Delta K}{E} \right)^2. \tag{11}$$

With $\sigma_{Y(C)}/E = 1/400$ and $E = 207 \times 10^6$ MPa $= 30 \times 10^6$ psi for steel, the empirical equations and the theoretically calculated growth rates according to the unzipping model are plotted in Fig. 9. The agreement is amazingly good in the intermediate ΔK region.

The measured crack growth data of various steels do not seem to vary much. It is well known that the cyclic yield strengths of steels do not vary much because the annealed soft steels usually cyclically harden, and hardened steels usually cyclically soften. The value of the ratio $\sigma_{Y(C)}/E$ does not vary much from one steel to another. If their crack growth is controlled by the deformation process, their growth rates should not vary much either.

Kobayashi et al. [27] measured the striation spacings of a variety of metals and alloys. The measured striation spacings are related to $\Delta J/\sigma_{\mathrm{flow}}$ as shown in Fig. 10. For a number of materials, $\sigma_{\mathrm{flow}} < 2\sigma_{Y(C)} < 2\sigma_{\mathrm{flow}}$. Two lines in Fig. 10 are calculated according to (7) with $\sigma_{Y(C)} = \frac{1}{2}\sigma_{\mathrm{flow}}$ for one line and $\sigma_{Y(C)} = \sigma_{\mathrm{flow}}$ for the other. Again the agreement is very good. The unzipping model and its comparison with the experimental data are given by Kuo and Liu [21], Liu, Yang, and Kuo [28], and Liu and Kobayashi [29]. This paper summarizes their earlier results.

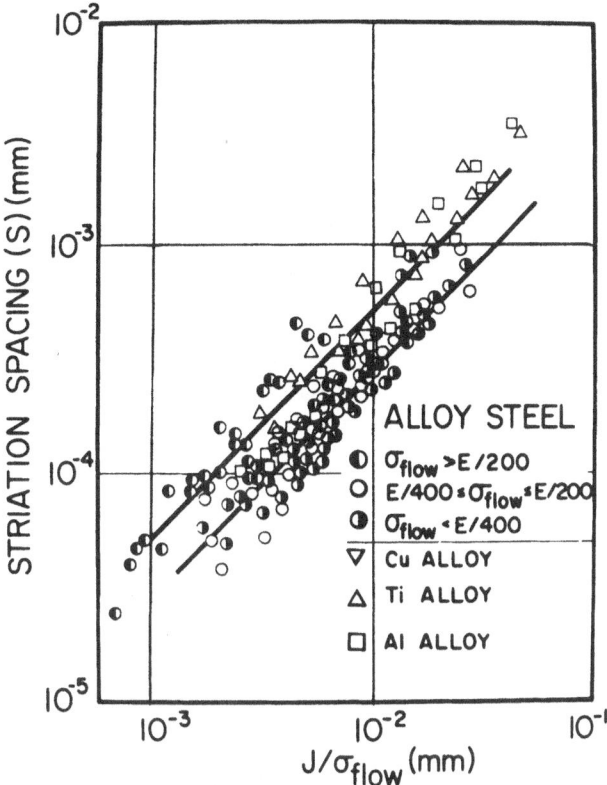

Fig. 10. The comparison of the calculated unzipping fatigue crack growth rate with the measured striation spacing.

6. Discussion

The unzipping model takes only the shear decohesion mechanism into consideration. It is a mechanism of plastic deformation. A crack may also grow by brittle fractures or by a combination of plastic deformation and brittle fracture processes.

Brittle particles and embrittled grain boundaries or interfaces may fracture at or below the applied K_{max} during a fatigue cycle. The basic crack growth mechanism could be plastic deformation, but the local brittle fractures will increase the overall growth rate. The unzipping model, based on the ductile deformation mechanism alone, gives only the baseline growth rate; any increased crack growth rate due to the local brittle fractures has to be added. The effect of the brittle fractures would be more severe in the high ΔK region, and local fractures will give an accelerated crack growth rate.

In the low ΔK region close to the threshold, ΔK_{th}, the fatigue crack growth rate drops sharply as shown schematically in Fig. 9 by the dash-dotted line. Near ΔK_{th}, the measured rates are much lower than the calculated unzipping growth rates.

The stress–strain relation of a single crystal is highly anisotropic. The "average" stress–strain relation of a polycrystal, averaged over a large enough number of randomly oriented grains, is isotropic. The finite element model of the unzipping process uses an isotropic stress–strain relation. Therefore, it is applicable only if the linear plastic zone size at a crack tip is more than 3 or 4 times the grain diameters so that enough grains are within the plastic zone and the isotropic average stress–strain relation becomes valid. Near ΔK_{th}, the crack tip plastic zone is only one to two times the grain diameter or less; thus the isotropic cyclic stress–strain relation and, thereby the finite element model of the unzipping process are no longer applicable.

Within a single grain, the shear slip is highly anisotropic. Near the threshold, when the crack tip plastic zone is smaller than the grain size, the shear decohesion process and the zigzag crack path follow the crystallographic planes closely. The length of each segment of the zigzag path is comparable to the grain diameter. In order to model the crack growth process accurately, the deviations from the assumed planar crack surface such as crack branching and zigzag crack path have to be taken into account.

On the other hand, when the crack tip plastic zone contains a large number of grains, the strong pattern of the crack tip plastic deformation for isotropic solids will force the zigzag path to turn more frequently. Hence, the length of each segment of the zigzag path is much smaller than the grain diameter, and the crack surface becomes more "planar". Therefore, the unzipping model applies well in the intermediate ΔK region, where the crack tip plastic zone is large enough so that the assumptions of isotropic deformation properties and planar crack surface are valid.

7. Summary

Plastic deformation is caused by dislocation motion and shear slip in a crystal. The slip process causes discontinuous displacements across the active slip plane and results in slip steps on a free surface. The "slip steps" at a crack tip will cause crack growth. Such crack growth mechanism is commonly referred to as crack tip shear decohesion mechanism.

The plastic deformation at a crack tip causes crack tip blunting and crack growth. Crack blunting is caused by the shear decohesion processes on all of the shear bands within the entire plastic zone, but crack growth is caused by the shear decohesion of the shear band at the very tip of a crack. The shear decohesions elsewhere cause blunting but not crack growth.

A fatigue crack grows primarily by the alternate shear decohesions on two intersecting shear bands. A *micro-mechanism based* finite element model is made for the unzipping process by the alternate crack tip shear decohesion mechanism to calculate the fatigue crack growth rate. The model takes the discontinuous displacement across the active shear band into account, and it singles out only the shear decohesions that contribute to fatigue crack growth. The calculated crack growth rate is

$$\frac{da}{dN} = 0.02(1 - v^2)\frac{\Delta K^2}{E\sigma_{Y(C)}} \cong 0.02\frac{\Delta J}{\sigma_{Y(C)}}. \tag{7}$$

The only undetermined constants of the equation are material constants: v, E, and $\sigma_{Y,(C)}$. Thus the unzipping model gives an explicit expression for fatigue crack growth rate which agrees amazingly well with the measured growth rates in a number of materials in the intermediate ΔK region.

References

1. D.S. Dugdale, *Journal of the Mechanics and Physics of Solids* 8 (1960) 100.
2. J.N. Goodier and F.A. Field, *Fracture of Solids*, John Wiley, New York (1963) 103.

3. B.A. Bilby, A.H. Cottrell and K.H. Swinden, *Proceedings Royal Society London* A272 (1963) 103.
4. J.R. Rise, *Fatigue Crack Propagation*, ASTM STP 415 (1967) 247–311.
5. N. Levy, P.V. Marcal, W.J. Ostergren, and J. Rice, *International Journal of Fracture Mechanics* 7 (1971) 143–156.
6. J.R. Rice and E.P. Sorensen, *Journal of the Mechanics and Physics of Solids* 26 (1978) 163.
7. R.M. McMeeking, *Journal of the Mechanics and Physics of Solids* 25 (1977) 357.
8. C.F. Shih, General Electric Company, TIS Report No. 79CR075, (April 1979).
9. F.A. McClintock, *Fatigue Crack Propagation*, ASTM STP 415 (1967) 170.
10. E. Orowan, *Report of Progress in Physics* 12 (1949) 185.
11. A.J. McEvily and R.C. Boettner, *Acta Metallurgica* 11 (1963) 725.
12. C. Laird, *Fatigue Crack Propagation*, ASTM STP 415 (1967) 131.
13. R.M.N. Pellous, *Transactions ASM* 62 (1969) 281.
14. B. Tomkins, *Philosophical Magazine* 18 (1968) 1041.
15. V.P. Neumann, *Z. Metallkde* 58 (1967) 780.
16. P. Neumann, *Acta Metallurgica* 22 (1974) 1155.
17. H. Vehoff and P. Neumann, *Acta Metallurgica* 27 (1979) 915.
18. R.D. Heidenreich and W. Shockley, in *Report of Conference on Strength of Solids*, Physical Society of London (1948) 57.
19. M.L. Williams, *Journal of Applied Mechanics* 24 (1957) 109–114.
20. G.R. Irwin, *Journal of Applied Mechanics* 24 (1957) 361.
21. A.S. Kuo and H.W. Liu, *Scripta Metallurgica* 10 (1976) 723.
22. A.S. Kuo, "An Experimental and FEM Study on Crack Opening Displacement and Its Application to Fatigue Crack Growth," Ph.D. dissertation, Syracuse University (1976).
23. C.Y. Yang, "Modelling of Crack Tip Deformation with Finite Element Method and Its Application," Ph.D. dissertation, Syracuse University (1979).
24. R.C. Bates and W.C. Clark, Jr., *Transactions ASM* 62 (1969) 380.
25. G.T. Hahn, R.C. Hoagland, and A.R. Rosenfield, Contract AF 33616-70-C-1630, Battelle Memorial Institute, Columbus, Ohio, August 1971.
26. J. Barsom, *Damage Tolerance in Aircraft Structures*, ASTM STP 486 (1971) 1–15.
27. H. Kobayashi, H. Nakamura, and H. Nakazawa, in *Proceedings 3rd International Conference on Mechanical Behavior of Materials*, Pergamon Press, 3 (1979) 529.
28. H.W. Liu, C.Y. Yang, and A.S. Kuo, in *Proceedings of the International Symposium on Fracture Mechanics*, George Washington University, Washington, DC, September 1978.
29. H.W. Liu and H. Kobayashi, *Scripta Metallurgica* 14 (1980) 525–530.

Résumé. La propagation d'une fissure de fatigue peut être due au mécanisme de déformation plastique cyclique de l'extrémité de la fissure, au mécanisme de rupture de portions fragiles et de frontières de grains fragilisées ou, souvent, à une combinaison de ces deux mécanismes.

Neumann et Vehoff ont procédé à des observations in situ des décohésions par cisaillement alterné dans deux bandes de glissement s'intersectant à l'extrémité d'une fissure, et ont décrit ce mécanisme par déformation plastique comme un mécanisme de base de la propagation d'une fissure de fatigue.

En vue de simuler le processus d'ouverture qui régit le mécanisme de décohésion par cisaillement à l'extrémité d'une fissure, on élabore un modèle par éléments finis basé sur un mécanisme à échelle microscopique. On trouve que les vitesses de propagation d'une fissure calculées grâce à ce modèle par éléments finis sont en très bon accord avec les vitesses mesurées dans la zone des ΔK intermédiaires et pour plusieurs matériaux.

International Journal of Fracture 39: 79–92 (1989)
© Kluwer Academic Publishers, Dordrecht

Use of parametric models in designing polymeric materials to specifications

E. VON MEERWALL[(1)] and F.N. KELLEY[(2)]
[(1)]*Department of Physics and Institute of Polymer Science,*
[(2)]*Institute of Polymer Science, University of Akron, Akron, Ohio 44325, USA*

Received 3 September 1987; accepted 1 April 1988

Abstract. In an effort to organize the design of polymeric materials, we have devised guidelines for analytically modeling the behavior of such materials as function of two independent variables such as rate and temperature. The parameters of successful models should have physical significance in terms of molecular theories or other measurements, and may be optimized by least-squares fits to appropriate data. Specimens may be compared by studying the variations of the fitted model parameters with sample constitution or preparation. We describe the implementation of this procedure on a computer graphics facility, and illustrate its use in the analysis of tear energy in filled polybutadiene networks. We envision an extension of this methodology to permit quantitative prediction of ingredients and treatment needed to produce materials of desired properties.

1. Introduction

The understanding of the properties of polymers provided by modern molecular theories is still relatively incomplete and must be supplemented and extended by observed correlations [1]. Simple polymer systems benefit in this regard from the equivalence [2] of changes in test rate to corresponding changes in temperature, a phenomenon referred to as thermorheological simplicity. The time-temperature superposition of the results of mechanical tests then produces a single master curve from whose interpolation the results of measurements at untested rate-temperature combinations may be reliably predicted.

The pronounced dependence of most physical properties of viscoelastic materials on temperature and rate demands a range of testing rates and/or temperatures for a proper understanding of the behavior of the material. But while the tradeoff between rate and temperature is well understood theoretically [3], the shape of the master curve itself still lacks quantitative prediction; mathematical models for it [4, 5] must be regarded as heuristic even when molecular theories are involved in their construction and parameterization. Unfortunately, this approach limits the accuracy of extrapolations of test results.

The usual complex, heterogeneous polymer materials often exhibit mechanical behaviors which cannot be reduced to a single master curve [6, 7, 8]. The question of what part of this failure is attributable to artifacts of the testing procedure at certain time-temperature combinations is, of course, of great interest to the understanding of the material at the molecular level. But this question is largely irrelevant in a macroscopic description of the material under any test which is designed to reproduce the conditions encountered in practical use. Thus a model description of mechanical tests as a function of temperature and rate must contain, in addition to the fully shiftable basic model, certain non-shiftable heuristic features. A representation of the results thus necessitates the use of two independent variables [8], the three-dimensional model constituting a curved surface which may be viewed in projection from a convenient angle.

We have produced a digital implementation of such a description of materials behavior on an engineering computer-graphics facility. Our models make extensive reference to molecular theories [9, 10, 11] and to the quantifiable aspects of materials preparation. With some of the parameters automatically adjustable to produce the best fit to the data at hand, the resulting values may be compared with those of an identical fit for a material of different constitution or preparation. This approach represents the initial stage of a procedure which envisions the prediction of materials and preparation required to produce polymer systems of desired physical properties. A more detailed description of the software has been presented [12]; the system has recently been used in several studies of tearing energy in filled polymer networks [13], and is currently being applied to stick-slip tearing.

2. System design

2.1. Desired attributes of the model

Any scalar mechanical measurement may be expressed as z, assumed to be a function of two independent variables, x and y. In our principal model, z represents the logarithm of the tear energy, x the absolute temperature, and y the logarithm of the rate of tearing. The function $z(x, y)$ has several fixed forms, one of which will be described below; it is driven by a set of 17 parameters p_j. Of these, n are adjustable; the remainder, whose values are known or which serve in a purely definitional capacity, remain fixed. Thus

$$z = F(x, y; p_1, \ldots, p_n). \tag{1}$$

The form of the function F is chosen to be able to represent a wide range of possible results for the test under consideration when its adjustable parameters are optimized. Often considerable freedom exists in the choice of various additive contributions to F and in the manner of parameterizing these. The mathematically desirable attributes of this parameterization are discussed in detail elsewhere [14]; a brief summary will suffice. The number of adjustable parameters must be appropriate: too few will not permit the data to be reproduced, while too many must result in conceptual ambiguities and mathematical difficulties. For meaningful results the number of data points should greatly exceed n. Whenever feasible the parameters should be designed to be approximately "orthogonal": each should affect the function in a manner not closely duplicated by any other. Parameters whose main effect on the function occurs in a region of x and y devoid of data cannot be determined at all unless data in the missing region become available.

The physical significance and intuitive meaning of the parameters are of principal interest. When possible, parameters should be simply relatable to materials properties and/or theoretical constructs (molecular theories; free-volume theory, scaling/reptation, related or complementary experiments, etc.) applicable to the test results. Since initial guesses for the adjustable parameters are required (see below), these need to be obtainable from a cursory inspection of the data. Baseline levels and slopes, as well as positions, heights and widths of the features to be described are much preferable to higher coefficients of polynomials (quite aside from the latter's obscure significance). Parameters representing the results of independent measurements (glass transition temperature, thermal expansivity, etc.), or reference temperatures, etc., may enter the model but are never adjusted.

Finally, the test and hence the model describing its results ought to be independent of any extensive, design-related dimensionality, and must concentrate on intensive materials properties. Characteristic or critical rates or temperatures, and asymptotic or extremal responses or their derivatives, are useful candidates as parameters, whether these describe responses to nondestructive applied forces and displacements or characterize some aspect of failure of the material.

In the model to be described later we have attempted to adhere to these precepts, arriving at suitable compromises among the often conflicting requirements and desiderata.

2.2. Adjusting the model to the data

For an unambiguous comparison between model and data, the latter must include experimental uncertainties, e.g., standard deviations of the results of repeated identical tests. Provided the results at a given x and y are available in the form $z \pm \Delta z$, the comparison can proceed on the preferred basis, the reduced chi-square statistical measure of goodness of fit [15]:

$$\chi_v^2 = \frac{1}{(N-n)} \sum_{i=1}^{N} \left[\frac{F(x_i, y_i; p_1, \ldots, p_n) - z_i}{\Delta z_i} \right]^2, \tag{2}$$

where N denotes the number of data points. When this index greatly exceeds unity, the fit is not acceptable. This may occur either because the parameters p_j have incorrect values or else because the model cannot be made to fit the data even when the parameters are allowed to adjust optimally. Because for a given comparison or fit the data are fixed, reduced chi-square is a (continuous) function solely of the adjustable parameters; optimization of these parameters is achieved by requiring the partial derivatives to vanish:

$$\partial \chi_v^2 / \partial p_j = 0, \quad j = 1, \ldots, n. \tag{3}$$

As was alluded to above, the form of F is subject to a large variety of requirements and preferences. Thus no flexibility can be sacrificed to ensure that the secular equations (3) are linear in the p_j, a set of severe mathematical constraints. Hence the standard procedures for dealing with linear simultaneous equations are not applicable here, leaving us obliged to employ the more difficult and temperamental nonlinear curvefitting methods; this subject is discussed in detail in the literature [15]. In order not to prejudice the range of convergence by locally linearizing chi-square or the model function, we have chosen the fully nonlinear Davidon (or Fletcher-Powell) variable metric minimization algorithm [16, 17] with a modification permitting the derivatives in (3) to be evaluated numerically [14, 18]. This avoids the need for incorporating code for all analytic derivatives $\partial F / \partial p_j$ into the program, and thereby facilitates experimentation to obtain the optimal forms for the contributions to F. Like all nonlinear fitting routines, ours requires initial guesses for all adjustable parameters to be supplied; sometimes the program is able to converge starting from a standard set of parameters contained within the program. The algorithm also checks that the obtained solution represents a minimum rather than a local maximum in chi-square, and attempts to establish that the minimum is universal rather than local.

Whenever an acceptable value of reduced chi-square indicates that the fit of the model to the data was successful, all significant information about the data is contained in the optimized set of parameters. The chosen form of the model together with the values of the parameters and their uncertainties constitute a complete and highly distilled description of the data irrespective of the number, coordinates, and precision of the data points. Thus except for the graphic presentation the original data have no further part in the analysis. It is this functional representation of the data which makes possible the extensive comparison among data sets, i.e., on the basis of variations in constitution or preparation, to be described below. The statistical uncertainties [15] in the fitted parameters p_j^* are important in deciding whether a given parameter changes significantly with specimen preparation, particularly large values indicating insensitivity of the model to that parameter given the available data. Fits producing excessive chi-square values are rejected; those with acceptable values (for a desired confidence level [15]) are associated with the data set and the model until they are replaced by the results of a better fit of the same model.

At present the program contains two models of which at least one is fitted to each set of data. A successful fit implies no unique status for the model in question; more than one model may adequately represent the data. Our experience shows that for well-constructed models a clear decision as to superior fit is usually not possible particularly when several sets of data are included in the comparison. Addition of further data to a set necessitates refitting, the principal changes typically being a reduction in the uncertainties Δp_j^*, and slight adjustments to the p_j^* themselves.

2.3. Correlation with preparation variables

The modeling described thus far is useful mainly as an aid to the interpolation of the behavior at values of x and y for which no test data has yet been taken. Of greater interest is the prediction of behavior of similar material differing in one or more aspects of constitution, preparation, or thermal and mechanical history. These attributes will be collectively referred to as preparation variables; each sample (in case the testing procedure is destructive, each batch of identical samples) is characterized by a set of preparation variables, whose numerical values are stored with the test data. Many preparation variables may take a continuous range of values (concentration of an ingredient, crosslink density, ageing time, etc.), while others have essentially integer, often binary, character (presence vs. absence of a trace ingredient, choice between two or more crosslinking agents, etc.). Even these may usually be transformed into other, molecular, variables having a continuous range.

The effect of changes in a given preparation variable on the test results may be studied by examining a series of samples differing systematically in (only or mainly) that variable, subjecting the results to the model fit. Our computerized procedure provides plots of any desired p_j for a given model as a function of any selected preparation variable, the data base including all specimens in the series of interest. This search, conducted interactively, reveals the parameters which are particularly sensitive to that preparation variable. To quantify that sensitivity, a low-order polynomial fit is automatically invoked and its coefficients are used as an aid in predicting p_j for as yet unexamined values of that preparation variable. This quantification may be repeated, under operator direction, for all p_j significantly sensitive to that preparation variable. This entire examination and analysis would need to be repeated for any other preparation variables whose effect on the materials properties is of interest.

The ultimate and still distant goal envisioned for this methodology is the prediction of the preparation variables required to produce a material of the desired properties, at least for materials within narrowly prescribed limits. The attainment of this goal will depend on the successful solution of three problems:

(a) The unambiguous disentanglement of the effects of a simultaneous variation of two or more preparation variables. The correlations thus engendered among the various p_j should be amenable to semiquantitative treatment by the interaction matrix method along the lines described by Kelley and Williams [1].
(b) The establishment of a quantitative connection between the p_j and the macroscopic properties of interest. The success of this step depends on the felicitous choice of the model and its parameterization, and may involve simple combinations of the p_j in the derivation of a given physical property.
(c) The prediction of properties from preparation variables, by inverting the forward correlations developed in (a) and (b). While this problem is almost entirely mathematical, conceptual difficulties may arise from multiple-valued solutions whenever the forward correlations are non-monotonic. In such cases the inverting algorithm must be able to detect all distinct non-trivial sets of preparation values resulting in the desired properties. The final choice of preparation is then likely to be made (initially off-line) on the basis of equally important but performance-unrelated criteria such as availability, time and expense, etc.

2.4. Details of principal model

The model to be described is regarded as the more appropriate of the two available in our present code for the representation of tear energy. In the earlier notation, $z = F(x, y)$, with F denoting the logarithms of the tear energy in Jm^{-2}, x being temperature (K), and y being the logarithm of the tear rate in ms^{-1}. The first requirement is to generate the time–temperature-superposable part F' of the function:

$$F'(x, y) = f(x_r, y') - \log \{x/[x_r(1 + \alpha(x - x_r))]\}. \tag{4}$$

Here x_r represents the reference temperature, which enters the shifted function f as well as the second term, the entropy-density correction [2]; the latter also involves the thermal expansivity α. The shifted rate is related to the original rate in the conventional way [2]:

$$y' = y + \log a_T, \tag{5}$$

with a_T calculated either from the Williams–Landel–Ferry (WLF) theory [3]:

$$\log_e a_T = -c_1^0(x - x_r)/(c_2^0 + x - x_r), \tag{6a}$$

or else from an Arrhenius-like expression:

$$\log_e a_T = A [(x - x_0)^{-1} - (x_r - x_0)^{-1}], \tag{6b}$$

with the choice between these determined by a binary parameter which is not fitted. The function f itself consists of a baseline z_0 plus a broadened edge feature [19] which terminates in a higher plateau of level z_p. If the approach to this plateau is more complex, a second, similar feature of height h may be optionally interposed [19]. Combining these, we write:

$$f(x_r, y') = z_0 + (z_p - z_0 - h)/\{1 + \exp[2(y' - y_a)/w_a]\} + h/\{1 + \exp[2(y' - y_b)/w_b]\}. \tag{7}$$

The midpoints along y of the edge features are given as y_a and y_b, while their y-widths are w_a and w_b, respectively. The shiftable part of F now being complete, a single non-shiftable Gaussian peak of height H, centered at coordinates x_c and y_c, with full widths at half height w_x and w_y in the respective directions, may be added:

$$F(x, y) = F'(x, y) + H \exp(-br^2), \tag{8}$$

where

$$b = 4 \ln 2,$$

$$r^2 = [(x - x_c)/w_x]^2 + [(y - y_c)/w_y]^2.$$

If h or H are set to zero, the respective expressions are not evaluated in order to save computational effort. In these cases, the other parameters contained in only these expressions become irrelevant and cannot be adjusted. The complete function may be seen to involve as many as 17 parameters, three of which are never fitted. The simplest cases call for $h = H = 0$, so that only four parameters (z_0, z_p, y_a, w_a) are subject to adjustment when the shift constants are known. A list of all parameters with some explanation is given in Table I.

Table I. Description of parameters for tear energy model

Parameter #	Symbol[a]	Explanation; comments
1	z_0	baseline, log of threshold tearing energy
2	z_p	upper plateau level of log tearing energy
3	y_a	y-midpoint of principal edge feature
4	w_a	y-width of principal edge
5	h	z-height of secondary edge feature
6	y_b	y-midpoint of secondary edge
7	w_b	y-width of secondary edge
8	c_1^0 or A	first time-temperature shift constant
9	c_2^0 or x_0	second shift constant; see parameter #17
10	x_c	x-position of centroid of peak feature
11	y_c	y-position of peak centroid
12	w_x	x-width of peak (full width at half height)
13	w_y	y-width of peak
14	H	z-height of peak
15[b]	α	thermal expansivity
16[b]	x_r	reference temperature (K)
17[b]	(0 or <0)	choice: WLF (0) or Arrhenius (<0) shifting

[a]Symbols refer to those used in (4) to (8).
[b]Parameter is not fitted.

The parameters have immediate physical significance, either by way of definition (x, and the WLF-Arrhenius choice of shift mechanism), or as results of other experiments (α as well as the shift constants), or else as quantities easily estimated by inspection of the data. The chosen form of F assures a reasonable degree of orthogonality among the parameters provided the available data suffices to determine all adjustable parameters.

3. Computer implementation

3.1. Flow of work

The procedures descibed have been implemented on the Engineering Computer Graphics Facility at the University of Akron, based on a Prime 850 time-sharing computer system, and are in the process of being adapted to DEC Microvax II and Sun 3/50 workstations. The program is designed to be used interactively from a high-resolution monochrome graphics terminal, but also produces optional output and plots on a line printer capable of a graphics mode, as well as one of several incremental pen plotters. Except for system calls, certain device drivers, and low-level graphics routines, all code is written in standard FORTRAN 77, and includes its own data base management facilities as well as provisions for data conversion, checking and, error recovery. The length of the source file is some 17 000 lines, over one-third of which is documentation and user directions.

After optionally entering new test data or modifying (e.g., adding to) the selected existing data file, the user enters commands directing the analysis: manual or automatic curve-fitting, comparison of model surfaces or model parameters (including automatic polynomial fits) among selected data sets, and generating plots, with viewing angles of 3-dimensional plots under user control. The curve-fitting stage permits the selection of the model to the applied, the setting and manual changing of model parameter values, the selection of the parameters which are to be optimized, and the multiparameter nonlinear optimization. The checking for uniqueness and reliability of the chi-square minimum is divided between automatic (random step) and manually directed features.

Association of optimal parameter combinations for that model with the data set is mandatory and automatic: a superior parameter combination (as evidenced by the lowest chi-square obtained so far), however obtained, supplants the previous best set. To help decide whether the inclusion of a particular feature of the model is "cost-effective" for the data set under consideration, the Gauss criterion may be applied [15]; for identical data, comparisons between fits of the same model with more or fewer parameters adjustable are made on the basis of reduced rather than standard chi-square.

3.2. Graphics output

The graphics routine concerned with representing the data as function of x and y together with the model first establishes a reference frame in the form of a rectangular box whose transparent surfaces represent the extremal values of x, y, and z; its edges are shown as lines. Axes drawn parallel to the three frontal edges are labeled and supplied with marks and corresponding numbers. The data points are represented as symbols viewed frontally; error bars are represented as vertical lines whenever they exceed the size of the parent symbol.

The model function using the current parameters is evaluated at the corners of a 30×30 regular rectangular grid in the xy plane, whose outer edges are coincident with the vertical planes of the box. Each point in this grid is connected by straight lines to its four nearest neighbors. The systems graphics routine which draws this surface permits the selection of a hidden-line algorithm; our experience shows that its use generally results in a less informative presentation than a surface fully transparent to the data points (but opaque to itself), provided that the user selects the most advantageous viewing angles.

The box and its contents are displayed in an orthographic projection, which lacks a vanishing point; it was felt that the modest reduction in visual ambiguity achieved by true perspective is not worth the extra computing effort. The operator needs to select only two viewing angles, the elevation above the base plane and the azimuth. Because the graphics terminals in use have two screen buffers, a view from another set of preprogrammed angles may be computed and transmitted to the terminal's undisplayed screen while the current display is being studied, until the new display is completely drawn. At that time the screens are logically interchanged, and the process is repeated for another set of viewing angles, thus simulating a rudimentary form of animation. When only two displays are to be studied (e.g., comparing two surfaces), these may be kept entirely in the terminal's display buffers and repeatedly and rapidly interchanged under program control without any transmission delay, permitting the convenient detection of small differences between the displays.

Model surfaces may be displayed with or without the corresponding data. Early experience with the system showed that considerable spatial ambiguity was possible in the presence of data, disguising the relative location of data points with respect to the surface. Experimentation showed that a vertical connection between each data point and the model surface above or below it (same x and y coordinates) effectively removed this ambiguity. This connection. was implemented as a slender wire-frame pyramid with a small square base coincident with the surface and sharing its slope, the apex located at the center of the data symbol. It is suppressed when the data point is within an error bar of the surface, leaving more distant points clearly referred to the surface.

A related set of routines shows two model surfaces simultaneously, usually with the hidden-line algorithm enabled; a display of the solid enclosed between the two surfaces is also possible. The hidden-line attribute is usually suppressed when the color pen plotter is used for hard-copy, where different colors or different line thicknesses can identify the surfaces. This advantage is denied to the line printer in its graphics mode, which closely resembles the terminal display. A conventional $x-y$ display of selected model parameters as function of an arbitary preparation variable, together with a polynomial fit, is also available.

4. Illustration: filled networks

As a simple illustration of the computerized data analysis procedure described above, we show excerpts of the modeling of a series of tear tests on a simulated solid rocket propellant, an industrial polybutadiene matrix (Arco 45M or 45HT) crosslinked and highly filled (70 vol. percent) with glassy polystyrene beads of $30\,\mu$m average diameter. In the data to be shown the beads were without surface reactivity, hence were not chemically linked to the matrix network. Tear energies were extracted from tear tests conducted on specimens in the trouser geometry, with results corrected for the energy required for bending. The results were not

SAMPLE PS-L-HC

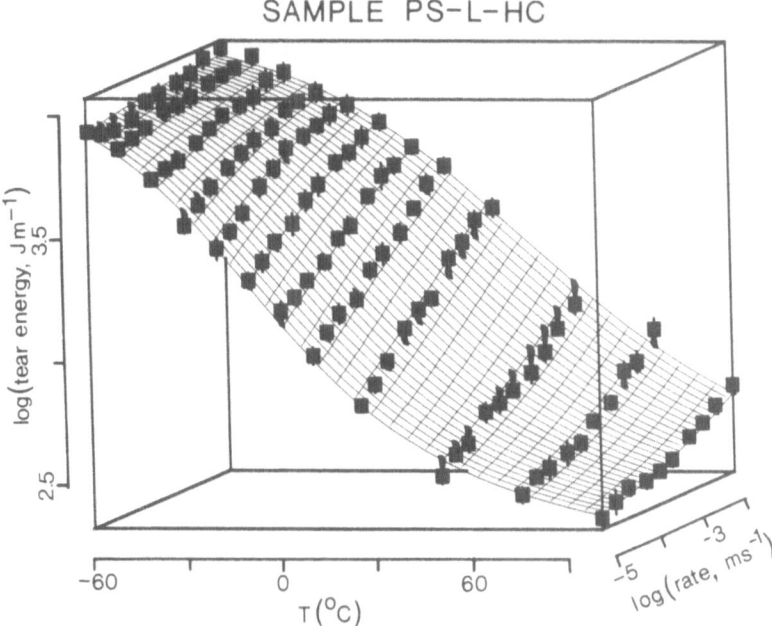

Fig. 1. Rate and temperature dependence of the tear energy in a filled polybutadiene network (sample PS-L-HC). Model surface incorporates a simple broadened edge between a threshold and an upper plateau in the rate-dependence, and is WLF-shifted in temperature. Plot shows experimental uncertainties in the data, and, where appropriate, a vertical connection between data points and surface.

further corrected for any small deviation of the tear path from linearity. Sample preparation and details of the tear test including data reduction and correction [20] are described elsewhere.

Tear energies were measured in a set of specimens, PS-L-HC, having a molecular weight between crosslinks $M_c \approx 4100$. A master curve was successfully prepared by empirically [2] shifting the data (rate-dependences at twelve temperatures between -60 and $+100°C$) to a reference temperature of 25°C. The computer program which performed this shifting then compared the extracted values of log a_T vs. $(x - x_r)$ with (6a), adjusting c_1^0 and c_2^0 to obtain the best fit. Since the shifting resulted in an acceptable master curve and was consistent with the WLF hypothesis, our three-dimensional treatment was expected to replicate these findings.

Figure 1 shows this set of data as an explicit function of both independent variables, together with a fitted model surface employing WLF shifting. Examination of the data suggested that the shape of the master curve is sufficiently simple to be successfully and economically described by a single broadened edge feature. Indeed, while the addition of the secondary edge slightly improved the fit, the three extra adjustable parameters actually raised chi-square per degree of freedom. Given that unshiftable features were not necessary, parameters number 5 and 14 were set to zero, making parameters number 6, 7, 10, 11, 12, and 13 irrelevant. With $p_{17} = 0$, $p_{16} = 298\,\mathrm{K}$, and $p_{15} \approx 4.8 \times 10^{-4}\,\mathrm{K}^{-1}$, the only adjustable parameters were number 1, 2, 3, 4, 8, and 9.

The initial parameter guesses needed as input for the automatic curve-fitting routine were arrived at in a brief visual inspection of the data displayed with the model surface generated with the current parameter estimates. An exception was the shifting parameters, for which the results of the master curve analysis were substituted. The reliability of the optimized parameters was found to be enhanced if the six parameters were fitted in two groups: alternately the edge feature (p_1 through p_4) and the shift constants (p_8 and p_9), and repeating once or twice before completing the process using the full six-parameter fit. This strategy minimized the possibility of being detained in a local chi-square minimum, and reduced the effort required to untangle the inevitable non-orthogonality (parameter correlation) between p_4 and p_8. The optimum obtained by the fitting routine was tested and slightly refined in five automatic random parameter displacements each followed by refitting. The value of reduced chi-square ($= 1.28$) characterizes the final fit as adequate.

The experiment was repeated using another set of specimens, PS-L-LC, identical to the earlier ones with the exception of its lower crosslink density ($M_c \approx 15\,000$). The model was fitted to the data from this experiment in a form and manner identical to the earlier case; the fit obtained was again satisfactory. The model surface obtained had a similar appearance, with many of the fitted parameters retaining their previous values within their combined respective uncertainties. Comparison between the surfaces, however, reveals significant differences as well. Figure 2 shows both surfaces displayed together, without the data.

Examination of the numerical results of the fit reveals that parameters p_1 and p_2 underwent the most significant changes as a result of the decrease in matrix crosslink density. The M_c-dependence of the tear energy asymptote p_1 is statistically significant, and admits of a direct molecular interpretation in terms of a concept proposed by Lake and Thomas [9]. The latter calls for a non-linear (square-root) dependence of p_1 on M_c, a fact not discernible here in the absence of data for three or more values of M_c. The application of the Lake–Thomas theory to tear data in unfilled specimens has been described elsewhere [4, 5].

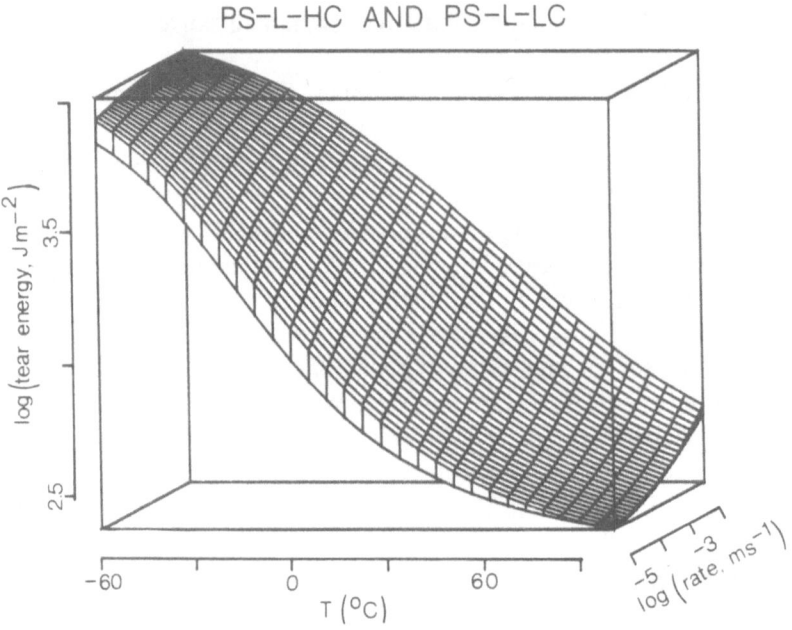

Fig. 2. Comparison of the model surface of Fig. 1 (top) with an identical fit to data in a set of similar specimens with lower crosslink density (see text). To reduce visual ambiguity (Moiré pattern), the surfaces are opaque and connected at the edges, simulating a solid object in projection.

R–45M–2–C and R–45M–2

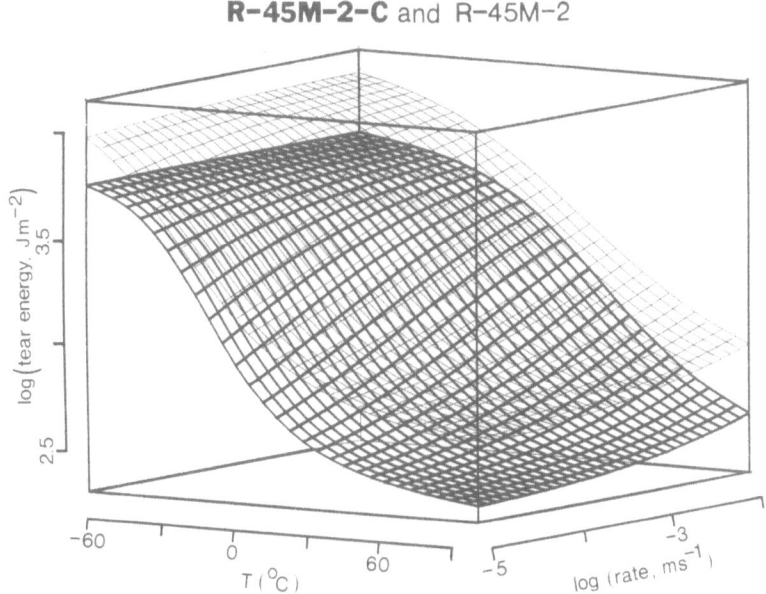

Fig. 3. Comparison of two model surfaces fitted to different versions of the same data, i.e., tearing energies in a filled PB network. The top surface (transparent; fainter lines) represents a fit of the model to the raw data; the lower surface describes a similar fit to the data after it was corrected for the bending energy. The two surfaces are not simply related. Data are suppressed for clarity.

The M_c-dependence of the plateau tear energy as expressed by p_2 is less precisely captured by the fit because fewer data points were available in the plateau region. Experience with these as well as various other materials has suggested that parameters p_2 and p_1 tend to scale together, perhaps because the height of the plateau above the threshold tearing energy is affected by molecular and morphological variables separate from those which determine the threshold. In that case the current parameterization is inappropriate in this respect (as well as undesirably "non-orthogonal" in the sense discussed above), and p_2 instead should express the logarithm of the ratio of plateau to threshold tear energy, analogous to other features whose vertical extents are referred to the baseline asymptote. This change in parameterization is currently being considered.

Figure 3 provides an illustration of the effect of artifacts of either the measurement or the initial data reduction. Here the model surfaces, as described above, are fitted to the same set of initial data (for another, similar, filled PB network) without and with a correction applied to the data for the work required to bend the trouser test specimen during tearing. The differences are highly significant and call for two observations. Firstly, the bending correction is larger in magnitude than the effects of sizable changes in sample constitution (cf. Fig. 2). Our experience shows that the domination of test artifacts over the usual differences in preparation variables is the rule rather than the exception, so that in extracting true materials properties the corrections for these artifacts must themselves be accurate to a high order. Secondly, since bending energy is viscoelastic in nature, any correction for it must depend on rate and temperature. Hence the differences between the two surfaces in Fig. 3 are non-trivial, i.e., not limited to a simple decrement of p_1 and p_2 by equal amounts. Comparison of the fitted parameters reveals significant changes to all four shape parameters p_1 through p_4, but essentially no changes to the shift constants.

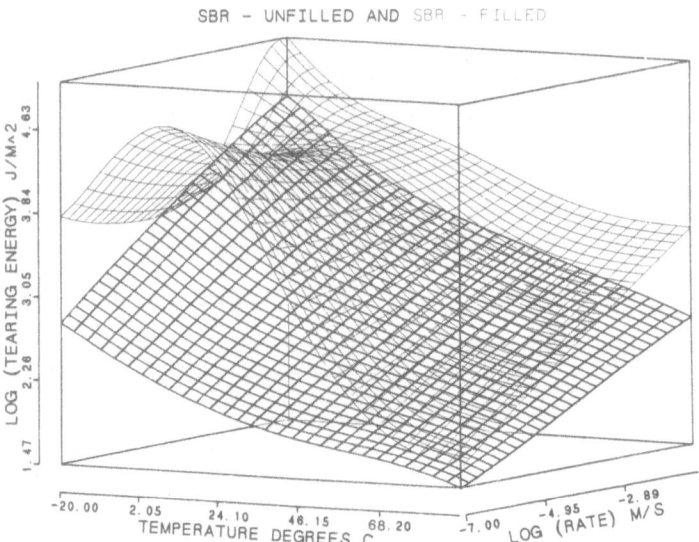

Fig. 4. Comparison of two model surfaces fitted to tearing data: (bottom) in unfilled amorphous SBR [13, 22] which tends to tear smoothly, and (top) in a similar SBR reinforced with 30 wt percent FT carbon black, extracted from a three-dimensional plot in [21]; this material exhibited knotty tearing in the region of the peak.

Three-dimensional representations of the temperature and rate dependences of tearing energy in polymers may be found in the literature as early as the 1950's. One particularly well-known example is Greensmith's [21] work in carbon-black reinforced styrene-butadiene rubber, which clearly defies time–temperature superposition. We extracted a set of values from the published grid and subjected it to a fit with the model described. To reproduce the pronounced rise in tear energies at low rates and intermediate temperatures it was necessary to invoke the Gaussian peak feature, and allowing parameters p_{10} through p_{14} to adjust. The result is shown in Fig. 4 (top); the resemblance to the original is satisfying but not exact. The features of the surface prior to the addition of the peak were obtained by performing a fully shiftable fit to data [22] in similar, unfilled SBR (bottom), then minimally modifying the results for filled SBR by allowing p_1 and p_2 to increase appropriately.

Related work of this laboratory [5, 13, 20, 22] has suggested that the peak in Fig. 4 coincides with the rate and temperature domain characterized by stick-slip (knotty) tearing. In that regime the tear is not straight but irregularly wavy, and often additionally displays elaborate branching, thus causing the length of the tear path greatly to exceed the macroscopic advance of the tear. It has been conjectured that materials undergoing knotty tearing derive (at least part of) their extra tear toughness not from any enhancement in microscopic strength or the efficacy of their strain-energy dissipation mechanisms, but simply from the increased length of the microscopic tear path. A semiquantitative estimate of this geometric effect is in accord with observations, but a more convincing demonstration was achieved by mechanically constraining the tear path to follow a nearly straight line through the use of flexible metallic shims of narrow separation on opposite sides of the trouser test specimens [22]. In these constrained tear tests the tear energies in the stick-slip regime for the filled SR closely resembled those in the unfilled, smoothly tearing SBR shown in Fig. 4; this marked reduction from the results (including Greensmith's [21]) for unconstrained tearing in very similar material supports the geometric origin of the extra energy required for knotty tearing. The availability of three-dimensional modeling programs, while only incidental to the discovery of such effects, is expected to aid materially in the quantitative understanding of their causes, such as the precise role played by the matrix-filler interaction.

5. Summary; further work

In this work we have elaborated a procedure which permits the accurate description of mechanical test results as function of two independent variables, and the comparison of these results among specimens with different ingredients or preparation. The procedure requires a computer graphics system with facilities for maintaining substantial data bases. The latter contain not only the results of the measurements, but also the parameters of the current best description in terms of one or more models proposed for the data.

Several requirements and desiderata govern the design of the models. They must, of course, be able to represent data from a given test on an entire family of materials under consideration, over the full range of both variables (e.g., temperature and rate) of interest. The models must also incorporate as much physical significance and theoretical insight as possible, both with respect to the form of the function as well as to its parameterization. Finally, the parameterization must permit the fitting procedure to obtain an unambiguous separation of the effects of the various adjustable parameters on the model function. All these requirements severely limit the choices available in constructing the model. Still, the success of any model is no assurance of its uniqueness as a valid description of the test data, and the final choice of the model to be used is likely to be made on practical or esthetic grounds.

The development of this facility has as its short-term objective the comparison of specimens differing in some aspect of preparation. While semiquantitative comparisons are possible simply through a visual examination of the fitted models, quantitative comparisons need to rely on the fitted parameters. From a study of the correlations of the parameters with the preparation variables it is possible to predict test results for rates and temperatures not yet explored. More important for the longer term, a methodology of this general nature is likely to be at the basis of any sophisticated prediction of the ingredients and preparation necessary to produce polymeric materials with desired properties. In the interim, the procedure described here is a very useful tool in the systematic quantification of structure-property relations in the context of academic materials science.

The methods described here are sufficiently general for application in a wide range of fields. The tested specimens need not be based on polymers, the tests are not restricted to the mechanical or viscoelastic variety, the number of independent variables may be one, two, or greater (with increasing difficulty of presenting a suitable visualization of the model), and the variables themselves need not include temperature or rate.

It may be equally profitable to use this methodology to study the testing procedures and data reduction itself. In several current efforts to measure tear energy in filled vulcanizates we are examining the effects of test specimen geometry and tear path constraint, as well as measures to prevent or correct for elastic or plastic specimen distortion during tearing. In these cases specimens are prepared in identical sets, and the comparisons between the results of different testing and data reduction procedures serve mainly to quantify the artifacts of each procedure, with a view to their reduction or elimination, or at least their standardization. Since the nature of the tear varies strongly with temperature and tear rate, variations in testing procedures will in general introduce non-trivial alterations into the data and the fitted model.

Acknowledgments

The authors express their sincere appreciation to D. Thompson for performing most of the development and debugging of the computer code. K.J. Min and R. Stacer kindly supplied the illustrative tear energy data prior to publication. We are grateful to Professor D. Plazek for several helpful conversations, and to L. Yanyo and R. Stacer for their assistance and helpful suggestions during the design phase. The generous financial support of this project by the Goodyear Tire and Rubber Company, Lord Corporation, the Morton-Thiokol Corporation, and the US Air Force Office of Scientific Research is gratefully acknowledged.

References

1. F.N. Kelley and M.L. Williams, *Rubber Chemistry and Technology* 42 (1969) 1175.
2. J.D. Ferry, *Viscoelastic Properties of Polymers*, 3rd ed., Wiley, New York (1980).
3. M.L. Williams, R.F. Landel, and J.D. Ferry, *Journal of the American Chemical Society* 77 (1955) 3701.
4. D.J. Plazek, I.-C. Choy, F.N. Kelley, E. von Meerwall, and L.-J. Su, *Rubber Chemistry and Technology* 56 (1983) 866.
5. L.-J. Su, Ph.D. dissertation (Polymer Science), University of Akron (1983).
6. H.W. Greensmith and A.G. Thomas, *Journal of Polymer Science* 18 (1955) 189.
7. L. Mullins, *Transactions Inst. Rubber Industry* 35 (1958) 213.
8. H.W. Greensmith, L. Mullins, and A.G. Thomas, *Transactions of the Society of Rheology* 4 (1960) 179.
9. G.J. Lake and A.G. Thomas, *Proceedings Royal Society London*, Series A, 300 (1967) 1460.
10. A.G. Thomas, *Rubber Chemistry and Technology* 48 (1975) 902.
11. A.N. Gent, in *The Science and Technology of Rubber*, F.R. Eirich (ed.), Academic Press, New York (1978) Chapter 10.
12. E. von Meerwall, D. Thompson, K.J. Min, and F.N. Kelley, *Journal of Materials Science* 21 1801; a preliminary report in *Bulletin of the American Physical Society* 31 (1986) 131.
13. R.G. Stacer, E. von Meerwall, and F.N. Kelley, *Rubber Chemistry and Technology* 58 (1985) 913.
14. E. von Meerwall, *Computer Physics Communications* 11 (1976) 211.
15. E.P. Bevington, *Data Reduction and Error Analysis for the Physical Sciences*, McGraw-Hill, New York (1969).
16. W.C. Davidon, U.S. Atomic Energy Commission R&D Report ANL-5990 (Rev.), Appendix (1959).
17. E. von Meerwall, *Computer Physics Communications* 9 (1975) 117.
18. E. von Meerwall, *Computer Physics Communications* 18 (1979) 411.
19. The modeling of the two-stage approach to an upper plateau is described in [4] and, in more detail, in [5].
20. K.J. Min, Ph.D. dissertation (Polymer Science), University of Akron (1987).
21. H.W. Greensmith, *Journal of Polymer Science* 21 (1956) 175.
22. R.G. Stacer, Ph.D. dissertation (Polymer Science), University of Akron (1986).

Résumé. Dans le but de systématiser la conception de matériaux polymères, on a élaboré une guidance pour la modélisation analytique du comportement de tels matériaux en fonction de deux variables indépendantes, telles que la vitesse et la température. Ainsi, les paramètres de modèles performants devraient avoir un sens physique du point de vue des théories moléculaires et d'autres critères, à charge de les optimiser par adjustements par moindre carrés sur des données pertinentes. Des comparaisons peuvent être faites entre échantillons en étudiant les variations des paramètres des modèles ajustés, par rapport à la préparation ou à la constitution de ces échantillons. On décrit l'introduction de cette procédure sur un terminal de calculs graphiques, et on illustre son utilisation à l'analyse de l'énergie d'arrachement dans les réseaux imprégnés de polybutadiène. On prévoit une extension de cette méthodologie à la prédiction quantitative des ingrédients et des traitements nécessaires à la production de matériaux à propriétés déterminées.

International Journal of Fracture 39: 93–102 (1989)
© Kluwer Academic Publishers, Dordrecht

Impulse viscoelasticity: a new method to study polymerization of thermosets

RICHARD J. FARRIS and MENAS S. VRATSANOS
Polymer Science and Engineering Department, University of Massachusetts, Amherst, MA 01003, USA

Received 20 September 1987; accepted 1 April 1988

Abstract. The technique of Impulse Viscoelasticity, in combination with incremental linear elasticity, is presented for monitoring the mechanical and rheological changes which occur during the solidification process. The technique is specifically applied to the curing of an epoxy resin. For such a system, the properties which can be measured as a function of the cure history are gelation time and temperature, equilibrium tensile modulus, cure stress and shrinkage, steady state elongational viscosity, mean relaxation time, thermal expansion coefficient, glass transition temperature and dynamic mechanical properties.

1. Introduction

The formation of a network in thermosetting resins is marked by large changes in mechanical behavior. Thus far, the most common mechanical characterizations of the *in situ* curing process in polymers have been dynamic mechanical methods. They are, however, limited in the amount of information which can be obtained. Recently, Farris [1] developed the technique of Impulse Viscoelasticity for following the solidification process. Impulse Viscoelasticity examines the mechanical response of an aging linear viscoelastic material subject to deformation. For a curing system, the following mechanical properties can be measured or calculated as a function of the polymerization history: gelation time and temperature, equilibrium modulus, stress and shrinkage due to cure, steady state elongational viscosity, mean relaxation time, thermal expansion coefficient, glass transition temperature and dynamic mechanical properties.

This paper highlights the mathematics of the Impulse Viscoelastic technique and incremental linear elasticity and its application to the curing of an epoxy resin. More complete mathematical treatments have been submitted for publication [2–5]. Experimental emphasis was placed on the measurement of the equilibrium tensile modulus and the stresses associated with cure.

2. Theory of impulse viscoelasticity

The Impulse Viscoelastic method involves the application of an arbitrary deformation to a material during its solidification. By deforming the material at periodic intervals during the polymerization, it is possible to monitor the mechanical changes as they occur. While the method does not depend upon the path of deformation, those which induce large stress relaxation (such as uniaxial step deformations) in a sample are best. In order to understand the changes during cure, we consider polymerization to be equivalent to an aging process.

As noted, the path of deformation is completely arbitrary except that the deformation must be of short duration. This satisfies the assumption that the aging process is slow in comparison to the time scale of the deformation. In order to simplify the mathematics it is required that the deformation return to its pre-deformation level. By virtue of the assumption of linear viscoelasticity, the stress must return to its pre-deformation value.

Equation (1) is a linear viscoelastic, non-aging constitutive equation written for uniaxial stress, where the elastic contributions (*Eeq*) are separate from the viscous contributions (*Er(t)*):

$$\sigma(t) \;=\; Eeq\,\varepsilon(t) + \int_0^t Er(t-t')\,\frac{\partial \varepsilon}{\partial t'}\,\mathrm{d}t',\tag{1}$$

where $\sigma(t)$ = stress response; $\varepsilon(t)$ = strain history; Eeq = equilibrium tensile modulus; $Er(t)$ = time-dependent tensile relaxation modulus, where $Er(t=\infty)=0$; t = time; t' = time parameter.

This equation is valid for an aging material when pulses of short duration are used. In this way, the mechanical properties can be considered constant during the pulse deformation. Later in the cure, another short duration deformation is applied to the sample. Although the mechanical properties have changed as a result of polymerization, they are again considered constant at their new values. Equation (1), using the new values of $\varepsilon(t)$, $\sigma(t)$, Eeq and $Er(t)$ is again applied to calculate the mechanical properties. This procedure is continued throughout the polymerization history.

By taking the Laplace transform of (1), and requiring the stress and strain response to return to their pre-deformation values, it can be shown that the equilibrium tensile modulus (*Eeq*) is given by (2):

$$Eeq \;=\; \int_0^\infty \sigma(t)\,\mathrm{d}t \Big/ \int_0^\infty \varepsilon(t)\,\mathrm{d}t\tag{2}$$

Equation (2) represents a simple criterion for gelation. By definition, a liquid has an *Eeq* of zero whereas a solid has a finite, non-zero equilibrium tensile modulus. Since the integral of a uniaxial box-strain deformation is always non-zero, the gel point can be established by determining the time at which the integral of the stress response ceases to be zero. This criterion for gelation is more fundamental than that based upon the equivalence of G' and G'' at the gel point [6]. Note that single frequency dynamic mechanical methods cannot differentiate liquid from solid behavior.

Having taken the Laplace transform of (1) it is possible to differentiate the transformed equation twice with respect to the Laplace parameter. The use of Laplace transform theorems introduces time-weighted integrals of the stress and strain responses. Such integrals make it possible to calculate the steady state elongational viscosity and mean relaxation time of a material during cure. These properties can be determined for both liquids and solids. Repeated differentiations of the transformed equation with respect to the Laplace parameter result in higher moments of the relaxation spectrum.

From the Fourier transform of (1) it is possible to calculate the dynamic mechanical properties at any operator selected frequencies [5]. These calculated properties were found to agree closely with those obtained by standard dynamic mechanical methods.

3. Theory of incremental linear elasticity

The calculation of such properties as the stress due to cure, shrinkage due to cure, thermal expansion coefficient and glass transition temperature (T_g) are also important in understanding the cure process. Although these properties are changing with time, it is possible to describe them using the equations of linear elasticity written in differential form. Equation (3) is written for aging linear elastic material in differential form:

$$Eeq[(d\varepsilon_{ij} - \delta_{ij}(\alpha\, dT - \dot{\gamma}\, dt)] = (1 + v)\, d\sigma_{ij} - v\delta_{ij}\, d\sigma_{kk}, \tag{3}$$

where Eeq, v, α and $\dot{\gamma}$ are dependent upon the cure history. σ_{ij} = stress tensor; ε_{ij} = strain tensor; Eeq = equilibrium tensile modulus; v = Poisson's ratio; α = linear thermal expansion coefficient; $\dot{\gamma}$ = linear rate of polymerization shrinkage; t = time; T = temperature; d = differential operator; δ_{ij} = Kronecker delta. Implicit in (3) are the assumptions of (a) incremental linear elastic behavior, (b) material isotropy, (c) time dependent moduli, (d) no viscoelastic behavior and (e) dimensional changes induced by pressure, temperature and polymerization.

In order to obtain information on the volumetric changes during polymerization the simplest geometry that can be used is uniaxial. Shear geometries are not sensitive to these dimensional changes. For the case of uniaxial stress, (3) reduces to (4):

$$d\sigma_{ij} = Eeq[d\varepsilon_{ij} - \delta_{ij}(\alpha\, dT - \dot{\gamma}\, dt)]. \tag{4}$$

Rewriting (4) for conditions of constant strain and temperature it is possible to calculate the one-dimensional stress due to cure using (5):

$$\sigma_{11}(t) = \int_0^t Eeq(t')\dot{\gamma}(t')\, dt'. \tag{5}$$

The cure stress is easily obtained experimentally by measuring the static load of an isothermally curing sample held at constant strain. In addition, (6) calculates the one-dimensional stress associated with a temperature change by solving (4) for completely polymerized samples held at constant strain:

$$\sigma_{11}(T) = -\int_{T_0}^T Eeq(T)\alpha(T)\, dT. \tag{6}$$

By monitoring the thermal stress and Eeq of a cured sample during cooling it is possible to calculate the thermal expansion coefficient. From the thermal stress data it is also possible to determine the T_g. T_g is taken as the temperature at which the thermal stress behavior changes from the rubbery to the glassy state.

4. Experimental

Epon 828, a DGEBA epoxy resin, and V-40, a polyamide curing agent, were chosen as the resin system. Both of these were provided by the Shell Chemical Company. V-40 is characterized by a flexible methylene backbone which contains both primary and secondary amines. In order to study the effect of the amine/epoxy ratio on network structure, three ratios of V-40/Epon 828 were chosen. Using equivalent weights of 190 g/mole for Epon 828 and 145 g/mole for V-40, amine/epoxy (A/E) ratios of 0.8, 1.0 and 1.2 were prepared. Cure temperatures of 55, 85, 115 and 165°C were studied.

In order to characterize the incipient stages of cure it was necessary to find a sample support which did not dominate the measured mechanical response. For this purpose, a soft rubber membrane in the shape of a tube was used to constrain the resin mixture. The resulting sample geometry was that of slender uniaxial cylinder with flared ends. Details regarding this sample configuration have been discussed by Vratsanos and Farris [7].

Samples were secured in the environmental chamber of the Dynastat mechanical spectrometer. A temperature ramp of 2.3°C/minute was used to heat the samples, which were held at constant strain, from room temperature to the isothermal cure temperature. A uniaxial box-strain pulse of five seconds duration was applied to the sample after which it was returned to its original length. The magnitude of the strain pulse was initially 5 percent and decreased as the epoxy cured so as to keep the material response in the linear viscoelastic range. The strain pulse was repeated every 100 seconds. Approximately 100 pulses were used to characterize each polymerization. Analog signals of the strain pulse, resulting stress response and sample temperature were digitized and stored on a Digital Professional 380 computer. Data were collected for fifty two seconds at a frequency of 10 Hz and began five seconds prior to the strain pulse. Data reductions were performed using software written specifically for the Impulse method.

5. Results

Since the emphasis of this study was determining the point of gelation and the stresses associated with the curing process, the data presented will be on the equilibrium tensile modulus (Eeq) and the stress due to cure as represented by the baseline stress level.

A typical pulse deformation indicating viscoelastic behavior taken from the curing of an epoxy is shown in Fig. 1. The data below were compiled by sequentially combining the information obtained from each individual pulse. The data have not been corrected for the contribution of the rubber membrane, which had an elastic modulus of 0.05 MPa over the entire temperature range studied. Such a correction was deemed unnecessary since the force required to deform the sample shortly after geletation was dominated by the sample. Figure 2(a) is a plot of Eeq during polymerization at the 1.0 A/E ratio for all four cure temperatures. For the same set of samples, Fig. 2(b) plots the baseline or shrinkage stress behavior during cure. As examples of the data that can be obtained during the cooling portions of these experiments, Figs. 3(a) and (b) plot Eeq and the shrinkage stress, respectively, for the samples cured at 165°C. From these two plots it is possible to obtain the T_g of each sample. Table I summarizes the mechanical properties at the end of cure in terms of cure time, gel time, gel temperature, Eeq, E' calculated at 0.1 Hz, baseline stress and T_g for each amine/epoxy ratio at all cure temperatures.

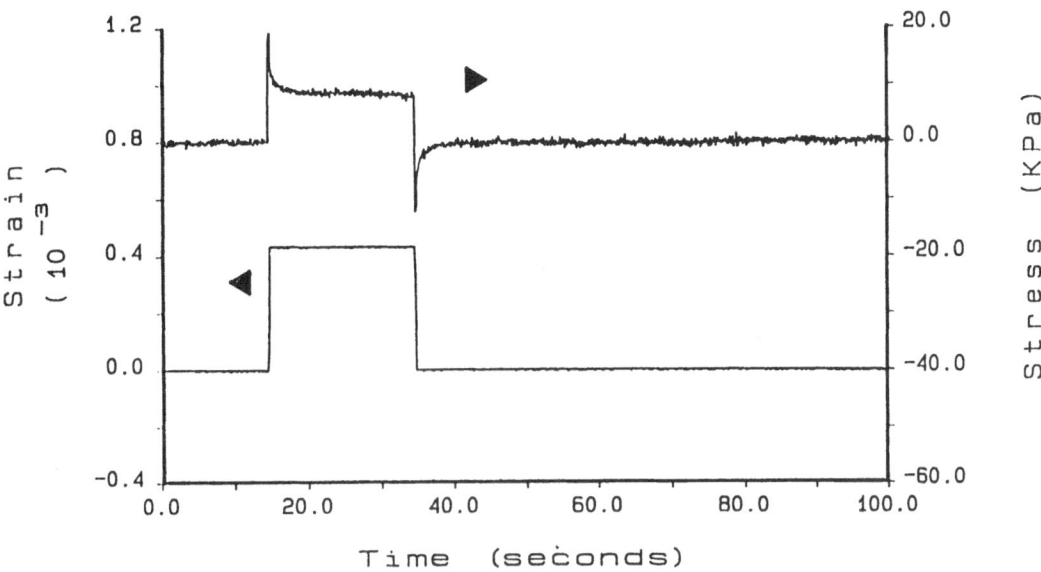

Fig. 1. Pulse deformation and the resulting stress response for a viscoelastic epoxy.

Table I. Summary of mechanical properties determined at the end of polymerization for V-40/Epon 828 epoxy resins

Cure temp [°C]	Amine/Epoxy ratio	Cure time [min]	Gel time [min]	Gel temp [°C]	*Eeq* [MPa]	*E'* [MPa]	Baseline stress [MPa]	T_g [°C]
55	0.8	160	78	55	20	180	0.02	45
55	1.0	155	78	55	15	150	0.01	45
55	1.2	155	76	55	5	30	0.01	45
85	0.8	100	37	84	320	800	0.06	76
85	1.0	130	37	84	60	270	0.05	72
85	1.2	210	36	83	12	15	0.02	57
115	0.8	130	32	82	26	26	−0.07	95
115	1.0	130	34	87	18	18	−0.05	78
115	1.2	125	34	90	13	13	−0.05	62
165	0.8	120	34	90	24	24	−0.65	90
165	1.0	135	32	90	19	19	−0.49	72
165	1.2	175	35	95	14	14	−0.38	60

6. Discussion

From Fig. 2(a) it is easy to find the sample gelation time despite the presence of the rubber membrane. The nearly equivalent gelation times for the samples cured at 85, 115 and 165°C are the result of the slow ramp in temperature that was used. The slow heating of these samples allowed sufficient time for gelation to occur prior to reaching the "isothermal"cure temperature. For these sets of samples the polymerization was essentially over before the

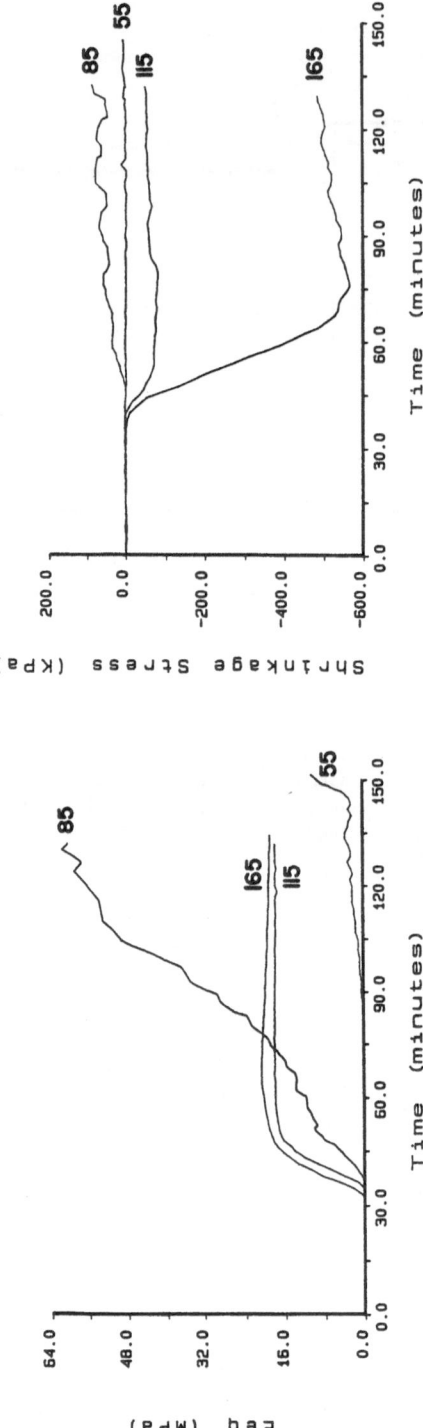

Fig. 2. (a) Equilibrium tensile modulus and (b) shrinkage stress as a function of cure time for the 1.0 amine/epoxy equivalence for the V-40/Epon 828 samples cured at 55, 85, 115 and 165°C.

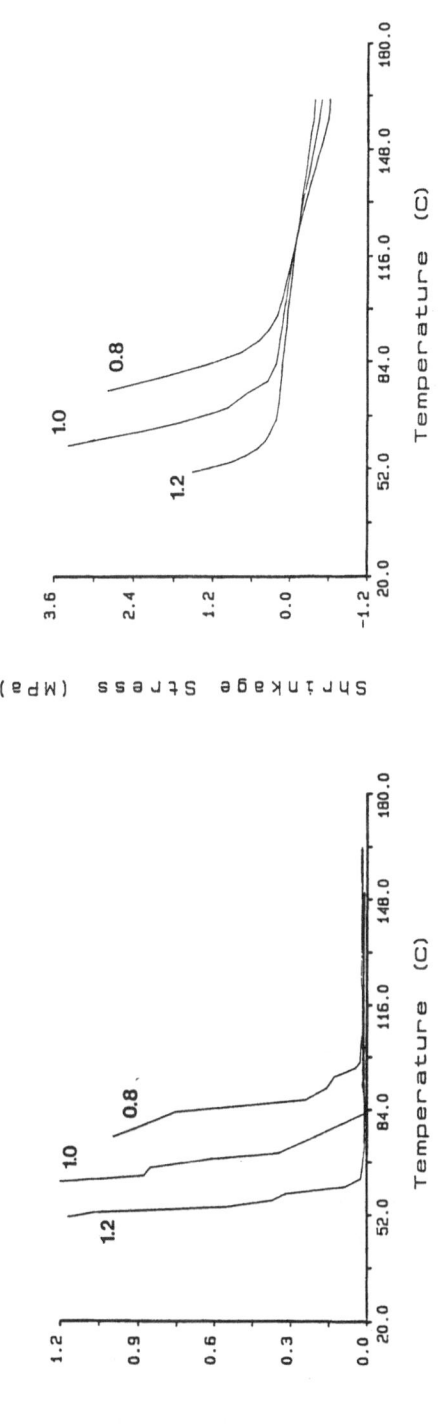

Fig. 3. (a) Equilibrium tensile modulus and (b) shrinkage stress as a function of temperature during cooling for the V-40/Epon 828 samples cured at 165°C for the three amine epoxy equivalences.

samples reached the isothermal cure temperature. Kinetic studies of the effect of cure temperature would require faster temperature ramps.

Some loss in *Eeq* was observed during polymerization at the 165°C cure temperature. This effect can be observed on Fig. 2a. The T_g's of these samples were lower than those obtained at the 115°C cure temperature. These results were attributed to thermal degradation. At all three equivalences, the 115 and 165°C cure temperatures yielded materials which behaved elastically over most of the polymerization. In addition, the values of E' and *Eeq* for these samples were very similar. This, coupled with the associated pulse shapes, suggests that the ratio of *Eeq*/E' be used as a measure of the time-dependency of a material. The limits of such a ratio are zero for a liquid and unity for an elastic solid.

The samples cured at 55°C were polymerized under isothermal conditions. This relatively low cure temperature resulted in a competition between vitrification and extent of reaction or cure. For this reason the T_g's of these samples were limited to 45°C. At all of the amine/epoxy ratios the pulse stress responses and dynamic data indicated that these samples cured viscoelastically. Similar findings were observed for the samples cured at 85°C. The effect of vitrification is to artificially raise *Eeq*. This is graphically evidenced on Fig. 2a by noting that the *Eeq* for the sample cured at 85°C was greater than that obtained at the 115°C cure temperature. In order to obtain values of *Eeq* more indicative of equilibrium, it is necessary to lengthen the pulse duration and data collection periods.

Effect of amine/epoxy ratio

It is particularly difficult to establish the role of the amine/epoxy ratio in network formation for epoxies cured with polyamides. Two structural factors hinder the understanding of the development of the network. The first concerns the structure of the V-40 molecule itself. Since primary and secondary amines are initially present, their relative reactivity towards epoxies will be important in determining the network structure. For simplicity, it will be assumed that the reactivities of primary and secondary amines are equivalent. The second factor concerns the purity of V-40. While the dominant isomeric structure present in polyamide curing agents is the dimerized species, it is well known that commercially available polyamides contain some trimerized and cyclic species [8, 9].

The samples cured at the 0.8 A/E ratio exhibited the highest dynamic and equilibrium tensile moduli for all cure temperatures. This is attributed to the formation of a more densely crosslinked network. Because of the relative deficiency in amine content, more epoxide-amine reactions will take place per molecule of V-40 and causes the network to have a lower molecular weight between crosslinks, Mc. An excess of amine requires fewer reactions per molecule of V-40 for complete conversion of the epoxide groups, and lowers the crosslink density. Also, since each V-40 molecule is not fully reacted, the network will contain more defects.

Cure stress behavior

It would be expected that as polymerization occurs the epoxy network increases in density. The densification should be proportional to the degree of crosslinking, i.e., the more tightly crosslinked networks should be of higher density. The stresses caused by this shrinkage and densification are tensile for a sample held at fixed length. From Table I, the samples cured at 55 and 85°C exhibited tensile baseline stresses of less than 0.06 MPa. Stresses of this order are small in comparison to load bearing stresses or the thermal stress associated with a few degrees temperature change. However, the samples cured at 115 and 165°C exhibited compressive baseline stresses at 0.05 and 0.5 MPa, respectively. This is due to the fact that these samples gelled before reaching their "isothermal" cure temperature, and exhibited the thermal expansion behavior of a solid upon further heating. This can be oberved on Fig. 2b. The apparent cure stress behavior of these samples is dominated by thermal stresses. For all samples the stresses associated with cure are negligible for the V-40/Epon 828 system. These stresses, due purely to cure shrinkage, can be dominated by a temperature change of a few degrees. This dominance of thermal stresses can also be determined by comparing the magnitude of the stresses induced during cooling (Fig. 3b) with those obtained during cure (Fig. 2b). Thus, the traditional prediction of tensile stresses can be incorrect in sign and magnitude. Similarly, the assumption of "stress free" conditions at the cure temperature may be wrong.

With regard to residual stress for non-isothermal cures, it is important to determine the gelation temperature. At gelation the baseline stress is just greater than zero. Upon further heating or extent of reaction the baseline stress usually ceases to be zero. While the stress at the cure temperature is the sum of the cure stress and the thermal stress, from an engineering perspective it is important to know the residual stress at the use temperature. The residual stress of a sample at its use temperature is the sum of the stress at the cure temperature plus the thermal stress associated with the change from the cure to the use temperature. For high performance thermosets which require high cure temperatures and post-curing, these residual stresses may be significant.

7. Summary and conclusions

The techniques of Impulse Viscoelasticity and incremental linear elasticity have been presented with emphasis on curing systems. With these techniques it is possible to measure the gel time and temperature, equilibrium tensile modulus, cure stress, cure shrinkage, steady state elongation viscosity, mean relaxation time, thermal expansion coefficient, glass transition temperature and dynamic mechanical properties as a function of the cure history.

The Impulse Viscoelastic method was applied to the curing of V-40/Epon 828 epoxy resins. Though the method is general with regard to the deformation shape, uniaxial box-strain deformations were used to characterize the curing process. From these simple deformations many of the relevant engineering properties associated with cure were measured. The most fundamental of these is the gel point, which was determined by the onset of a non-zero equilibrium tensile modulus.

Of the three amine/epoxy ratios investigated, the 0.8 ratio yielded the largest dynamic and equilibrium moduli at all cure temperatures. The 115°C cure temperature was the most efficient in terms of the time needed for complete cure. The static or baseline stress of the samples at the cure temperature can be tensile or compressive depending on when gelation occurred. For non-isothermal cures the bulk of this stress could be attributed to thermal stress as a result of the slow temperature ramp used to heat the samples. For the samples cured at 115 and 165°C thermal stresses dominated the apparent cure stress behavior. The magnitude of the one-dimensional tensile stress associated with cure shrinkage is about 0.05 MPa.

Acknowledgement

The authors would like to thank the Center for UMass-Industry Research on Polymers (CUMIRP) for financial support.

References

1. R.J. Farris, *Journal of Rheology* 28 (1984) 347.
2. M.S. Vratsanos and R.J. Farris, in *Composite Interfaces*, H. Ishida and J.L. Koenig (eds.), Elsevier, New York (1986) 71.
3. M.S. Vratsanos and R.J. Farris, "Impulse Viscoelasticity: A New Technique for Measuring Relevant Mechanical Properties during Cure, Part I: Theory of Impulse Viscoelasticity", *Journal of Polymer Science: Polymer Physics Ed.*, submitted.
4. ibid, "Part II: Theory of Incremental Linear Elasticity".
5. M.S. Vratsanos and R.J. Farris, *Journal of Applied Polymer Science* 36 (1988) 403.
6. C.Y. Tung and P.J. Dynes, *Journal of Applied Polymer Science* 27 (1982) 59.
7. M.S. Vratsanos and R.J. Farris, *Journal of Applied Polymer Science* 33 (1986) 915.
8. D.E. Floyd, D.E. Peerman and H. Wittcoff, *Journal of Applied Chemistry* 7 (1957) 250.
9. D.E. Peerman, W. Tolberg and H. Wittcoff, *Journal of the American Chemical Society* 76 (1954) 6085.

Résumé. On présente une technique basée sur la visco-élasticité sous impulsion qui, en combinaison avec l'élasticité linéaire incrémentielle, permet de surveiller les changements mécaniques et rhéologiques qui surviennent lors d'un processus de solidification. On applique en particulier cette technique à la cuisson d'une résine epoxy. Pour un tel système, diverses propriétés peuvent être mesurées en fonction du progrès de la polymérisation: durée et température de gélification, module de traction à l'équilbre, contraintes et bridages associés à la polymérisation, viscosité de l'allongement en conditions stationaires, temps moyen de relaxation, coefficient de dilatation thermique, température de transition à l'état vitreux et propriétés mécaniques en conditions dynamiques.

International Journal of Fracture 39: 103–110 (1989)
© Kluwer Academic Publishers, Dordrecht

Effect of crosslink type on the fracture of natural rubber vulcanizates

LYNN C. YANYO
Lord Corporation, 405 Gregson Drive, Cary, North Carolina 27512, USA

Received 22 September 1987; accepted 1 April 1988

Abstract. The effect of the chemical nature of the crosslinks on the fatigue crack growth behavior of filled natural rubber has been investigated. By varying the ratio of sulfur to accelerator, the relative amounts of polysulfidic to monosulfidic crosslinks was controlled. Carbon–carbon crosslinking was introduced via peroxide cure. All elastomers tested were prepared at the same number average crosslink density as confirmed by equilibrium swelling and modulus measurements. At the same crosslink density, polysulfidic crosslinks were most resistant to fatigue over the range of tearing energies investigated. Vulcanizates with primarily monosulfidic crosslinks exhibited lower cut growth rates than peroxide cured specimens, although the monosulfidic network strength may have been enhanced by the presence of some polysulfidic crosslinks.

Introduction

The strength of a particular elastomer is determined by its molecular structure including the crosslink density and type, and filler type and loading. This study investigates the effect of the chemical crosslink which transforms a viscous polymeric fluid into an elastomeric solid. In natural rubber, the crosslink can be made of many different chemical types including, commonly, polysulfidic (chains of sulfur atoms between 3 and 8 atoms in length), disulfidic, monosulfidic and carbon-carbon linkages.

Many researchers have investigated the effect of crosslink type on the threshold tear energy [1, 2], tear energy master surfaces [3], and tensile strength [4–7]. The effect of crosslink type on the strength properties of elastomers is still under debate.

Lake and Thomas [1] have reported that under threshold tearing conditions (where energy dissipative processes are minimized) an elastomer with mainly polysulfidic crosslinks has a higher threshold tear strength than one with predominately monosulfidic crosslinks. This result has been confirmed by other workers [2] and leads to the conclusion that enhanced energy dissipation cannot account for the increase in strength due to polysulfidic crosslinking. The basic network strength (as quantified by the threshold tearing energy) is directly affected by the type of crosslink. Interestingly, Lake and Thomas's theory of threshold strength does not include a dependence on the crosslink type, however different crosslink distributions can be incorporated [1, 8].

Early work by Greensmith and co-workers [7] reported sulfur cured systems having a greater tensile strength than peroxide cured systems which again was higher than radiation crosslinked rubber. Bateman and co-workers [5] proposed that high strength in rubber again was obtained with weakly covalent bonded crosslinks, having observed Greensmith's work and the fact that carbon-carbon bonds are stronger than monosulfidic which are stronger than polysulfidic bonds. They argued that weaker crosslinks would be able to slip under the localized high-stress concentrations, thereby relieving the stress and allowing the network as a whole to bear more load. Later work by Lal [6] disagreed completely with Greensmith [7] by showing that the tensile strength in natural rubber did not depend at all on the type of sulfur crosslinks.

Theories have been advanced that different strengths in natural rubber were not due to the nature of the crosslink but to the effect of different crosslink structures on strain induced crystallization [9]. As support for this argument, several researchers have shown non-strain-crystallizing elastomers to have no change in tensile strength with different crosslink structures [10, 11]. A recent study by Kok and Yee [4] disagrees and concludes, based on extensive testing, that changes in strength are not due to the crystallizability of the rubber. Furthermore, strength properties depended mainly on the type of crosslink with rubbers vulcanized by sulfur giving higher tensile strengths than those vulcanized by peroxide. Over a broad range of crosslink density, the tensile and tear strengths followed the order: conventional (high polysulfidic content) > efficient (low polysulfidic content) > peroxide (carbon-carbon crosslinks) [4].

Rubber is commonly used in applications where the elastomer is subjected to dynamic stresses. Under cyclic deformation, natural rubber is often observed to fail by a crack growth mechanism [7]. Depending upon the severity of the applied stresses, the rubber lifetime can range from an infinite period of time without failure to catastrophic rupture. The author is unaware of any systematic investigation of the effect of crosslink type on the fatigue crack growth in elastomers.

Some of the errors in investigating the effect of molecular structure on the strength of elastomers may be the result of applying test procedures (such as tensile tests) which are not truly testing the material behavior but also include the effect of the geometry of the test (the way in which the strains or loads were applied).

The amount of crack growth in an elastomer is determined by the magnitude of the strain energy which can be released by creating new surface area [12–14]. The tearing energy, i.e., the energy available for crack growth, is a determining factor in the amount of crack growth per cycle. The tearing energy, τ, can be calculated from the externally applied forces or strains. For a given tearing energy, the amount of cut growth per cycle which occurs in a sample is independent of the type of test piece used [15–18].

The pure shear test configuration allows one to determine the cut growth per cycle of the material irrespective of how the forces are applied (geometry). The cut growth rate is related to the energy input to the sample, not the strain or stress or the way in which the forces are applied.

Pure shear in the rectangular elastomer sample is achieved by pulling it in its width direction as illustrated in Fig. 1. The elastomer becomes wider at the expense of the thickness, yet its length does not change appreciably. The sample is in pure shear since one of its principal strains, that parallel to its length, is zero. For this configuration [19] the tearing energy is given simply as the product of the strain energy density, W, and the unstrained height, h,

$$\tau = Wh. \tag{1}$$

The most fortuitous aspect of (1) is that the tearing energy in pure shear does not depend on the crack length. The strain energy density is evaluated separately from the unloading portion of stress-strain loops of the samples before a cut is initiated [20].

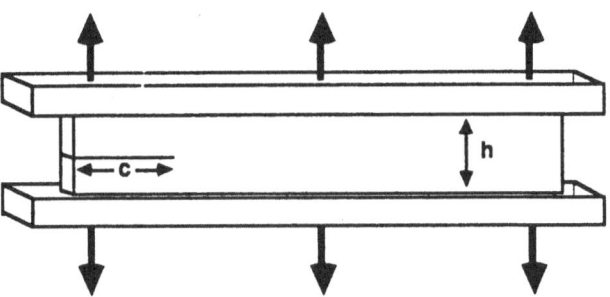

Fig. 1. Pure shear fatigue test specimen.

Under relaxing conditions (where the elastomer is allowed to return to zero load during each cycle), there is a relationship between the amount of crack growth per cycle, dc/dn, and the tearing energy calculated for the maximum of deformation of the cycle. Below some threshold tearing energy, the very slow crack growth is due to chemical rather than mechanical effects [21–23]. In vacuo the threshold tearing energy can be three times as large as in an oxygen environment [21]. A theoretical derivation for the threshold tearing energy has been given and verified [1, 24, 25].

As the tearing energy in a specimen is increased, e.g. the maximum applied strain is increased, the crack grows at a more rapid pace. At some tearing energy the crack propagates catastrophically. This catastrophic tearing energy is approximately the tearing energy obtained in conventional tear tests using, for example, the trouser tear test piece [23].

Between the threshold and catastrophic tearing energies the rate of crack growth has been represented by a power law expression [21, 22, 26, 27].

The influence of hysteresis on fatigue has been theoretically investigated by Payne and Whittaker [27]. Hysteresis can enhance the fatigue life of materials by dissipating energy which would otherwise be used in crack growth. However, hysteresis also leads to an increase in sample temperature due to heat build-up. Under cyclic loading, because of the hysteresis in the material, fatigue failure can be enhanced by the rise in temperature in the elastomer. In this study the temperature rise was avoided by maintaining the samples at a temperature higher than what would be the steady state temperature under the given cyclic conditions.

2. Experimental

Materials

Three vulcanizates of filled natural rubber were used. The formulations are listed in Table I.

When a comparison between different types of crosslinks is made it is necessary to maintain the same crosslink density. In this study the level of accelerator (CBS) to sulfur was adjusted to alter the ratio of monosulfidic to polysulfidic crosslinks [3]. A peroxide cure was used to generate carbon-carbon crosslinks. The amounts of the curvatives and the cure times at 140°C were adjusted so that the amount of swelling in benzene after 7 days was identical in all three networks [3].

These formulations also maintain the molecular weight of the polymer (by controlling the energy input during mixing) and additive ingredients. It is desirable to maintain the same

106   *L.C. Yanyo*

Table I. Elastomer formulations

	C-Poly	C-Mono	C-CC
Natural Rubber (SMR CV)	100 pphr	100 pphr	100 pphr
Zinc Oxide	5	5	5
Stearic Acid	2	2	2
Agerite Resin D	1	1	1
HAF Carbon Black	50	50	50
Sulfur	2.8	0.6	–
CBS (N-cyclohexyl-2-benzothiazolesulfenamide)	0.6	9.0	–
Dicumyl Peroxide	–	–	3

content of other ingredients, such as zinc oxide and stearic acid, for the different cure systems in order to get comparable results, although this would not be customary compounding practice. For example, zinc oxide is not necessary for peroxide vulcanization.

Compounding was carried out using a two stage mixing process. The masterbatches of rubber, carbon black, zinc oxide, stearic acid and Agerite resin D were mixed in a banbury at 80°C. The curatives were added on a 2-roll mill at 60°C. For the accelerated sulfur systems, the compounds were cured to an optimum state at 140°C as determined by the Monsanto Rheometer curves. The cure time used was 24 minutes for both the polysulfidic crosslinked (C-Poly) and monosulfidic crosslinked (C-Mono) elastomers. Since the peroxide vulcanization exhibits a marching modulus, the cure time was selected to maintain the same level of crosslinking as in the sulfur cured samples. The peroxide cured samples (C-CC) were molded for 100 minutes at 140°C.

Determination of crosslink density and distribution

Equilibrium swelling in benzene was used to determine the crosslink density of the three elastomers. Samples of each testpiece were swollen in benzene for seven days at 25°C. Assuming 2 crosslinks for each elastically effective network chain, the crosslink density was calculated from the equation [28],

$$v = \frac{-\ln(1 - v_r) + v_r + \mu v_r^2}{2V_0\varrho_r(v_r^{1/3} - v_r/2)}, \qquad (1)$$

where v is the moles of crosslinks per gram of rubber, v_r is the volume fraction of crosslinked rubber in the swollen sample, ϱ_r is the density of rubber (0.911 g/cm³), V_0 is the molar volume of solvent (88.82 cm³/mole), and μ is the polymer-solvent interaction parameter which depends on the volume fraction as $\mu = 0.44 + 0.18v_r$ [29].

The volume fraction of rubber in the swollen network is calculated by

$$v_r = \frac{(W_d - W_f)/\varrho_r}{(W_d - W_f)/\varrho_r + (W_s - W_d)\varrho_s}, \qquad (2)$$

where W_d is the weight of the sample after swelling and drying, W_f is the weight of the filler in the sample, W_s is the weight of the swollen sample, and ϱ_s is the density of the solvent (0.879 g/cm³).

The proportion of polysulfidic crosslinks was determined using a chemical probe of piperidine-propane-2-thiol which is capable of cleaving trisulfide and higher bonds [30].

Table II. Crosslink density and distribution

Network	Crosslink density $(\times 10^4)$ mole/g	% Polysulfidic	% Disulfidic	% Monosulfidic/ carbon–carbon
C-Poly	0.74	77.5	3.4	19.1
C-Mono	0.73	10.0	31.6	58.4
C-CC	0.75	4.6	0.0	95.4

Benzene extracted samples, 2 mm thick, were placed under nitrogen in a 0.2 M solution of propane-2-thiol dissolved in piperidine for 6 h. Samples of each network were swollen in benzene for 6 days after probe treatment. A comparison of the original crosslink density to the value after probe treatment provides the concentration of polysulfidic crosslinks.

The polysulfidic and disulfidic crosslinks in the networks can be broken using lithium aluminum hydride [31–33]. The probe solution was a fresh solution of 5g of lithium aluminum hydride in 200 ml of benzene-tetrahydrofuran (1 : 1) mixed solvent. Specimens 2 mm thick were placed in the probe solution under nitrogen for 3 hours. After treatment the samples were deswollen in acetone (containing a small amount of water) for a few minutes, rinsed with benzene for a few minutes and reswollen in benzene for 7 days. The vacuum dried samples were boiled in water to remove the reaction residuals of the probe to determine the deswollen weight of the specimens. In this way, the number of mono-sulfidic or carbon–carbon crosslinks was determined by comparison with the untreated samples.

The crosslink densities and distributions of the three networks are listed in Table II. As an indication of the inherent error in these studies it is noted that the peroxide cured samples exhibited a positive polysulfidic crosslink content of about five percent.

Cut growth fatigue testing

Fatigue testing was done in a pure shear geometry as illustrated in Fig. 1. Each compound was compression molded into testpieces 13 cm long, 1.2 cm high and 1 mm thick. The samples were bonded during cure along their length to metal plates which were held by the grips in the hydraulically driven fatigue tester. In this way, the forces wree transmitted equally along the length of the specimen.

The appropriate dynamic strains to be imposed on the sample to generate a given tearing energy were determined from the unloading curve of a stress-strain plot on the original test piece. The modulus of each of the networks was also calculated from these stress-strain plots and indicated that each of the samples had the same stiffness, about $1700 \, \text{kN/m}^2$ (250 psi), implying the networks had the same crosslink density since the same filler type and amount was used in each of the elastomers.

Testing was done at a constant temperature of 50°C to eliminate discrepancies in cut growth rates due to temperature rises from dynamic heat build-up. A constant frequency of 10 Hz was used. The applied tearing energy varied from approximately 50 to 250 N/m. The crack growth rate dc/dn, was determined from the slope of a plot of crack length versus the number of cycles for each sample at a given input tearing energy.

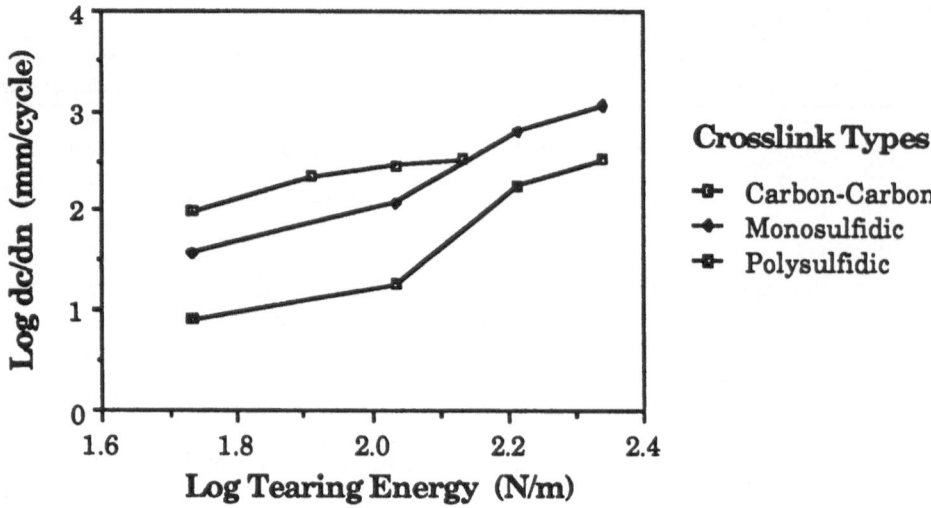

Fig. 2. Fatigue crack propagation data for the three elastomers.

3. Results and discussion

The results from the fatigue testing are shown in Fig. 2. Over the range of tearing energies tested the polysulfidic network (C-Poly) exhibited the slowest cut growth rate. The sample C-Mono, which consisted of mostly monosulfidic crosslinks, appears more resistant to crack growth than the peroxide cured specimen, C-CC. It may be that the presence of some disulfidic and polysulfidic crosslinks enhances the fatigue performance of C-Mono above that of a purely monosulfidic network.

At first thought, it may be that the hysteresis in the polysulfidic samples is higher than that of the other two networks. Payne [27] has shown that increased hysteresis correlates with increased fatigue life. However other researchers [1, 2] have shown that in the absence of energy dissipation, a difference in tear strength between polysulfidic, monosulfidic and carbon-carbon crosslinked elastomers is still apparent. Therefore, although hysteresis differences (if they exist between polysulfidic and monosulfidic networks) may account for a portion of the fatigue life differences under the conditions tested here, the intrinsic non-hysteretic strength of the network as quantified by the threshold tearing energy is still a major factor.

Under cyclic deformation, network chains are alternately extended and relaxed. Free mobility of chain segments depends on the structure of the crosslinks, whether carbon–carbon, monosulfidic, disulfidic or polysulfidic. Using an argument similar to that advanced by Bateman et al. [5], the longer the crosslinks, the easier it is for individual chains to move (rather than break) when the elastomers are subjected to mechanical stress. The polysulfidic crosslinks may provide this flexibility and hence improve fatigue life.

Although the previous discussion is plausible the author would like to add another possibility to the continuing debate. Depending upon the length of the polysulfidic chains, the crosslinks themselves may act as short polymer segments. It has been experimentally demonstrated [34, 35] and theoretically predicted [8] that mixtures of short network chains with relatively long chains can increase the tensile and tear strengths of elastomers by increasing the number of network chains available for deformation without severely diminishing the average lengths of the chains which must be stretched to their full length before rupture can occur.

Crosslink distribution is also a factor. Depending upon the assumed crosslink distribution, different results for the threshold tearing strength are predicted by the theory of Lake and Thomas [1]. Although random crosslinking is usually assumed for both sulfur and peroxide curing, it is possible that the distributions may be different. With distributions approaching bimodality, enhancement in strength is predicted [8].

References

1. G.J. Lake and A.G. Thomas, *Proceedings Royal Society (London)* A300 (1967) 108.
2. A.K. Bhowmick, A.N. Gent and C.T.R. Pulford, *Rubber Chemistry and Technology* 56 (1983) 226.
3. M.J. Wang and F.N. Kelley, unpublished report (1986).
4. C.M. Kok and V.H. Yee, *European Polymer Journal* 22 (1986) 341.
5. L.C. Bateman et al. in *The Chemistry and Physics of Rubber-Like Substances*, John Wiley and Sons, New York (1963) 715.
6. J. Lal, *Rubber Chemistry and Technology* 43 (1970) 664.
7. H.W. Greensmith et al., in *The Chemistry and Physics of Rubber-Like Substances*, John Wiley and Sons, New York (1963) 249.
8. L.C. Yanyo and F.N. Kelley, *Rubber Chemistry and Technology* 60 (1987) 78.
9. J.A. Brydson, *Rubber Chemistry*, Applied Sciences, London (1978).
10. D.S. Pearson and G.G.A. Bohm, *Rubber Chemistry and Technology* 45 (1972) 193.
11. R.F. Fedors and R.F. Landel, *Transactions Society of Rheology* 9.1 (1965) 195.
12. A.A. Griffith, *Philosophical Transactions Royal Society (London)* A221 (1921) 163.
13. A.A. Griffith, *Proceedings International Congress of Applied Mechanics* (1924) 55.
14. R.S. Rivlin and A.G. Thomas, *Journal of Polymer Science* 10 (1953) 291.
15. P.B. Lindley and S.C. Teo, *Plastics and Rubber: Materials and Applications* (1979) 29.
16. G.J. Lake and P.B. Lindley, in *Physical Basis of Yield and Fracture: Conference Proceedings*, The Institute of Physics and The Physical Society (1966) 176.
17. A.N. Gent, P.B. Lindley, and A.G. Thomas, *Journal of Applied Polymer Science* 8 (1964) 455.
18. A. Stevenson, *Rubber Chemistry and Technology* 59 (1986) 208.
19. A. Ahagon, A.N. Gent, H.J. Kim and Y. Kumagai, *Rubber Chemistry and Technology* 48 (1975) 896.
20. E.H. Andrews, *Journal of the Mechanics and Physics of Solids* 11 (1963) 231.
21. G.J. Lake and P.B. Lindley, *Journal of Applied Polymer Science* 9 (1965) 1233.
22. G.J. Lake and P.B. Lindley, *Journal of Applied Polymer Science* 10 (1966) 343.
23. P.B. Lindley, *International Journal of Fracture* 9 (1973) 449–462.
24. A.N. Gent and R.H. Tobias, *Journal of Polymer Science: Polymer Physics Edition* 20 (1982) 2051.
25. L.C. Yanyo and F.N. Kelley, *Rubber Chemistry and Technology* 60 (1987) 78.
26. G.J. Lake and P.B. Lindley, *Journal of Applied Polymer Science* 8 (1964) 707.
27. A.R. Payne and R.E. Whittaker, *Journal of Applied Polymer Science* 15 (1971) 1941.
28. P.J. Flory, in *Principles of Polymer Chemistry*, Cornell University Press, Ithaca, NY (1953) 579.
29. G. Kraus, *Rubber World* 135 (1956) 67.
30. B. Saville and A.A. Watson, *Rubber Chemistry and Technology* 40 (1967) 100.
31. M.L. Studebaker and L.G. Nabor, *Rubber Chemistry and Technology* 40 (1967) 100.
32. M.L. Studebaker and L.G. Nabor, in *Proceedings International Rubber Conference*, Washington (1959) 237.
33. A.Y. Coran, *Rubber Chemistry and Technology* 37 (1964) 668.
34. J.E. Mark and M.Y. Tang, *Journal of Polymer Science: Polymer Physics Edition* 22 (1984) 1849.
35. J.E. Mark, *Polymer Journal* 17 (1985) 265.

Résumé. On a étudié l'effet de la nature chimique des liaisons dans le caoutchouc naturel sur son comportement à la propagation des fissures de fatigue. On contrôle la quantité relative de liaisons polysulfurées par rapport aux liaisons monosulfurées en faisant varier le rapport soufre-accélérateur de réaction. Par une vulcanisation sous peroxyde, on peut introduire des liaisons C-C. Tous les élastomères soumis à essais ont été préparés une même valeur moyenne de densité de liaisons, ce qui est confirmé par le gonflement à l'équilibre et par des mesures de module. A même densité de liaison, les liaisons polysulfurées se révèlent les plus résistantes en fatigue, sur la gamme des énergies d'arrachement étudiée. Des composants vulcanisés à liaisons principalement monosulfurées ont montré des vitesses de croissance d'une entaille plus faibles que des éprouvettes vulcanisées sous peroxyde, bien que la résistance du réseau mono-sulfuré puisse avoir été accrue par la présence de quelques liaisons polysulfurées.

International Journal of Fracture 39: 111–120 (1989)
© Kluwer Academic Publishers, Dordrecht

The conceptual physical framework of stochastic fracture kinetics

A.S. KRAUSZ* and K. KRAUSZ
*Faculty of Engineering, University of Ottawa, Canada

Received 20 September 1987; accepted in revised form 1 April 1988

Abstract. Physically based, rational, constitutive laws are developed by analysis or synthesis: the second represents the maturity of deformation and fracture science. The basic processes are thermally activated discrete steps, each controlled by the elementary rate constant derived rigorously by rate theory considerations of statistical mechanics. The rate constants are combined by the laws of kinetics, leading to the constitutive relations which express the rate of the process as a function of the applied load, component geometry, temperature, and microstructure. The physical process is stochastic and the appropriate mathematics is expressed by the Markov chain or by the Fokker–Planck transport differential equation: both are the appropriate forms of the random walk process. The constitutive laws can lead to direct engineering applications and have critical significance at the threshold conditions.

1. Constitutive laws: analysis or synthesis?

Without wanting to be rigid in classification, it is convenient to consider that the constitutive laws of crack growth (and plastic deformation) can be derived through either a process of analysis or synthesis. This is a convenient grouping; more often than not, actual research is a combination of both, but with one or the other dominating.

The *analysis* approach starts with observed experimental results and seeks to interpret these by using descriptive forms, empirical formulations, or, if possible, and as much as possible, physically based laws. The investigator is forced by the scope and character of the project to approximate the observed behavior and describe it as best as possible recognizing that the target, the observation, has to be met; often this can be achieved only by proposing specially designed mechanisms, by phenomenology, or only by empirical representation. These can be valuable: some of the analytical approaches turned out to be critical leads to physical understanding. Just to mention two: the well-known Arrhenius equation [1] although an empirical expression, paved the way to the all-important reaction rate theory of Eyring [2, 3] that has a fundamental role in time and temperature dependent crack growth descriptions; another that comes easily to mind is the logarithmic formula of Ludwik [4] and the sinh form [5]. Because of its very character, analysis provides guidance to "ongoing" problems and is, therefore, often given in simple mathematical form that lends itself well to engineering design practice. It is, however, severely limited in scope, and by its very nature and goals usually burdened by empiricism. Its statements are prone to revisions; its predictions limited in validity; its conclusions less than firm and frequently contested.

Synthesis starts from established, basic laws of physics (and chemistry) and derives conclusions using rigorous techniques to describe crack growth behavior. To do so, a model is constructed that represents conditions of crack growth having as wide interest as compatible with keeping the model simple. The model is constructed so that its behavior is described physically rigorously; the behavior of this model is then investigated. It provides a valid description of crack growth processes that are within the scope of the conditions of the model. The corresponding constitutive laws have permanent validity; their predictions are solid; they are not prone to be superceded by other proposed mechanisms, only expanded. The synthesis approach, however, requires a "global" viewpoint, usually not starting with a posed practical problem but has to work toward it; the physically based constitutive law may be quite complex because nature is rather subtle. Synthesis provides, however, the most mature and powerful means to the derivation of the constitutive laws of crack growth. In the following, some of the thoughts are presented on the synthesis approach to the understanding of time and temperature dependent crack growth. Their application to environment assisted cracking will also be indicated.

2. The stochastic synthesis model of crack growth

Environment assisted crack growth is thermally activated. Atomic bonds are broken in sequence when the sum of the mechanical energy, W, and the thermal energy, $\Delta G^*(W)$, is equal to the bond energy ΔG^*. The thermal energy varies in space and time randomly; the mechanical energy delivered to the crack tip depends on the microstructure and is also a random variable; the bond energy, or more precisely the energy needed to rearrange the atomic configuration that corresponds to a crack growth step is also a microstructural term and hence random even in an ideally homogeneous material. Hence, the rate of crack growth steps, ℓ, is a stochastic quantity. Figure 1 gives an illustration of this [6].

It was shown that a physically rigorous model can be constructed from this basic principle using the theory of rate processes and the kinetics of thermal activation. There are two techiques particularly powerful for synthesis modelling.

It is recognized that the stochastic process of crack growth is a random walk mechanism [7]. The process can then be described wit the Markov-chain mathematics of the typically stochastic behavior, and with the Fokker–Planck equation of transport processes [8–10]. Both were applied to environmental crack growth in stress corrosion cracking and corrosion fatigue [11–13].

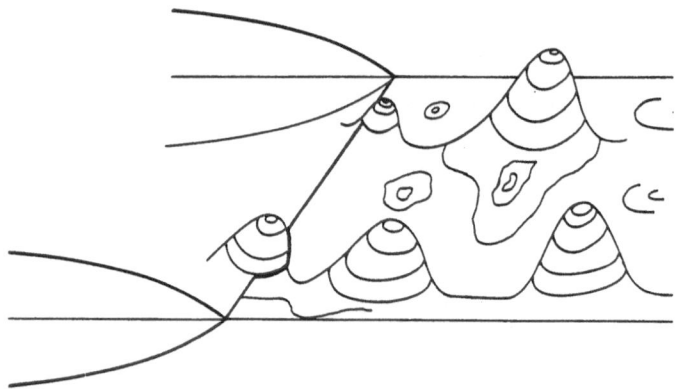

Fig. 1. A schematic representation of the random thermal energy in the plane of the crack.

The random walk, and one of its mathematical formulations the Markov-chain, is not a common concept. It represents, however, the actual discrete and stochastic behavior of energy transport in solids, and, therefore, the crack growth process. Once the initial strangeness wears off and becomes a familiar, comfortable concept, it provides powerful insights into the real process and, because of that, a simple and widely valid framework – the true physical tool of great efficiency. The random walk concept is essentially the expression of the stochastic "wandering" of the energy in the solid. On the arrival of an energy peak to the crack tip a bond breaking step, and thus a crack growth step, results: crack growth is controlled by stochastic energy transport.

"Much of transport of energy in the universe takes place by wave motion of some sort or another. The only competing mode of energy transport involves the bodily motion of matter, and even in this case, the motion of matter is described at a fundamental level by wave (quantum) mechanics" [14].

There are three basic descriptions (conditions, equations):

(1) kinematic;
(2) conservation (of mass or energy);
(3) dynamic.

(i) There are two types of these:

(a) Eulerian: describes what takes place at a point, as particles move by;
(b) Lagrangian: a volume element (particle) and its development, (motion) is followed up.

Their equivalent is expressed by

- what happens to an infinite number of cracks during the period t to $t + \Delta t$;
- what happens to one crack over an infinitely long period;
- one particle doing random motion over a long time;
- many particles doing random motion over the elementary period (this being the time unit that corresponds to the particle as a unit in the first case.)

That is: the fraction of cracks moving over the barrier during Δt is the same as the probability that one crack moves over the barriers in Δt time.

(ii) The conservation law = equation of continuity (in fluid flow, diffusion, etc.)

$$\frac{\partial \varrho}{\partial t} = - \operatorname{div} \varrho \bar{v} = - \nabla \cdot \varrho \bar{v},$$

where ϱ is the number of discrete units, t is the time, $\bar{v}$ is the velocity.

For these the variational principle applies and the Euler equation

$$\frac{\mathrm{d}}{\mathrm{d}\sigma} \left(\frac{\partial F}{\partial g'} \right) - \frac{\partial F}{\partial g} = 0$$

leads to the appropriate differential equations which, in the case of mass, or energy transfer, is the Fokker–Planck equation. The above equation is the necessary and sufficient condition

114   A.S. Krausz and K. Krausz

that the definite integral

$$I(g) = \int_{\sigma_1}^{\sigma_2} F \, d\sigma$$

be an extremum, where

$$F \equiv F[\sigma, g(\sigma), g'(\sigma)]$$

σ = independent variable;
g = unknown function of σ;
F = a given function of σ, g, g'; and
g has prescribed values at σ_1, σ_2.

It is easy to show that from this [15–17] the Fokker–Planck equation of typical transport process, stochastic conditions follow. For crack growth it was shown to be

$$\frac{\partial \varrho}{\partial t} = \tfrac{1}{2}a^2(\ell_b + \ell_h)\frac{\partial^2 \varrho}{\partial x^2} - a(\ell_b - \ell_h)\frac{\partial \varrho}{\partial x},$$

where ϱ is the number of cracks at the coordinate x and time t; a is the interatomic distance. The rate of bond breaking is

$$\ell_b = \frac{kT}{h}\exp\left(-\frac{\Delta G_b^* - W_b}{kT}\right),$$

and the rate of bond healing is

$$\ell_h = \frac{kT}{h}\exp\left(-\frac{\Delta G_h^* + W_h}{kT}\right),$$

where k and h is the Boltzmann and Planck constant respectively; T is the absolute temperature; and W is the function of the crack driving force such as K, ΔK, J-integral, C^*, etc. The solution of this differential equation, together with the boundary conditions, is the constitutive law of crack growth. It expresses the distribution of crack sizes as a function of the load and geometrical configuration, temperature, microstructure, and the time evolution. It provides a full, physically based rigorous description in quantitative terms.

When the physical process of crack growth is considered, as it should be, as a stochastic transport process, it is firmly established as one in the family of chemical reactions, diffusion, plastic deformation, heat transfer by conduction. All are controlled by thermal energy transfer; they form the same family and are interrelated. This is of particular advantage in the understanding of environment assisted crack growth, a major form of fracture.

Illustrative applications

Demonstrations of the concept expressed here are given in the following applications.

(a) Consider the threshold condition classically expressed as the design stress intensity K_{th}

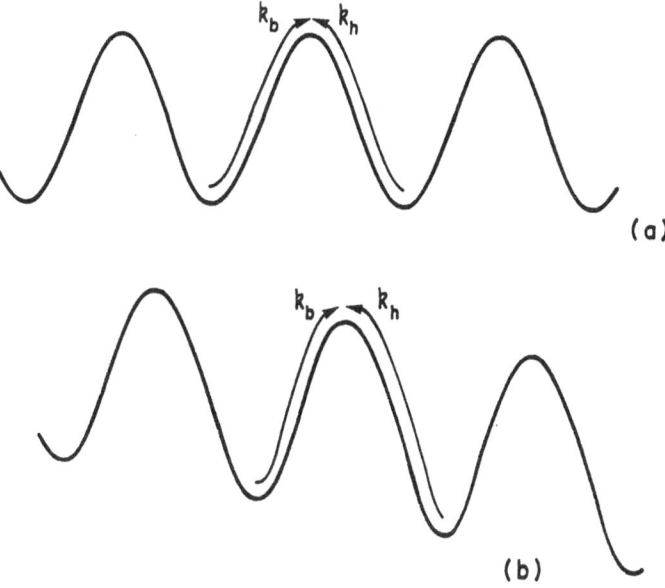

Fig. 2. The energy-barrier system: (a) equilibrium, threshold, condition, $k_b = k_h = k$; (b) slow crack growth, $k_b > k_h$.

at which cracks do not grow. The energy barrier in the absence of a driving force is shown in Fig. 2(a). At increased applied load the breaking and healing rate constants change as demonstrated in Fig. 2(b). When

$$k_b \ = \ k_h \ = \ k,$$

the conventional concept of crack velocity, v, expects that

$$v \ = \ L(k_b - k_h) \ = \ L(k - k) \ = \ 0$$

and no crack growth occurs. The stochastic character of the physical process, however, leads to the understanding that a distribution of crack sizes exists. At $K < K_{th}$ equilibrium consideration shows that the crack size, a, distribution is described as

$$\varrho(a)k_b \ = \ \varrho(a + a)k_h,$$

where $\varrho(a)$ and $\varrho(a + a)$ is the number of cracks of size a and $a + a$, with a the interatomic distance. Hence, the distribution is

$$\varrho(a + a) \ = \ \varrho(a) \exp\left[- \frac{\Delta G_b^*(W) - G_h^*(W)}{kT} \right].$$

This is recognized as the typical Boltzmann distribution.

When $k_b = k_h = k$ the stochastic model is represented by the differential equation

$$\frac{\partial \varrho}{\partial t} \ = \ a^2 k \frac{\partial^2 \varrho}{\partial x^2}.$$

The solution is

$$\varrho = \frac{\varrho_{\text{total}}}{2(\pi a^2 k t)^{1/2}} \exp\left(-\frac{x^2}{4a^2 k t}\right).$$

It is recognized that $4a^2 k t$ is the variance. Note that in the random walk concept the variance is defined as the root mean square of displacement of the crack tip.

The probability of crack of x size is, (when the crack can shrink as well as grow from the initial size)

$$P(x) = \frac{1}{(4a^2 \pi k t)^{1/2}} \exp\left(-\frac{x^2}{4a^2 k t}\right).$$

The mean square displacement is $\overline{\Delta x^2}$: this is the quantity that has to be determined because the positive and negative displacements are equally probable, $k_b = k_h = k$. By definition

$$\overline{\Delta x^2} = \int_{-\infty}^{+\infty} x^2 P(x)\, dx.$$

Substitution of $\S^2 = x^2/(4a^2 k t)$ leads through

$$\S = \frac{x}{(4a^2 k t)^{1/2}}; \quad d\S = (4a^2 k t)^{-1/2}\, dx; \quad \text{and}$$

$$P(x) = \frac{1}{(4a^2 \pi k t)^{1/2}} \exp(-\S^2), \text{ to}$$

$$\overline{\Delta x^2} = 2a^2 k t = \text{standard deviation.}$$

It is then clear that at the classical threshold design condition

$$k_b = k_h$$

cracks grow, even to fracture, in obvious contradiction to the usual expectations. The stochastic physical condition can be demonstrated by the following example. Consider the apparent activation energy of

$$\Delta G^+(W) = 0.75\,\text{eV},$$

a very realistic value for structural metals at threshold load, and the operating temperature of 600 K. The rate constant is

$$k = \frac{kT}{h} \exp\left[-\frac{\Delta G^+(W)}{kT}\right]$$

$$\simeq 0.6 \times 10^{13}\,\text{sec}^{-1} \exp\left(-\frac{0.75\,\text{eV}}{0.05\,\text{eV}}\right)$$

$$= 2 \times 10^5\,\text{sec}^{-1}.$$

About 32 percent of the cracks, which is the standard deviation, will be of $2^{1/2} a(\ell t)^{1/2}$ size. During the test program that measures the threshold condition, for say one week period ($\simeq 6 \times 10^5$ sec), one standard deviation is about 2×10^{-8} cm(2×10^5 sec$^{-1} \times 6 \times 10^5$ sec)$^{1/2} \simeq 7 \times 10^{-3}$ cm, a size growth below the sensitivity of detection techniques. In service, however, conditions are very different. In two years, cracks of one millimeter standard deviation size will appear. These will be in numbers that correspond to the 32 percent of defects that can grow. There will be, of course, ~ 2 percent (that is, the equivalent of three standard deviation) of cracks that reached 3 mm size and so large enough to have grown even further because the stress intensity factor increased substantially during this period due to the increase in crack size by an order of magnitude. It follows that many of the cracks that appear in structures over their service life are not caused by defective production, or testing, or the Weibull effect, but are the inevitable consequence of the stochastic character of the physical process itself. For valid design as well as for maintenance scheduling this has to be taken into consideration. Through this realization, rational, physically based designs can be developed.

A case of exactly this type occurred just recently: Zirconium alloy heat exchanger tubes developed observable cracks after about 15 years of service, at about half of the expected service life. No accident occurred, but the tubes had to be replaced [18]; a fracture kinetics based research and development program was successfully applied to prevent further operational difficulties.

(b) Previous studies in stress corrosion cracking and corrosion fatigue have shown that crack growth steps are thermally activated. The rate of these steps is described by the elementary rate constant as

$$\ell = \frac{kT}{h} \exp\left(-\frac{\Delta G^+ - W}{kT}\right),$$

where W is a function of the stress intensity factor K

$$W = \alpha K.$$

The crack velocity is then

$$v = L \frac{kT}{h} \exp\left(-\frac{\Delta G^+ - \alpha K}{kT}\right).$$

The crack grows during a fatigue cycle as

$$\frac{da}{dN} = \int_{t_1}^{t_2} L \frac{kT}{h} \exp\left[-\frac{\Delta G^+ - \alpha K(t)}{kT}\right] dt.$$

In positive saw-tooth loading, or when the load-time relation can be approximated with this pattern, the growth per cycle is

$$\frac{da}{dN} = L \frac{kT}{h} \exp\left(-\frac{\Delta G^+ - \alpha K_0}{kT}\right) \int_0^\tau \exp \frac{\alpha}{kT} \frac{\Delta K}{\tau} t \, dt,$$

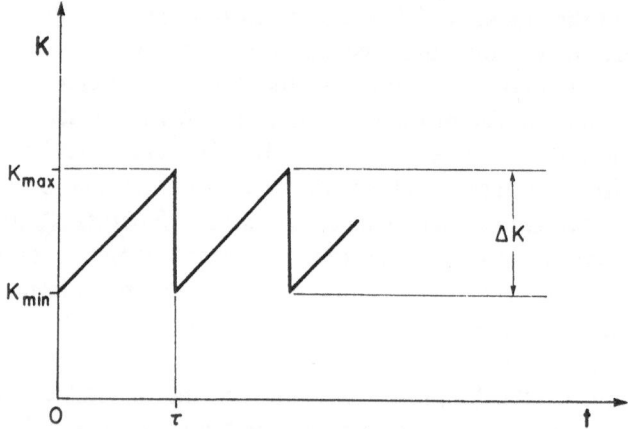

Fig. 3. The load vs. time relation for positive saw-tooth loading. The stress intensity varies as $K = K_0 + (\Delta K/\tau)t$.

where τ is the cycle time (Fig. 3). Accordingly,

$$\frac{\mathrm{d}a}{\mathrm{d}N} = L\frac{(kT)^2}{h}\frac{\tau}{\alpha\Delta K}\exp\left(-\frac{\Delta G^* - \alpha K_0}{kT}\right)\left(\exp\frac{\alpha}{kT}\Delta K - 1\right).$$

The pre-exponential ΔK factor has a much weaker effect than in the exponent and its variation over the region above the inflexion can be neglected. Because

$$\exp\frac{\alpha}{kT}\Delta K \gg 1$$

the growth rate is

$$\frac{\mathrm{d}a}{\mathrm{d}N} = A\ell_0 \exp\frac{\alpha}{kT}\Delta K.$$

It is of interest to note that this fracture kinetics development demonstrates the physical reason for the dependence of fatigue rate on the stress intensity range.

Acknowledgement

One of the authors (A.S. Krausz) gratefully acknowledges the financial assistance of the Natural Sciences and Engineering Research Council of Canada.

Appendix

A numerical example to illustrate the effect of the pre-exponential ΔK

The expression for crack growth rate is

$$\frac{\mathrm{d}a}{\mathrm{d}N} = A'\frac{kT}{\alpha\Delta K}\left[\exp\left(\frac{\alpha}{kT}\Delta K\right) - 1\right], \tag{1}$$

where

$$A' = L \frac{kT}{h} \tau \exp\left(-\frac{\Delta G^* - \alpha K_0}{kT}\right),$$

and it is considered that an approximation

$$\frac{\mathrm{d}a}{\mathrm{d}N} \simeq A \exp\left(\frac{\alpha}{kT} \Delta K\right) \tag{2}$$

can be used. The following illustrates that this is valid.

For (2), it was determined [11] that

$$A = 3.8 \times 10^{-9} \, \mathrm{m/cycle}$$

and

$$\frac{\alpha}{kT} = 0.3 \, \mathrm{MN}^{-1} \, \mathrm{m}^{3/2}$$

Consideration of (1), without the -1 term, leads to

$$A = 9.6 \times 10^{-9} \, \mathrm{m/cycle}$$

and

$$\frac{\alpha}{kT} = 0.36 \, \mathrm{MN}^{-1} \, \mathrm{m}^{3/2}.$$

When the -1 term is also considered, i.e. the full (1), the slope controlling α/kT remains unchanged, only A varies. With $\alpha/kT = 0.36 \, \mathrm{MN}^{-1} \, \mathrm{m}^{3/2}$ the value of A is now

$$A = 9.68 \times 10^{-9} \, \mathrm{m/cycle},$$

an obviously negligible, ~ 1 percent, difference.

References

1. S. Arrhenius, *Zeitschrift für Physikalische Chemie* 4 (1889) 226.
2. S. Glasstone, K.J. Laidler, and H. Eyring, *The Theory of Rate Processes*, McGraw-Hill, New York (1941).
3. A.S. Krausz and H. Eyring, *Deformation Kinetics*, Wiley-Interscience, New York (1975).
4. P. Ludwik, *Elemente der Technologischen Mechanik*, Springer, Berlin (1909).
5. A. Nadai, *Theory of Flow and Fracture of Solids*, McGraw-Hill, New York (1950).
6. A.S. Krausz and K. Krausz, *Fracture Kinetics of Crack Growth*, Kluwer Academic, Dordrecht (1988).
7. A.S. Krausz, *Engineering Fracture Mechanics* 12 (1979) 499–504.
8. S. Chandrasekhar, *Reviews of Modern Physics*, 15 (1943) 1–89.

9. A.S. Krausz and K. Krausz, *International Journal of Engineering Science* 22 (1984) 1075–1081.

10. A.S. Krausz, K. Krausz, and D.-S. Necsulescu, in *Proceedings of the International Conference on Numerical Methods for Non-Linear Problems*, Vol. 2, Pineridge Press, Swansea, U.K. (1984).

11. A. S. Krausz and K. Krausz, in *Proceedings of the ASTM 20th National Symposium on Fracture Mechanics: Perspectives and Directions*, Lehigh University (1987) STP publication in press.

12. A.S. Krausz, K. Krausz, and D.-S. Necsulescu, *Zeitschrift für Naturforschung* 38a (1983) 497–502.

13. A.S. Krausz and K. Krausz, *Transactions ASME, Journal of Mechanical Design*, Special Issue 104 (1982) 666–670.

14. W.C. Elmore and M.A. Heald, in *Physics of Waves*, Dover, New York (1985) 45.

15. A.S. Krausz, J. Mshana, and K. Krausz, *Engineering Fracture Mechanics* 13 (1980) 759–766.

16. A.S. Krausz, *International Journal of Fracture* 12 (1976) 239–242.

17. A.S. Krausz and K. Krausz, in *Materials Science Monographs*, 38B, Proceedings of the World Congress on High Tech Ceramics, P. Vicenzini (ed), Elsevier, Amsterdam (1987) 1239–1245.

18. E.C.W. Perryman, *Nuclear Engineering* 17 (1978) 95–105.

Résumé. L'analyse ou la synthèse constituent deux voies pour développer des lois constitutives, rationnelles et basées sur la Physique. La seconde voie est représentative de la maturité de la science des déformations et de la rupture. Les processus fondamentaux relèvent d'étapes discrètes activées par effet thermique, chacune étant contrôlée par une constante de vitesse élémentaire, déduite de considérations sur la théorie de la vitesse en mécanique statistique. Combinées aux lois de la cinétique, les constantes de vitesse conduisent aux expressions constitutives de la vitesse d'un processus en fonction de la charge appliquée, de la géométrie du composant, de la température et de la microstructure. Le processus physique a un caractère stochastique et son expression mathématique est obtenue par une chaîne de Markov ou par une équation différentielle de transfert de Fokker-Planck. Ces deux approches sont les formes adéquates pour un processus itératif aléatoire. Les lois constitutives peuvent mener à des applications directes en construction, et ont une signification critique aux conditions limites.

International Journal of Fracture 39: 121–127 (1989)
© Kluwer Academic Publishers, Dordrecht

Viscoplastic flow due to penetration: a free boundary value problem

DANG DINH ANG, TIM FOLIAS, FRITZ KEINERT and FRANK STENGER
Department of Mathematics, University of Utah, Salt Lake City, Utah 84112, USA

Received 10 December 1987; accepted in revised form 1 April 1988

Abstract. Under the action of a pressure gradient, a solid body B penetrates into another body. Body B is assumed to be of an incompressible, viscoplastic, Bingham material. As a first model, the problem may be treated one-dimensionally in the space variable x as well as the time variable t.

By utilizing the Green's function, the location of the moving boundary $s(t)$, i.e., the boundary between the region of viscoplastic flow and the core, is expressed in terms of an integral equation, the solution of which may then be sought numerically.

1. Formulation of the problem

A solid body B of width $2H$ and under the action of a pressure gradient, penetrates into another body, in an action similar to that of a bullet entering an object. We assume that body B is an incompressible viscoplastic Bingham body, that is, it satisfies Bingham's law

$$\tau - \tau_0 = \pm \mu \frac{\partial u}{\partial x}, \tag{1}$$

where τ_0 is the yield stress, μ the coefficient of viscosity and u the velocity in the y-direction. The movement is in the y-direction only and is assumed to be independent of z and symmetric about the plane $x = H$ (see Fig. 1).

The body B is divided into two parts

$$B_1 = \{x: |x| < s(t) \quad \text{or} \quad |x| > 2H - s(t)\}$$

$$B_2 = \{x: s(t) \leq |x| \leq 2H - s(t)\}.$$

In B_1 (resp. B_2) the tangential stress is larger (resp. smaller) than the yield stress τ_0. We call B_1 the zone of viscoplastic flow and B_2 the core.

In the zone of viscoplastic flow, the velocity $u(x, t)$ satisfies the diffusion equation* (see Rubinstein [2], Chapter 4)

$$k^2 \frac{\partial u}{\partial t} + \frac{1}{v\varrho} \frac{\partial p}{\partial y} = \frac{\partial^2 u}{\partial x^2}, \quad k^2 = v^{-1}, \tag{2}$$

---

* The reader should notice that at the interface $x = s(t)$, $\partial u/\partial x = 0$.

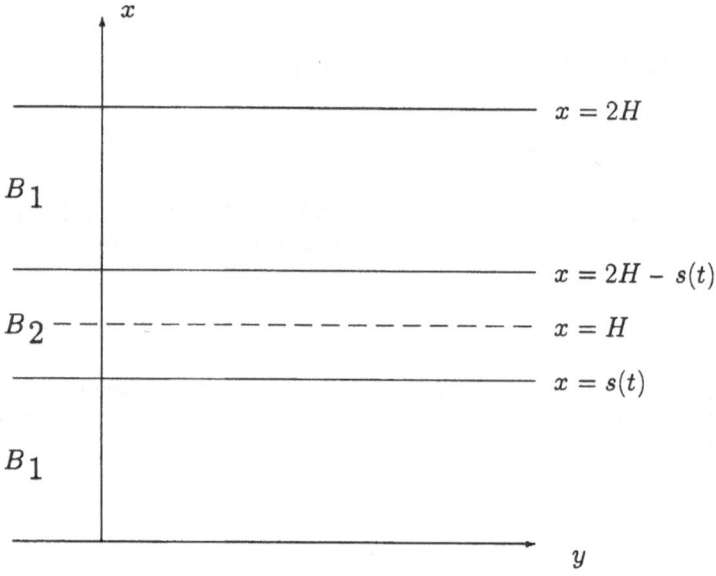

Fig. 1. Geometrical configuration.

where $\partial p/\partial y$ is the pressure gradient in the y-direction and may be interpreted as a driving force, ϱ the (constant) density and v the kinematic viscosity. Due to symmetry, it is sufficient to consider (2) in the domain $0 < x < s(t)$, and furthermore we assume that

$$u(0, t) = f(t), \tag{3}$$

$$\frac{1}{\varrho} \frac{\partial p}{\partial y} = g(t), \tag{4}$$

where $f(t)$ and $g(t)$ are given functions. Since the core is rigid, the velocity in it is

$$u = u_0(t) = u(s(t), t), \tag{5}$$

where it is assumed that $u_0(0) \neq 0$. At the interface $x = s(t)$, the tangential stress is equal to the yield stress, and hence by (1) we must have

$$\frac{\partial u}{\partial x} = 0 \quad \text{at} \quad x = s(t). \tag{6}$$

The problem therefore is to determine $s(t)$ from the above conditions as well as the following

$$s(0) = b > 0 \tag{7}$$

$$u(x, 0) = \phi(x), \quad \phi(0) = f(0) \tag{8}$$

$$\dot{u}_0(t) = -g(t) - \frac{\tau_0}{\varrho(H - s(t))}, \tag{9}$$

where the dot on top of a function indicates differentiation and (9) follows from a consideration of the forces acting on the core (e.g., see [2]).

Since by (6)

$$\dot{u}_0(t) \;=\; \frac{\mathrm{d}}{\mathrm{d}t}\,[u(s(t),\,t)] \;=\; u_x(s(t),\,t)\dot{s}(t) \,+\, u_t(s(t),\,t) \;=\; u_t(s(t),\,t),$$

we obtain from (2), (4), by letting $x \to s(t)$, that

$$\dot{u}_0(t) \;=\; -g(t) \,+\, vu_{xx}(s(t),\,t), \tag{10}$$

which upon comparing it with (9) gives

$$u_{xx}(s(t),\,t) \;=\; -\,\frac{\tau_0}{v\varrho(H - s(t))}. \tag{11}$$

Moreover, in order to be compatible with our previous assumptions, at $t = 0$ we must require that

$$\ddot{\phi}(b) \;=\; -\,\frac{\tau_0}{v\varrho(H - b)}. \tag{12}$$

Notice that in this analysis we have for simplicity assumed that $s(0) > 0$. The case $s(0) = 0$ requires some special mathematical rigor which for the sake of brevity we will omit. Perhaps it is appropriate at this point to comment on the difference between the present problem and the classical Stefan problem. For the classical Stefan problem, the location of the moving boundary, $x = s(t)$, is governed by the velocity u as well as its derivative with respect to x, whereas in the present problem it is also governed by the time derivative, i.e., an additional constraint which makes the solution even more difficult.

For the solution of the problem, we shall use the method of Green's functions. However, before engaging in the details of the construction of the solution, we first define the concept of a solution to our problem. By a solution to our problem, henceforth called the FBP, is meant an ordered pair $u(x, t)$, $s(t)$ of functions, $u(x, t)$ defined on $0 \leqslant x \leqslant s(t), 0 \leqslant t \leqslant \sigma$, $s(t)$ defined on $0 \leqslant t \leqslant \sigma$, for some $\sigma > 0$, such that

    (i) u_{xx}, u_t are continuous in $0 \leqslant x \leqslant s(t)$ for $0 < t < \sigma$
    (ii) u and u_x are continuous for $0 \leqslant x \leqslant s(t), 0 \leqslant t \leqslant \sigma$
    (iii) u satisfies (2) in $0 < x < s(t), 0 < t \leqslant \sigma$
    (iv) conditions (3)–(9) are satisfied
    (v) is Lipschitzian on $(0, \sigma]$.

2. Method of solution

We shall formulate the problem in terms of an integral equation, and for this purpose we shall require some regularity conditions on the initial and boundary data:

(vi) $f(t)$ is continuous, $g(t)$ is C^1 on $t \geq 0$

(vii) $\phi(x)$ is C^2 on $(0, b)$, and the left-hand derivative $\dot\phi(b)$ exists

(viii) $s(t)$ is C^1 for $t \geq 0$.

We now define

$$K(x, t; \xi, \tau) = \frac{k}{2\sqrt{\pi}} \frac{1}{\sqrt{t - \tau}} \exp\left(-\frac{k^2(x - \xi)^2}{4(t - \tau)}\right),$$

and

$$G(x, t; \xi, \tau) = K(x, t; \xi, \tau) - K(x, t; -\xi, \tau)$$

$$N(x, t; \xi, \tau) = K(x, t; \xi, \tau) + K(x, t; -\xi, \tau)$$

$$0 < x < s(t), \quad 0 < \xi < s(\tau), \quad 0 < \tau < t.$$

These are the Green's functions we shall use. For their various properties the reader is referred to Friedman [1] (chapter on free boundary value problems) or to Rubinstein [2]. We shall use them freely in our subsequent analysis without explicit mention of the references.

Thus, let $u(x, t)$, $s(t)$ be a solution of our FBP. Integrating the identity

$$(Gu_\xi - G_\xi u)_\xi - k^2(Gu)_\tau = k^2 Gg \tag{13}$$

over the region $\{(\xi, \tau): 0 \leq \xi \leq s(\tau), \varepsilon \leq \tau \leq t - \varepsilon\}$, applying Green's identity and letting $\varepsilon \to 0$, we obtain

$$u(x, t) = \int_0^b \phi(\xi)G(x, t; \xi, 0)d\xi - \frac{1}{k^2}\int_0^t u_0(\tau)G_\xi(x, t; s(\tau), \tau)d\tau$$

$$+ \int_0^t u_0(\tau)G(x, t; s(\tau), \tau)\dot s(\tau)d\tau + \frac{1}{k^2}\int_0^t f(\tau)G_\xi(x, t; 0, \tau)\,d\tau$$

$$- \int_0^t \int_0^{s(\tau)} G(x, t; \xi, \tau)d\xi g(\tau)d\tau \quad 0 < x < s(t), \quad t > 0. \tag{14}$$

We now differentiate both sides of (14) with respect to x for $0 < x < s(t)$

$$u_x(x, t) = \int_0^b \phi(\xi)G_x(x, t; \xi, 0)d\xi - \int_0^t u_0(\tau)N_\tau(x, t; s(\tau), \tau)d\tau$$

$$+ \int_0^t u_0(\tau)G_x(x, t; s(\tau), \tau)\dot s(\tau)d\tau + \int_0^t f(\tau)N_\tau(x, t; 0, \tau)d\tau$$

$$- \int_0^t \int_0^{s(\tau)} G_x(x, t; \xi, \tau)d\xi g(\tau)d\tau, \quad 0 < x < s(t), \quad t > 0, \tag{15}$$

where we have made use of the identity

$$G_{x\xi} = k^2 N_\tau.$$

Integrating by parts, we have

$$\int_0^b \phi(\xi)G_x(x, t; \xi, 0)d\xi = -\int_0^b \phi(\xi)N_\xi(x, t; \xi, 0)d\xi$$

$$= \phi(0)N(x, t; 0, 0) - \phi(b)N(x, t; b, 0)$$

$$+ \int_0^b \phi'(\xi)N(x, t; \xi, 0)d\xi \tag{16}$$

$$-\int_0^t u_0(\tau)N_\tau(x, t; s(\tau), \tau)d\tau$$

$$= \int_0^t u_0(\tau)\left[N_\xi(x, t; s(\tau), \tau)\dot{s}(\tau) - \frac{d}{d\tau}N(x, t; s(\tau), \tau) \right]d\tau$$

$$= -\int_0^t u_0(\tau)G_x(x, t; s(\tau), \tau)\dot{s}(\tau)d\tau - u_0(t)N(x, t, s(t), t)$$

$$+ u_0(0)N(x, t; b, 0) + \int_0^t \dot{u}_0(\tau)N(x, t; s(\tau), \tau)d\tau \tag{17}$$

$$\int_0^t f(\tau)N_\tau(x, t; 0, \tau)d\tau = -f(0)N(x, t; 0, 0) - \int_0^t \dot{f}(\tau)N(x, t; 0, \tau)d\tau \tag{18}$$

$$\int_0^{s(\tau)} G_x(x, t; \xi, \tau)d\xi = -\int_0^{s(\tau)} N_\xi(x, t; \xi, \tau)d\xi$$

$$= -N(x, t; s(\tau), \tau) + N(x, t; 0, \tau). \tag{19}$$

Substituting (16)–(19) and (9) into (15) and simplifying yields

$$u_x(x, t) = \int_0^b \phi'(\xi)N(x, t; \xi, 0)d\xi - \frac{\tau_0}{\varrho}\int_0^t \frac{1}{H - s(\tau)}N(x, t; s(\tau), \tau)d\tau$$

$$- \int_0^t [\dot{f}(\tau) + g(\tau)]N(x, t; 0, \tau)d\tau \tag{20}$$

and upon letting $x \nearrow s(t)$, by (6)

$$0 = \int_0^b \phi'(\xi)N(s(t), t; \xi, 0)d\xi - \frac{\tau_0}{\varrho}\int_0^t \frac{1}{H - s(\tau)}N(s(t), t; s(\tau), \tau)d\tau$$

$$- \int_0^t [\dot{f}(\tau) + g(\tau)]N(s(t), t; 0, \tau)d\tau. \tag{21}$$

Consider what happens in (21) as $t \to 0$. Splitting up the first integral by using $N(s(t), t; \xi, 0) = K(s(t), t; \xi, 0) + K(s(t), t; -\xi, 0)$ and likewise the second one, we observe that three of the resulting five integrals tend to zero exponentially. The other two have leading terms of $\sqrt{t}$, which of course have to cancel. Considering only the leading terms, we find

$$s(t) - s(\tau) \approx \dot{s}(t)(t - \tau)$$

$$K(s(t), t; s(\tau), \tau) \approx \frac{k}{2\sqrt{\pi}} (t - \tau)^{-1/2}$$

so

$$\frac{\tau_0}{\varrho} \int_0^t \frac{1}{H - s(\tau)} K(s(t), t; s(\tau), \tau) d\tau \approx \frac{\tau_0}{\varrho} \frac{k}{\sqrt{\pi}} \frac{1}{H - s(t)} t^{1/2}. \tag{22}$$

Also,

$$\int_0^b K(s(t), t; \xi, 0) d\xi = \frac{1}{\sqrt{\pi}} \int_{k(s(t)-b)/2\sqrt{t}}^{ks(t)/2\sqrt{t}} e^{-x^2} dx$$

$$\approx 1/2 - \frac{k\dot{s}(0)}{2\sqrt{\pi}} t^{1/2}$$

and

$$\int_0^b (\xi - s(t)) K(s(t), t; \xi, 0) d\xi \approx \frac{\sqrt{t}}{k\sqrt{\pi}} [e^{-k^2 s(t)^2/4t} - e^{-k^2 \dot{s}(0)^2 t}]$$

$$\approx - \frac{1}{k\sqrt{\pi}} t^{1/2}.$$

As $t \to 0$, $K(s(t), t; \xi, 0)$ behaves like an approximate δ-function with peak near b. Thus, if h is a C^1-function, then

$$\int_0^b h(\xi) K(s(t), t; \xi, 0) d\xi \approx \int_0^b [h(b) + \dot{h}(b)(\xi - s(t) + \dot{s}(t)t)] K(s(t), t; \xi, 0) d\xi$$

$$\approx \tfrac{1}{2} h(b) - \frac{1}{\sqrt{\pi}k} \left[\frac{k^2 \dot{s}(0)}{2} h(b) - \dot{h}(b) \right] t^{1/2}. \tag{23}$$

In particular, if $h = \phi$, we find

$$\int_0^b \phi(\xi) K(s(t), t; \xi, 0) d\xi \approx \frac{\ddot{\phi}(b)}{k\sqrt{\pi}} t^{1/2}. \tag{24}$$

Thus, as mentioned above, the consistency of (21) as t approaches zero requires condition (12). The rough estimates above can be made more rigorous and lead to the conclusion that if the given functions f, g, ϕ are smooth and (12) is violated, s cannot be Lipschitzian at $t = 0$. A sharp corner in the moving boundary is expected in this case.

3. Numerical method

Assuming that (12) is satisfied and $s(t)$ is smooth, we find

$$N(s(t), t; s(\tau), \tau) = \frac{k}{2\sqrt{\pi}} (t - \tau)^{-1/2} + F(s(t), t; s(\tau), \tau),$$

where F is smooth for $\tau < t$, and

$$F(s(t), t; s(\tau), \tau) = O((t - \tau)^{1/2}) \quad \text{as} \quad \tau \to t.$$

Thus we have

$$\frac{k}{2\sqrt{\pi}} \frac{\tau_0}{\varrho} \int_0^t \frac{1}{H - s(\tau)} (t - \tau)^{-1/2} d\tau = \int_0^b \phi(\xi) N(s(t), t; \xi, 0) d\xi$$

$$- \frac{\tau_0}{\varrho} \int_0^t \frac{1}{H - s(\tau)} F(s(t), t; s(\tau), \tau) d\tau$$

$$- \int_0^t [\dot{f}(\tau) + g(\tau)] N(s(t), t; 0, \tau) d\tau, \tag{25}$$

The integral equation may now be solved by an iteration scheme. Starting with an initial guess $s^{(0)}(t)$ for the moving boundary, for example $s^{(0)}(t) \equiv b$, we can substitute the ith iterate $s^{(i)}(t)$ into the right-hand side of (25) and calculate a new approximation $s^{(i+1)}(t)$.

We are now in the process of testing this method and plan to publish the results in a subsequent paper.

Acknowledgements

Part of this research was supported by U.S. Army research contract No. DAAL03-87-K-0008. The authors gratefully acknowledge this support.

References

1. A. Friedman, *Partial Differential Equations of Parabolic Type*, Prentice-Hall, Englewood Cliffs (1964).
2. L.I. Rubinstein, *The Stefan Problem*, Translations of Mathematical Monographs, 27, American Mathematical Society, Providence (1970).

Résumé. On traite le cas d'un solide B, supposé incompressible, viscoplastique et en matériau de Bingham, dans un autre corps sous l'effet d'un gradient de pression. En première analyse, le problème peut être traité suivant une dimension, sur une variable d'espace x ou de temps t.

En recourant à une fonction de Green, on exprime sous forme d'une équation intégrale la position de la frontière en mouvement $s(t)$, à savoir la frontière entre la région d'écoulement viscoplastique et la portion dure. La solution de cette équation peut être trouvée par voie numérique.

International Journal of Fracture 39: 129–146 (1989)
© Kluwer Academic Publishers, Dordrecht

The three-dimensional stress field around a cylindrical inclusion in a plate of arbitrary thickness

F.E. PENADO[1] and E.S. FOLIAS[2]

[1]*Division of Design and Analysis, Hercules Aerospace Co., Magna, Utah 84044, USA;* [2]*Department of Mechanical Engineering, University of Utah, Salt Lake City, Utah 84112, USA*

Received 1 November 1987; accepted in revised form 1 April 1988

Abstract. The three-dimensional Navier's equations are solved analytically for the case of a cylindrical inclusion of radius "a" which is embedded in a plate of arbitrary thickness $2h$. Both the plate and the inclusion are assumed to be of homogeneous and isotropic materials with different material properties. Perfect bonding is assumed to prevail at the interface. As to loading, a uniform tension is applied in the plane of the plate at points remote from the inclusion.

The analysis shows all stresses including the octahedral shear stress to be sensitive to the radius to half thickness ratio (a/h) as well as the material properties. In the limit, as (μ_2/μ_1) → 0 and as (μ_2/μ_1) → 1 (where μ_2 and μ_1 are, respectively, the shear moduli of the inclusion and of the plate) the results for a cylindrical hole and a continuous plate are recovered. Similarly as (a/h) → ∞ (very thin plate) the plane stress solution is recovered. Moreover, for (μ_2/μ_1) > 1.0 the presence of a stress singularity near the point of intersection of the inclusion and the free surface of the plate is confirmed by the numerical results.

1. Introduction

The three-dimensional stress field around a cylindrical inclusion which is embedded in a plate is of considerable importance to the field of fracture mechanics. For example, solutions of this type can help us to understand better the failure mechanism in fiber-reinforced materials. Although an analytical solution to the title problem does not exist, related three-dimensional solutions can be found in the literature. These include an axially loaded rod partially embedded in an elastic half space for the cases of the elastic [1] or rigid [2] rod. The absence of a solution to the problem under consideration is not due to a lack of interest, but rather to the mathematical complexities encountered in solving this kind of three-dimensional problem. Thus, the purpose of this paper is twofold: first, to provide an analytical model for an isolated fiber in a matrix under uniaxial transverse tension; second to lay out the mathematical foundations that will allow the solution of similar three-dimensional problems in elasticity.

Two-dimensional solutions (plane stress or plane strain) for plates with perfectly bonded circular inclusions can be found in the literature for single [3] as well as multiple inclusions [4]. Reference [4] is an extension of the single inclusion case obtained by using the Schwarz method of successive approximations. Other related two-dimensional solutions involving a smooth (frictionless) circular inclusion that does not separate from its surrounding plate are found in [5] and [6] and correpond, respectively, to cases where the diameter of the inclusion is the same or larger than that of the hole in the plate. It was assumed in [5] and [6] that the plate and inclusion were of the same material. The case when a smooth circular inclusion separates from a plate of different material properties is discussed in [7].

Two-dimensional asymptotic solutions for the stresses in two bonded wedges of dissimilar materials in the neighborhood of the intersection of the free and bonded edges are found in [8] for orthogonal wedges, and in [9] for wedges of an arbitrary angle. The analysis shows that the stresses are proportional to $\varrho^{-\alpha}$, where α depends on the shear moduli ratio and the two Poisson's ratios. The largest value of α found in [8] was 0.311 and occurred for the case of one material being rigid and the other incompressible. Recently, Folias [10] used a three-dimensional analysis to investigate the asymptotic behavior of the stresses in the neighborhood of the intersection of a cylindrical inclusion and the free surface of a plate. The analysis shows the stresses to be singular for $\mu_2/\mu_1 > 1$ (where μ_2 and μ_1 are, respectively, the shear moduli of the inclusion and of the plate), with the strength of the singularity increasing as the ratio μ_2/μ_1 increases.

2. Formulation of the problem

Consider the equilibrium of a body which occupies the space $|x| < \infty, |y| < \infty, |z| \leqslant h$ and contains two regions of different elastic properties. Their common boundary consists of a through-the-thickness cylindrical surface of radius $r = a$, whose generators are parallel to the z-axis (see Fig. 1). The regions $r \geqslant a$ and $r \leqslant a$ are called, respectively, plate and inclusion and are denoted by the superscripts (1) and (2). Both the plate and the inclusion are considered to be made of homogeneous, isotropic and linearly elastic materials. At the

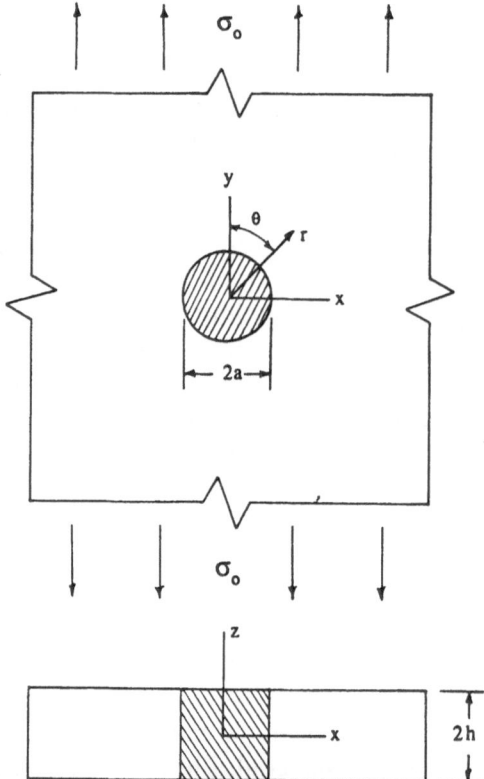

Fig. 1. Infinite plate of arbitrary thickness with cylindrical inclusion.

interface ($r = a$) perfect bonding is assumed to prevail. As to loading, a uniform tension σ_0 is applied at the boundary of the plate at points remote from the inclusion. For both regions, the surfaces $|z| = h$ are assumed to be free of stresses and constraints.

In the absence of body forces, the coupled differential equations governing the displacement functions $u^{(i)}$, $v^{(i)}$ and $w^{(i)}$ ($i = 1, 2$) are

$$\frac{m_i}{m_i - 2} \frac{\partial e^{(i)}}{\partial x_k} + \nabla^2 u_k^{(i)} = 0; \quad i = 1, 2; k = 1, 2, 3, \tag{1}$$

where ∇^2 is the 3D Laplacian operator, $m_i \equiv 1/v_i$, v_i is Poisson's ratio and

$$e^{(i)} \equiv \frac{\partial u_k^{(i)}}{\partial x_k}; \quad k = 1, 2, 3. \tag{2}$$

The stress-displacement relations are given by Hooke's law as:

$$\sigma_{kl}^{(i)} = 2\mu_i \left\{ \frac{1}{m_i - 2} e_{jj}^{(i)} \delta_{lk} + e_{lk}^{(i)} \right\}; \quad k, l = 1, 2, 3, \tag{3}$$

where μ_i are the respective shear moduli.

As to the boundary conditions, one must require that

$$\text{as } |x| \to \infty: \sigma_{xx}^{(1)} = \tau_{xy}^{(1)} = \tau_{xz}^{(1)} = 0 \tag{4}$$

$$\text{as } |y| \to \infty: \tau_{xy}^{(1)} = \tau_{yz}^{(1)} = 0, \sigma_{yy}^{(1)} = \sigma_0 \tag{5}$$

$$\text{at } |z| = h: \tau_{xz}^{(i)} = \tau_{yz}^{(i)} = \sigma_{zz}^{(i)} = 0; \quad i = 1, 2 \tag{6}$$

$$\text{at } r = a: \sigma_{rr}^{(1)} - \sigma_{rr}^{(2)} = \tau_{r\theta}^{(1)} - \tau_{r\theta}^{(2)} = \tau_{rz}^{(1)} - \tau_{rz}^{(2)} = 0 \tag{7}$$

$$u_r^{(1)} - u_r^{(2)} = u_\theta^{(1)} - u_\theta^{(2)} = u_z^{(1)} - u_z^{(2)} = 0. \tag{8}$$

Finally, at $r = 0$ we must require that all stresses and displacements be bounded.

It is found convenient at this stage to seek the solution to (1) in the form:

$$u^{(i)} = u^{(p)(i)} + u^{(c)(i)} \tag{9}$$

$$v^{(i)} = v^{(p)(i)} + v^{(c)(i)}; \quad i = 1, 2, \tag{10}$$

$$w^{(i)} = w^{(p)(i)} + w^{(c)(i)} \tag{11}$$

where the component with the superscript (p) represents the particular solution, and the component with the superscript (c) the complementary solution.

The particular solution in cylindrical coordinates is:
(i) for the plate:

$$u_{rr}^{(p)(1)} = \frac{\sigma_0 r}{4\mu_1} \left[\frac{1 - v_1}{1 + v_1} + \cos(2\theta) \right] \tag{12}$$

$$u_\theta^{(p)(1)} = -\frac{\sigma_0 r}{4\mu_1} \sin(2\theta) \tag{13}$$

$$u_{zz}^{(p)(1)} = -\frac{\sigma_0}{2\mu_1} \frac{\nu_1}{1 + \nu_1} z \tag{14}$$

$$\sigma_{rr}^{(p)(1)} = \tfrac{1}{2}\sigma_0[1 + \cos(2\theta)] \tag{15}$$

$$\sigma_{\theta\theta}^{(p)(1)} = \tfrac{1}{2}\sigma_0[1 - \cos(2\theta)] \tag{16}$$

$$\tau_{r\theta}^{(p)(1)} = -\tfrac{1}{2}\sigma_0 \sin(2\theta) \tag{17}$$

$$\tau_{rz}^{(p)(1)} = \tau_{\theta z}^{(p)(1)} = \sigma_{zz}^{(p)(1)} = 0 \tag{18}$$

(ii) for the inclusion:

$$u_{rr}^{(p)(2)} = C_1 r + C_2 r \cos(2\theta) \tag{19}$$

$$u_{\theta\theta}^{(p)(2)} = -C_2 r \sin(2\theta) \tag{20}$$

$$u_{zz}^{(p)(2)} = -2\frac{\nu_2}{1 - \nu_2} C_1 z \tag{21}$$

$$\sigma_{rr}^{(p)(2)} = 2\mu_2 \left[\frac{1 + \nu_2}{1 - \nu_2} C_1 + C_2 \cos(2\theta) \right] \tag{22}$$

$$\sigma_{\theta\theta}^{(p)(2)} = 2\mu_2 \left[\frac{1 + \nu_2}{1 - \nu_2} C_1 - C_2 \cos(2\theta) \right] \tag{23}$$

$$\tau_{r\theta}^{(p)(2)} = -2\mu_2 C_2 \sin(2\theta) \tag{24}$$

$$\tau_{rz}^{(p)(2)} = \tau_{\theta z}^{(p)(2)} = \sigma_{zz}^{(p)(2)} = 0, \tag{25}$$

where C_1 and C_2 are constants to be determined later from the boundary conditions at $r = a$. Note that the above particular solution for the inclusion satisfies the continuity conditions at $r = 0$.

In view of the particular solution, one needs to find six complementary displacements, i.e., $u^{(c)(i)}$, $v^{(c)(i)}$, $w^{(c)(i)}$ ($i = 1, 2$), such that they satisfy the partial differential equation (1) and the following boundary conditions:

$$\text{at } |z| = h: \tau_{xz}^{(c)(i)} = \tau_{yz}^{(c)(i)} = \sigma_{zz}^{(c)(i)} = 0 \tag{26}$$

$$\text{at } r = a: \sigma_{rr}^{(c)(1)} - \sigma_{rr}^{(c)(2)} = -\sigma_{rr}^{(p)(1)} + \sigma_{rr}^{(p)(2)} \tag{27}$$

$$\tau_{r\theta}^{(c)(1)} - \tau_{r\theta}^{(c)(2)} = -\tau_{r\theta}^{(p)(1)} + \tau_{r\theta}^{(p)(2)} \tag{28}$$

$$\tau_{rz}^{(c)(1)} - \tau_{rz}^{(c)(2)} = -\tau_{rz}^{(p)(1)} + \tau_{rz}^{(p)(2)} = 0 \tag{29}$$

$$u_{rr}^{(c)(1)} - u_{rr}^{(c)(2)} = -u_{rr}^{(p)(1)} + u_{rr}^{(p)(2)} \tag{30}$$

$$u_{\theta\theta}^{(c)(1)} - u_{\theta\theta}^{(c)(2)} = -u_{\theta\theta}^{(p)(1)} + u_{\theta\theta}^{(p)(2)} \tag{31}$$

$$u_{zz}^{(c)(1)} - u_{zz}^{(c)(2)} = -u_{zz}^{(p)(1)} + u_{zz}^{(p)(2)}. \tag{32}$$

Moreover, in order to complete the formulation of the complementary problem we must require that:

as $r \to \infty$: all complementary displacements and stresses for the plate must vanish $\tag{33}$

and the continuity condition:

at $r = 0$: all complementary displacements and stresses for the inclusion must be bounded. $\tag{34}$

3. Method of solution

A general method for constructing solutions for some three-dimensional mixed boundary-value problems which arise in elastostatics was developed by Folias [11] who illustrated the method by applying it to the problem of a uniform extension of an infinite plate containing a through the thickness line crack. Later, Folias and Wang [12] specialized the general solution to the case of a plate of an arbitrary thickness containing a cylindrical hole. Based on these results, one can deduce that the general form of the solution to system (1) which automatically satisfies the boundary conditions at the plate faces, i.e. (26) is[†]:

$$u^{(c)(i)} = \frac{1}{m_i - 2} \sum_{v=1}^{\infty} \frac{\partial^2 H_v^{(i)}}{\partial x^2} \{2(m_i - 1)f_1(\beta_v z) + m_i f_2(\beta_v z)\}$$

$$+ \sum_{n=1}^{\infty} \left\{ -\frac{\partial^2 H_n^{(i)}}{\partial x^2} + \alpha_n^2 H_n^{(i)} \right\} \cos(\alpha_n h) \cos(\alpha_n z) \tag{35}$$

$$+ \lambda_1^{(i)} - y \frac{\partial \lambda_3^{(i)}}{\partial x} + \frac{1}{m_i + 1} z^2 \frac{\partial^2 \lambda_3^{(i)}}{\partial x \partial y}$$

---

† Note that because of symmetry in the present problem, one needs only to consider the region $0 \leqslant \theta \leqslant \pi/2$.

$$v^{(c)(i)} = \frac{1}{m_i - 2} \sum_{v=1}^{\infty} \frac{\partial^2 H_v^{(i)}}{\partial x \partial y} \{2(m_i - 1)f_1(\beta_v z) + m_i f_2(\beta_v z)\}$$

$$- \sum_{n=1}^{\infty} \frac{\partial^2 H_n^{(i)}}{\partial x \partial y} \cos(\alpha_n h) \cos(\alpha_n z) \tag{36}$$

$$+ \frac{3m_i - 1}{m_i + 1} \lambda_3^{(i)} + \lambda_2^{(i)} - y \frac{\partial \lambda_3^{(i)}}{\partial y} - \frac{1}{m_i + 1} z^2 \frac{\partial^2 \lambda_3^{(i)}}{\partial x^2}$$

$$w^{(c)(i)} = \frac{1}{m_i - 2} \sum_{v=1}^{\infty} \frac{\partial H_v^{(i)}}{\partial x} \beta_v \{(m_i - 2)f_3(\beta_v z) - m_i f_4(\beta_v z)\} - \frac{2}{m_i + 1} z \frac{\partial \lambda_3^{(i)}}{\partial y}. \tag{37}$$

Furthermore, the stresses are given by:

$$\frac{1}{2\mu_i} \sigma_{xx}^{(c)(i)} = \frac{1}{m_i - 2} \sum_{v=1}^{\infty} \left\{ 2\beta_v^2 \frac{\partial H_v^{(i)}}{\partial x} f_1(\beta_v z) \right.$$

$$+ \frac{\partial^3 H_v^{(i)}}{\partial x^3} [2(m_i - 1)f_1(\beta_v z) + m_i f_2(\beta_v z)] \Bigg\}$$

$$+ \sum_{n=1}^{\infty} \left\{ -\frac{\partial^3 H_n^{(i)}}{\partial x^3} + \alpha_n^2 \frac{\partial H_n^{(i)}}{\partial x} \right\} \cos(\alpha_n h) \cos(\alpha_n z) \tag{38}$$

$$+ \frac{\partial \lambda_1^{(i)}}{\partial x} - y \frac{\partial^2 \lambda_3^{(i)}}{\partial x^2} + \frac{2}{m_i + 1} \frac{\partial \lambda_3^{(i)}}{\partial y} + \frac{1}{m_i + 1} z^2 \frac{\partial^3 \lambda_3^{(i)}}{\partial x^2 \partial y}$$

$$\frac{1}{2\mu_i} \sigma_{yy}^{(c)(i)} = \frac{1}{m_i - 2} \sum_{v=1}^{\infty} \left\{ 2\beta_v^2 \frac{\partial H_v^{(i)}}{\partial x} f_1(B_v z) \right.$$

$$- \left(\frac{\partial^3 H_v^{(i)}}{\partial x^3} - \beta_v^2 \frac{\partial H_v^{(i)}}{\partial x} \right) [2(m_i - 1)f_1(\beta_v z) + m_i f_2(\beta_v z)] \Bigg\}$$

$$+ \sum_{n=1}^{\infty} \left\{ \frac{\partial^3 H_n^{(i)}}{\partial x^3} - \alpha_n^2 \frac{\partial H_n^{(i)}}{\partial x} \right\} \cos(\alpha_n h) \cos(\alpha_n z) \tag{39}$$

$$+ \frac{2m_i}{m_i + 1} \frac{\partial \lambda_3^{(i)}}{\partial y} - \frac{\partial \lambda_1^{(i)}}{\partial x} + y \frac{\partial^2 \lambda_3^{(i)}}{\partial x^2} - \frac{1}{m_i + 1} z^2 \frac{\partial^3 \lambda_3^{(i)}}{\partial x^2 \partial y}$$

$$\frac{1}{2\mu_i} \sigma_{zz}^{(c)(i)} = -\frac{m_i}{m_i - 2} \sum_{v=1}^{\infty} \frac{\partial H_v^{(i)}}{\partial x} \beta_v^2 f_2(\beta_v z) \tag{40}$$

$$\frac{1}{2\mu_i} \tau_{xy}^{(c)(i)} = \frac{1}{m_i - 2} \sum_{v=1}^{\infty} \frac{\partial^3 H_v^{(i)}}{\partial x^2 \partial y} \{2(m_i - 1)f_1(\beta_v z) + m_i f_2(\beta_v z)\}$$

$$- \sum_{n=1}^{\infty} \left\{ \frac{\partial^3 H_n^{(i)}}{\partial x^2 \partial y} - \frac{1}{2} \alpha_n^2 \frac{\partial H_n^{(i)}}{\partial y} \right\} \cos(\alpha_n h) \cos(\alpha_n z) \tag{41}$$

$$+ \frac{m_i - 1}{m_i + 1} \frac{\partial \lambda_3^{(i)}}{\partial x} + \frac{\partial \lambda_2^{(i)}}{\partial x} - y \frac{\partial^2 \lambda_3^{(i)}}{\partial x \partial y} - \frac{1}{m_i + 1} z^2 \frac{\partial^3 \lambda_3^{(i)}}{\partial x^3}$$

$$\frac{1}{2\mu_i}\tau_{yz}^{(c)(i)} = -\frac{m_i}{m_i-2}\sum_{v=1}^{\infty}\frac{\partial^2 H_v^{(i)}}{\partial x \partial y}\beta_v\{f_3(\beta_v z)+f_4(\beta_v z)\}$$

$$+\frac{1}{2}\sum_{n=1}^{\infty}\alpha_n\frac{\partial^2 H_n^{(i)}}{\partial x \partial y}\cos(\alpha_n h)\sin(\alpha_n z) \tag{42}$$

$$\frac{1}{2\mu_i}\tau_{xz}^{(c)(i)} = -\frac{m_i}{m_i-1}\sum_{v=1}^{\infty}\frac{\partial^2 H_v^{(i)}}{\partial x^2}\beta_v\{f_3(\beta_v z)+f_4(\beta_v z)\}$$

$$+\frac{1}{2}\sum_{n=1}^{\infty}\left[\frac{\partial^2 H_n^{(i)}}{\partial x^2}-\alpha_n^2 H_n^{(i)}\right]\alpha_n\cos(\alpha_n h)\sin(\alpha_n z), \tag{43}$$

where

$$\alpha_n = \frac{n\pi}{h}, \quad n = 1, 2, 3, \ldots, \tag{44}$$

β_v are the roots of the equation

$$\sin(2\beta_v h) = -(2\beta_v h), \tag{45}$$

$H_v^{(i)}$ and $H_n^{(i)}$ are functions of x and y which satisfy the reduced wave equation:

$$\left(\frac{\partial^2}{\partial x^2}+\frac{\partial^2}{\partial y^2}-\beta_v^2\right)\frac{\partial H_v^{(i)}}{\partial x} = 0 \tag{46}$$

$$\left(\frac{\partial^2}{\partial x^2}+\frac{\partial^2}{\partial y^2}-\alpha_n^2\right)\frac{\partial H_n^{(i)}}{\partial y} = 0, \tag{47}$$

$\lambda_1^{(i)}$, $\lambda_2^{(i)}$ and $\lambda_3^{(i)}$ are two dimensional harmonic functions, and

$$f_1(\beta_v z) \equiv \cos(\beta_v h)\cos(\beta_v z) \tag{48}$$

$$f_2(\beta_v z) \equiv \beta_v h \sin(\beta_v h)\cos(\beta_v z) - \beta_v z \cos(\beta_v h)\sin(\beta_v z) \tag{49}$$

$$f_3(\beta_v z) \equiv \cos(\beta_v h)\sin(\beta_v z) \tag{50}$$

$$f_4(\beta_v z) \equiv \beta_v h \sin(\beta_v h)\sin(\beta_v z) + \beta_v z \cos(\beta_v h)\cos(\beta_v z). \tag{51}$$

By virtue of its construction, the complementary solution automatically satisfies the boundary conditions at the plate faces $|z| = h$. It remains next to satisfy the boundary conditions on the surface of the inclusion.

Utilizing the appropriate coordinate transformations from rectangular to cylindrical coordinates, (27)–(32) can be written in the form:

$$\sin^2\theta(\sigma_{xx}^{(c)(1)}-\sigma_{xx}^{(c)(2)}) + \cos^2\theta(\sigma_{yy}^{(c)(1)}-\sigma_{yy}^{(c)(2)}) + \sin(2\theta)(\tau_{xy}^{(c)(1)}-\tau_{xy}^{(c)(2)})$$

$$= -\left[\frac{1}{2}\sigma_0 - 2\mu_2\frac{1+\nu_2}{1-\nu_2}C_1\right] - [\tfrac{1}{2}\sigma_0 - 2\mu_2 C_2]\cos(2\theta) \tag{52}$$

$$\tfrac{1}{2} \sin (2\theta)(\sigma_{xx}^{(c)(1)} - \sigma_{xx}^{(c)(2)}) - \tfrac{1}{2} \sin (2\theta)(\sigma_{yy}^{(c)(1)} - \sigma_{yy}^{(c)(2)}) + \cos (2\theta)(\tau_{xy}^{(c)(1)} - \tau_{xy}^{(c)(2)})$$

$$= [\tfrac{1}{2}\sigma_0 - 2\mu_2 C_2] \sin (2\theta) \tag{53}$$

$$\sin \theta(\tau_{xy}^{(c)(1)} - \tau_{xy}^{(c)(2)}) + \cos \theta(\tau_{yz}^{(c)(1)} - \tau_{yz}^{(c)(2)}) = 0 \tag{54}$$

$$\sin \theta(u^{(c)(1)} - u^{(c)(2)}) + \cos \theta(v^{(c)(1)} - v^{(c)(2)})$$

$$= -\left[\frac{\sigma_0 a}{4\mu_1} \frac{1 - \nu_1}{1 + \nu_1} - C_1 a\right] - \left[\frac{\sigma_0 a}{4\mu_1} - C_2 a\right] \cos (2\theta) \tag{55}$$

$$\cos \theta(u^{(c)(1)} - u^{(c)(2)}) - \sin \theta(v^{(c)(1)} - v^{(c)(2)}) = \left[\frac{\sigma_0 a}{4\mu_1} - C_2 a\right] \sin (2\theta) \tag{56}$$

$$w^{(c)(1)} - w^{(c)(2)} = \left[\frac{\sigma_0}{2\mu_1} \frac{\nu_1}{1 + \nu_1} - \frac{2\nu_2}{1 - \nu_2} C_1\right] z. \tag{57}$$

Examining the nature of (52)–(57), one notices that the θ-dependency can be eliminated by considering the following forms of the solution to (46)–(47):

$$\frac{\partial H_v^{(1)}}{\partial x} = c_{1v} K_0(\beta_v r) + c_{2v} K_2(\beta_v r) \cos (2\theta) \tag{58}$$

$$\frac{\partial H_v^{(2)}}{\partial x} = c_{3v} I_0(\beta_v r) + c_{4v} I_2(\beta_v r) \cos (2\theta) \tag{59}$$

$$\frac{\partial H_n^{(1)}}{\partial y} = c_{1n} K_0(\alpha_n r) + c_{2n} K_2(\alpha_n r) \sin (2\theta) \tag{60}$$

$$\frac{\partial H_n^{(2)}}{\partial y} = c_{3n} I_0(\alpha_n r) + c_{4n} I_2(\alpha_n r) \sin (2\theta) \tag{61}$$

$$\lambda_1^{(1)} = \frac{A}{r} \sin \theta - \frac{2Ba^2}{r^3} \sin (3\theta) \tag{62}$$

$$\lambda_2^{(1)} = \frac{A}{r} \cos \theta - \frac{2Ba^2}{r^3} \cos (3\theta) \tag{63}$$

$$\lambda_3^{(1)} = \frac{D}{r} \cos \theta \tag{64}$$

$$\lambda_1^{(2)} = E r \sin \theta \tag{65}$$

$$\lambda_2^{(2)} = -E r \cos \theta \tag{66}$$

$$\lambda_3^{(2)} = G r \cos \theta, \tag{67}$$

$$\frac{\mu_2}{\mu_1} = 2.0, \quad v_1 = v_2 = 0.33, \quad a/h = 0.05$$

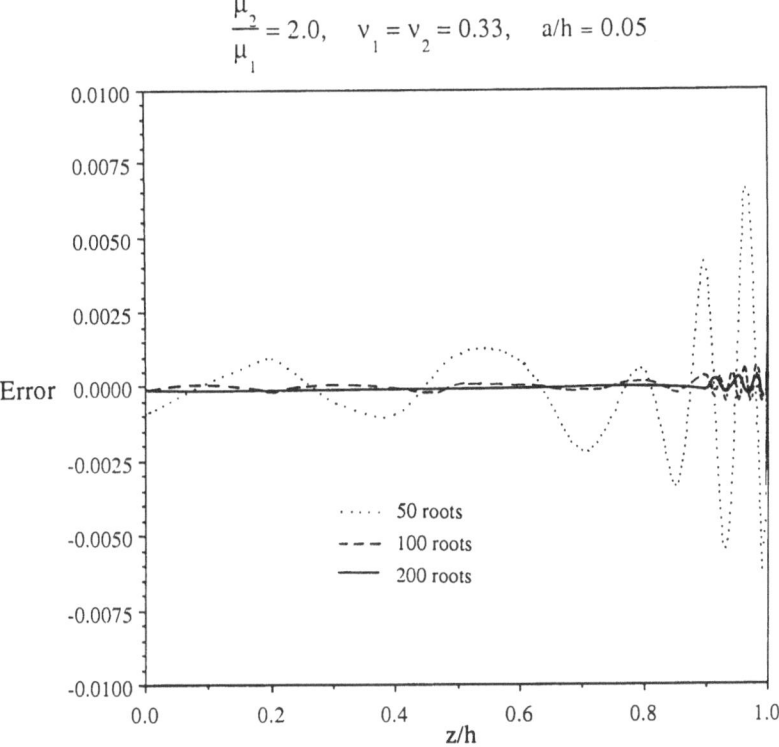

Fig. 2. Boundary condition across the thickness at $r = a$ for the stress $\tau_{r\theta}$ for $\mu_2/\mu_1 = 2.0$, $v_1 = v_2 = 0.33$ and $a/h = 0.05$.

where I_m and K_m ($m = 0, 2$) are, respectively, the modified Bessel functions of the first and second kind of order m, and c_{kv}, c_{kn} ($k = 1, 2, 3, 4$; $v, n = 1, 2, 3, \ldots$), A, B, D, E, G are arbitrary constants.

Finally, substituting (58)–(67) into (52)–(57) and letting $r = a$ one arrives at a system of six equations involving series in z. The system may then be solved numerically, by the method of [13], for the unknown coefficients. Details of the numerical solution can be found in [14]. Although the system is extremely sensitive to small changes in the coefficients, the method does furnish a solution which converges as the number of characteristic roots increases. The rate of convergence may be seen in Fig. 2 where the results for the boundary condition $\tau_{r\theta}$ for 50, 100 and 200 roots are plotted. This boundary condition was chosen because it is the most difficult one to satisfy. The reader should also notice that the little oscillation at the end is the result of the stress singularity which is present, for $\mu_2/\mu_1 = 2$, in the neighborhood of the point $z = h$ (see [10]).

As a check, the following three limiting cases will be examined:

(i) *Continuous plate.* If $v_1 = v_2$ and $\mu_2/\mu_1 = 1$, then the solutions for the plate and the inclusion reduce to the particular solution for the plate, i.e. (12)–(18).

(ii) *Thin plate.* If $h/a \to 0$, then one recovers precisely the plane stress solution given by Goodier [3].

(iii) *Cylindrical hole.* In this case $\mu_2/\mu_1 \to 0$. The stress $\sigma_{\theta\theta}^{(1)}/\sigma_0$ was evaluated numerically for $\theta = \pi/2$, $\mu_2/\mu_1 = 0.00001$, $v_1 = v_2 = 0.33$ and different values of a/h. The results are

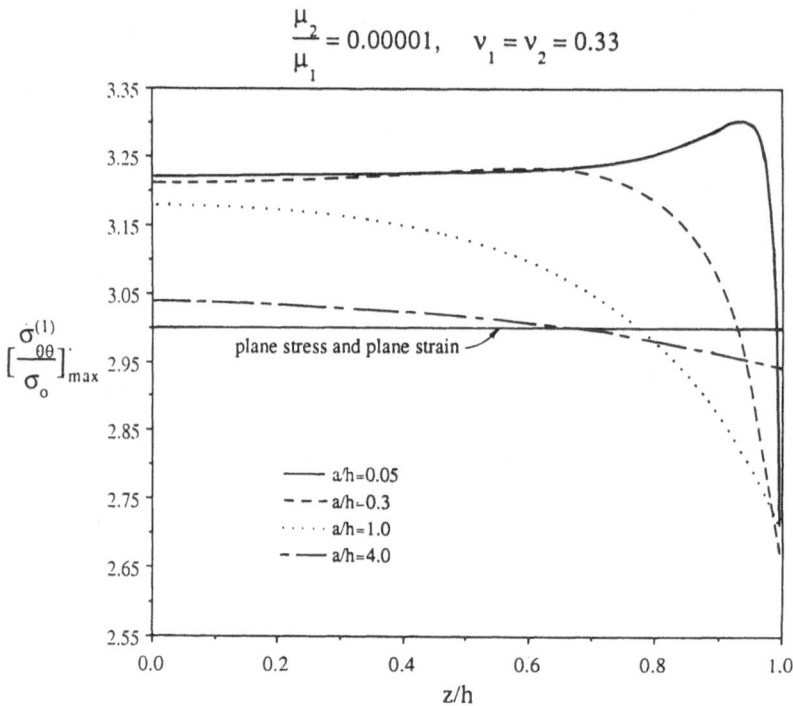

Fig. 3. Maximum normal stress for the plate across the thickness for $\mu_2/\mu_1 = 0.00001$ (cylindrical hole), $\nu_1 = \nu_2 = 0.33$ and different values of a/h.

shown in Fig. 3 where they are compared with those obtained by Folias and Wang [12]. The agreement is excellent indicating, therefore, that in the limit as $\mu_2/\mu_1 \to 0$ the present results tend to those obtained for a cylindrical hole.

4. Numerical results

Once the coefficients have been determined, the stresses and displacements may then be calculated at any point in the body. More specifically, we will calculate the stresses $\sigma_{\theta\theta}$, σ_{zz}, τ_{oct} and the out of the plane displacement u_{zz} and we will compare them with the results for plane stress and plane strain obtained in [3].

To quantify the effect that the applied load has upon failure, we choose as a suitable parameter the octahedral shear stress for it is directly related to the von Mises criterion, i.e.

$$\tau_{oct} = \frac{\sqrt{2}}{3} \sigma_Y, \tag{68}$$

where σ_Y represents the yield stress in simple tension.

Equation (68) describes the locus of the points in the composite plate where τ_{oct} attains a high value which may lead to failure initiation. Figures 4–7 show the maximum octahedral shear stress for the plate as a function of z/h, for different values of the thickness parameter, a/h, and the shear moduli ratio, μ_2/μ_1. The maximum value of τ_{oct} occurs at $r = a$ and $\theta = 0$

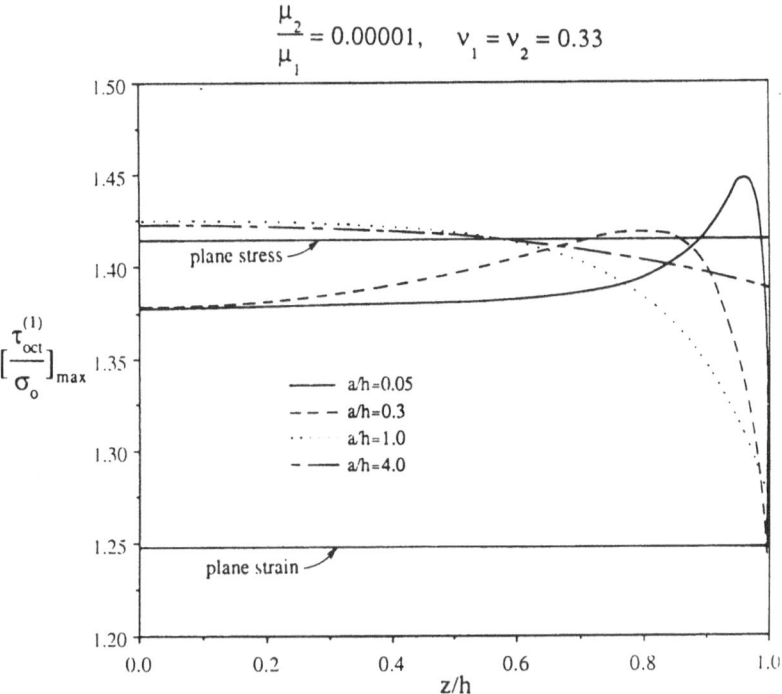

Fig. 4. Maximum octahedral shear stress for the plate across the thickness for $\mu_2/\mu_1 = 0.00001$ (cylindrical hole), $v_1 = v_2 = 0.33$ and different values of a/h.

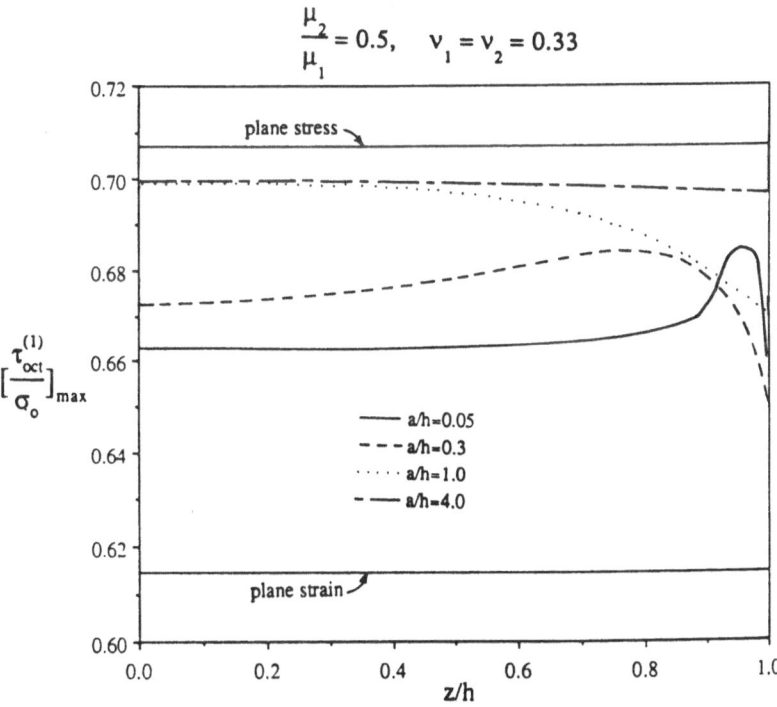

Fig. 5. Maximum octahedral shear stress for the plate across the thickness for $\mu_2/\mu_1 = 0.5$, $v_1 = v_2 = 0.33$ and different values of a/h.

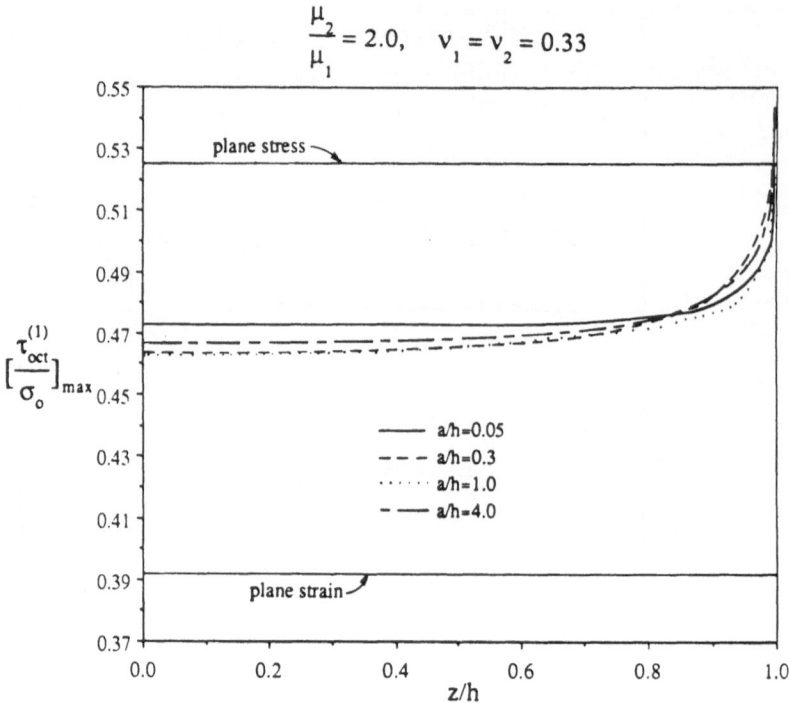

Fig. 6. Maximum octahedral shear stress for the plate across the thickness for $\mu_2/\mu_1 = 2.0$, $\nu_1 = \nu_2 = 0.33$ and different values of a/h.

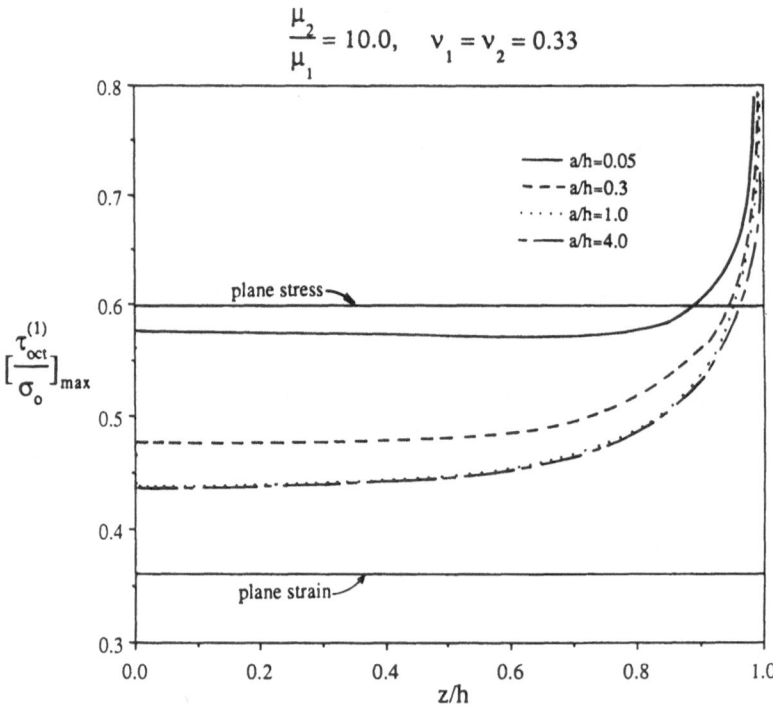

Fig. 7. Maximum octahedral shear stress for the plate across the thickness for $\mu_2/\mu_1 = 10.0$, $\nu_1 = \nu_2 = 0.33$ and different values of a/h.

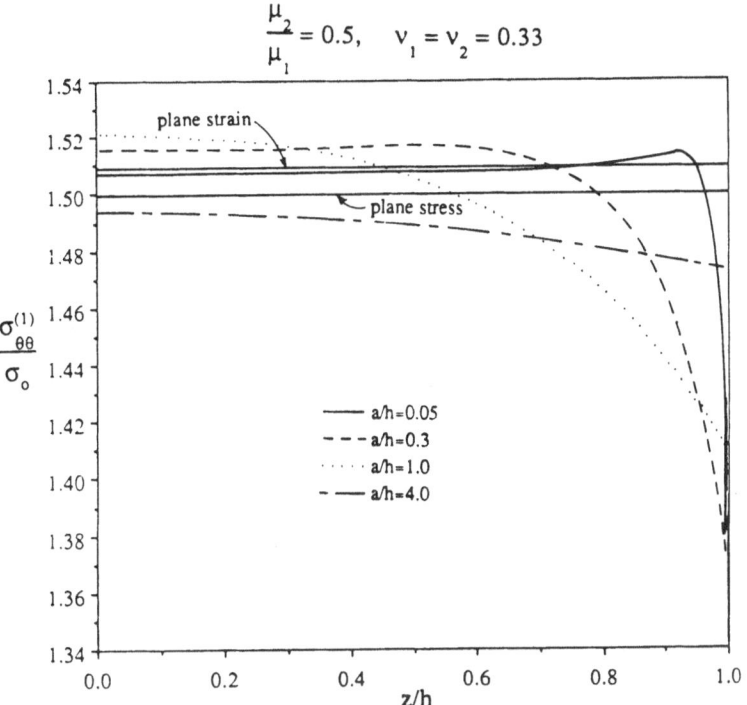

$$\frac{\mu_2}{\mu_1} = 0.5, \quad \nu_1 = \nu_2 = 0.33$$

Fig. 8. Stress $\sigma_{\theta\theta}$ for the plate across the thickness at $r = a$, $\theta = \pi/2$ for $\mu_2/\mu_1 = 0.5$, $\nu_1 = \nu_2$ 0.33 and different values of a/h.

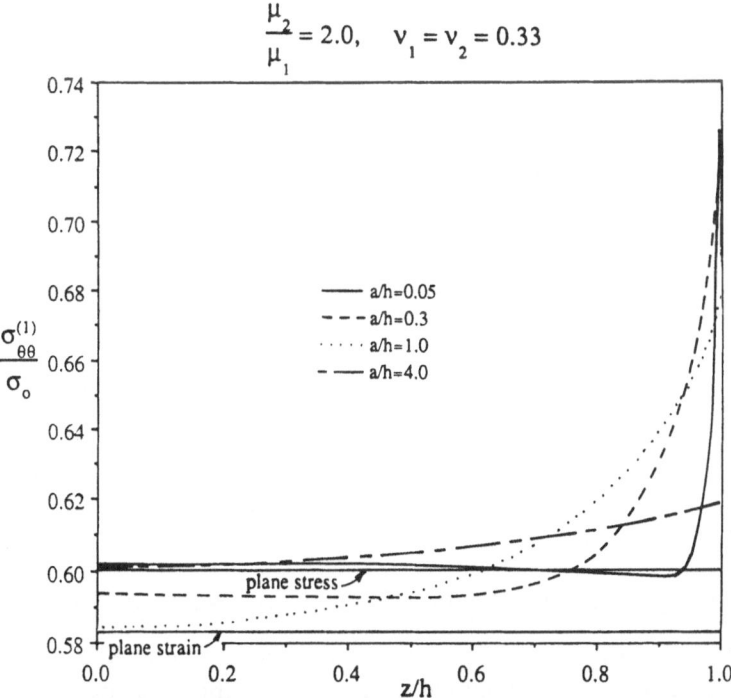

$$\frac{\mu_2}{\mu_1} = 2.0, \quad \nu_1 = \nu_2 = 0.33$$

Fig. 9. Stress $\sigma_{\theta\theta}$ for the plate across the thickness at $r = a$, $\theta = \pi/2$ for $\mu_2/\mu_1 = 2.0$, $\nu_1 = \nu_2 = 0.33$ and different values of a/h.

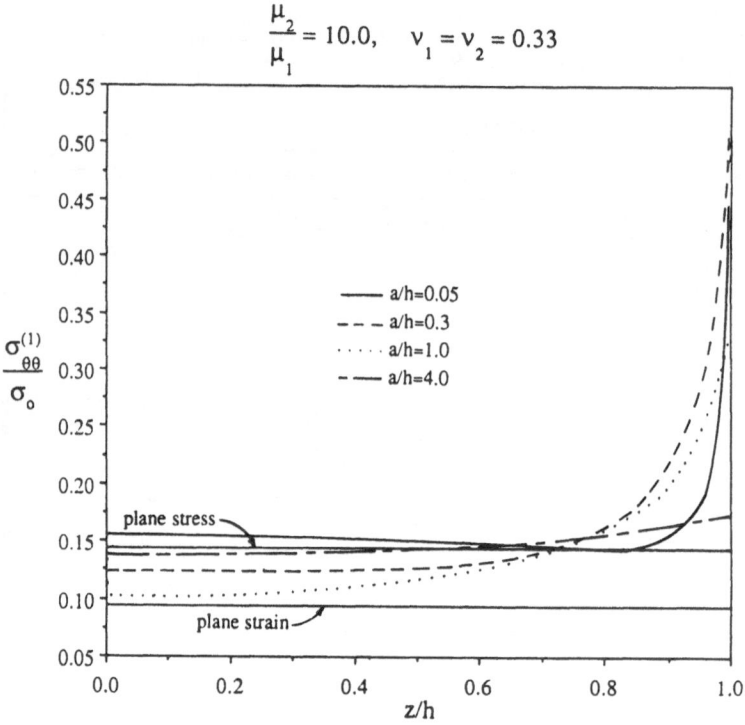

$$\frac{\mu_2}{\mu_1} = 10.0, \quad \nu_1 = \nu_2 = 0.33$$

Fig. 10. Stress $\sigma_{\theta\theta}$ for the plate across the thickness at $r = a$, $\theta = \pi/2$ for $\mu_2/\mu_1 = 10.0$, $\nu_1 = \nu_2 = 0.33$ and different values of a/h.

if the inclusion is of a softer material, or at $\theta = \pi/2$ if the inclusion is of a stiffer material. Three important observations can be made from these figures. First, for thick plates ($a/h = 0.05$) the plane strain solution always gives nonconservative approximations to the three-dimensional solution in the interior of the plate. Second, in the neighborhood of $z/h = 1.0$ and for $\mu_2/\mu_1 = 2.0, 10.0$, the stresses become high, which is compatible with the stress singularities predicted in [10] for $\mu_2/\mu_1 > 1.0$. It should be noted that the plane strain solution fails to predict these singularities for it neglects the three-dimensional effects. Third, as a/h becomes large (thin plates) the three-dimensional solution tends to the plane stress solution.

Figures 8–13 show the stresses $\sigma_{\theta\theta}$ and σ_{zz} for the plate at $r = a$, $\theta = \pi/2$ as a function of z/h and illustrate how individual stresses vary across the plate thickness. Finally, Figs. 14 and 15 show the displacement u_{zz} at the surface of the plate ($z = h$) along the radial line $\theta = \pi/2$.

5. Conclusions

In view of the foregoing results, the following conclusions may be drawn.

(1) Thickness as well as the material properties play a fundamental role on the failure mechanism of a plate with a cylindrical inclusion. The thickness parameter, a/h, and the shear moduli ratio, μ_2/μ_1, control not only the maximum load that can be applied to the plate, but also the location where failure initiation may occur.

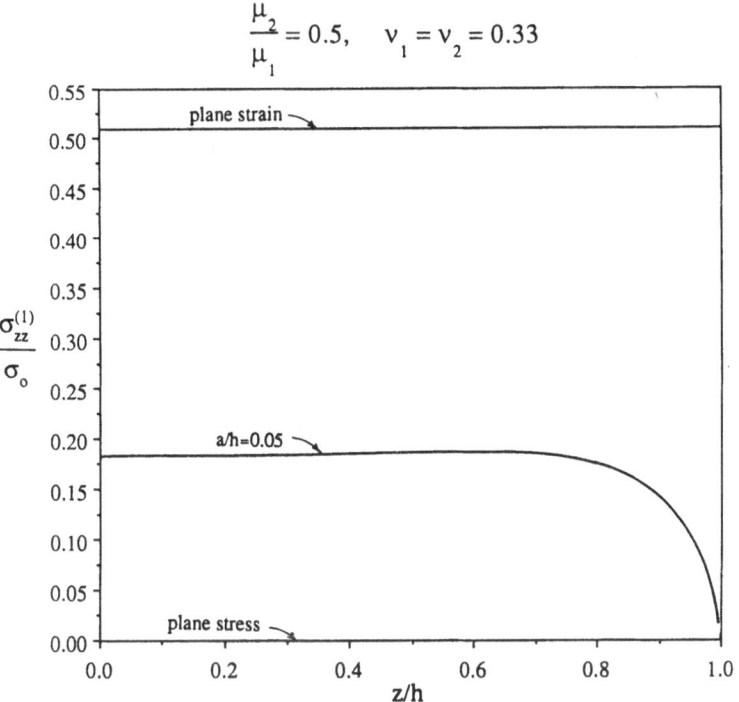

Fig. 11. Stress σ_{zz} for the plate across the thickness at $r = a$, $\theta = \pi/2$ for $\mu_2/\mu_1 = 0.5$, $\nu_1 = \nu_2 = 0.33$ and $a/h = 0.05$.

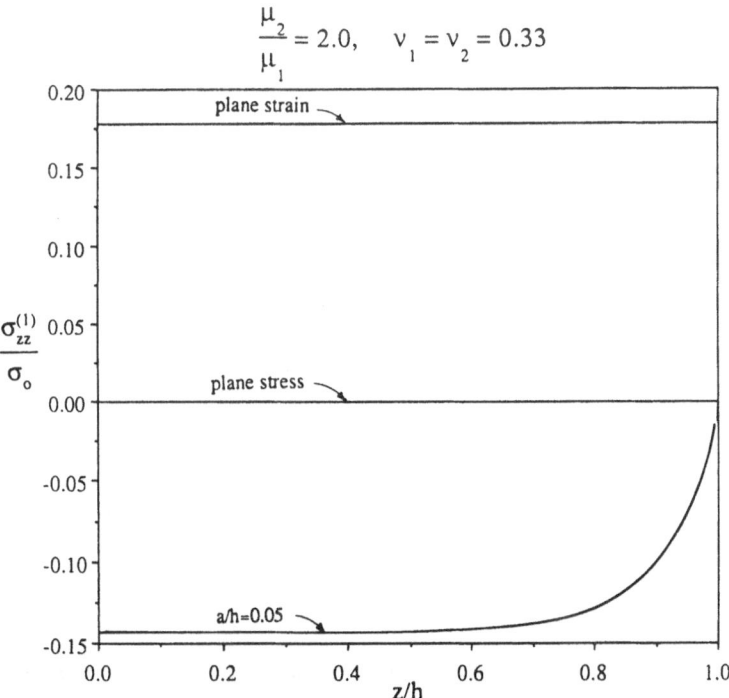

Fig. 12. Stress σ_{zz} for the plate across the thickness at $r = a$, $\theta = \pi/2$ for $\mu_2/\mu_1 = 2.0$, $\nu_1 = \nu_2 = 0.33$ and $a/h = 0.05$.

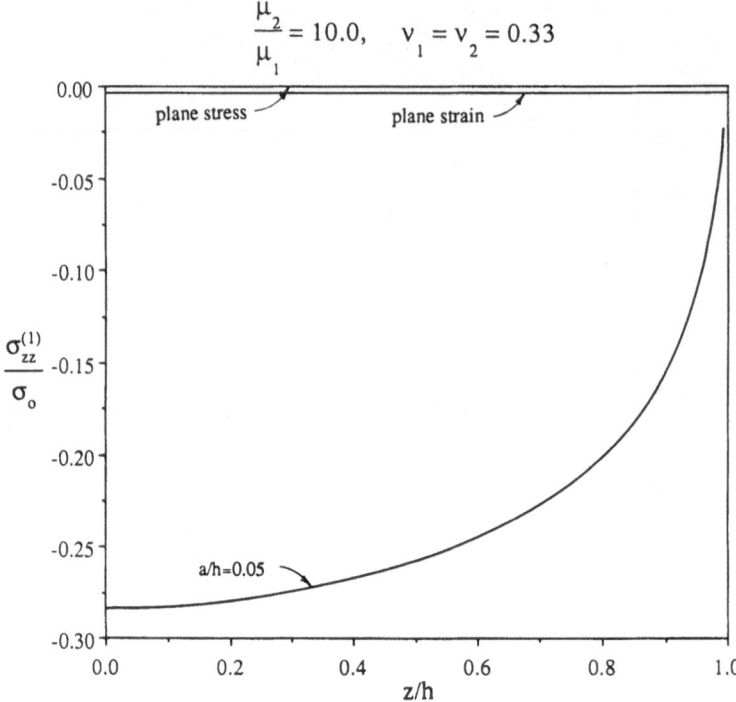

Fig. 13. Stress σ_{zz} for the plate across the thickness at $r = a$, $\theta = \pi/2$ for $\mu_2/\mu_1 = 10.0$, $\nu_1 = \nu_2 = 0.33$ and $a/h = 0.05$.

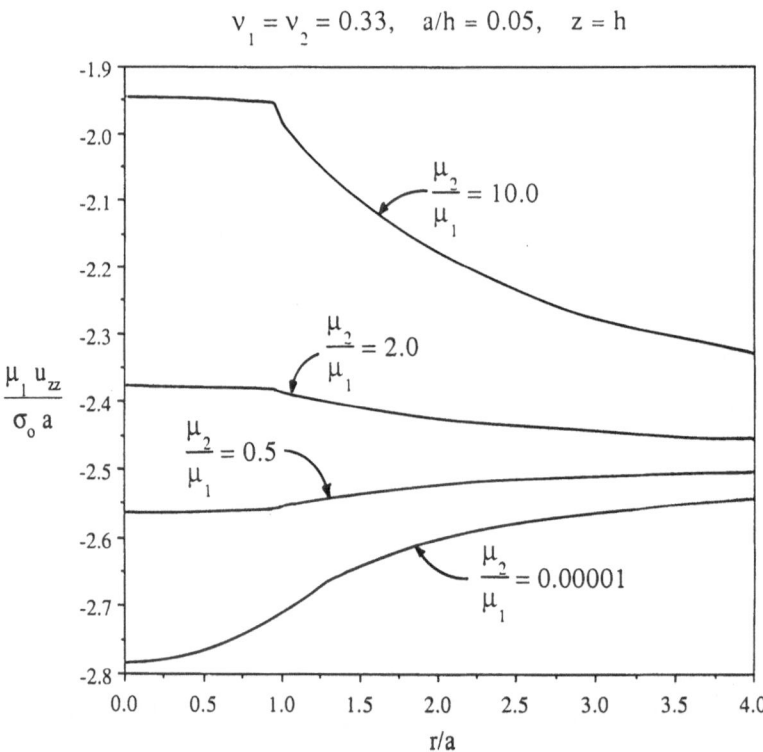

Fig. 14. Displacement u_{zz} as a function of r at $z = h$, $\theta = \pi/2$ for $a/h = 0.05$, $\nu_1 = \nu_2 = 0.33$ and different values of μ_2/μ_1.

$$\nu_1 = \nu_2 = 0.33, \quad a/h = 1.0, \quad z = h$$

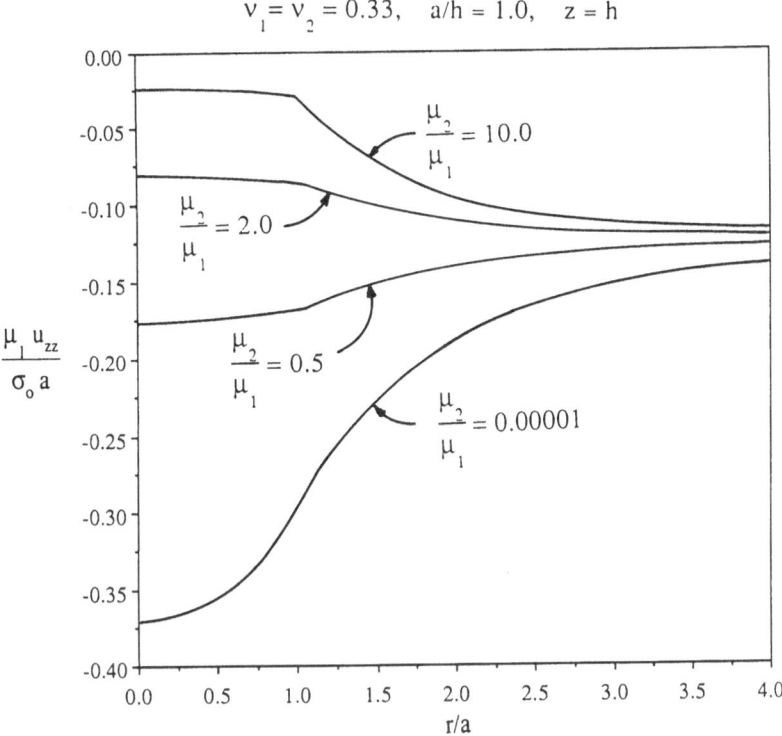

Fig. 15. Displacement u_{zz} as a function of r at $z = h$, $\theta = \pi/2$ for $a/h = 1.0$, $v_1 = v_2 = 0.33$ and different values of μ_2/μ_1.

(2) It is observed that for relatively thick plates and shear moduli ratios of $\mu_2/\mu_1 > 1$ the stresses are singular at the intersection of the cylindrical inclusion and the free surface of the plate. The strength of these singularities, however, is best if it is extracted by analytical means (see [10]).

(3) For relatively thick plates, the maximum octahedral shear stress predicted by the plane strain solution is always lower than the corresponding three-dimensional solution.

(4) As $h \to 0$, the plane stress solution is recovered.

(5) Finally, the plane stress solution is a good approximation for plates with ratios of $(a/h) \geqslant 4$.

Acknowledgments

This work was supported in part by the Air Force Office of Scientific Research Grant No. AFOSR-87-0204. The authors would like to thank Lt. Col. G. Haritos for this support. Also, the authors would like to acknowledge the time on the CRAY supercomputer at San Diego donated by the National Science Foundation.

References

1. R. Muki and E. Sternberg, *International Journal of Solids and Structures* 6 (1970) 69–90.
2. V.K. Luk and L.M. Keer, *International Journal of Solids and Structures* 15 (1979) 805–827.

3. J.N. Goodier, *Journal of Applied Mechanics, Transactions ASME* 55 (1933) 39–44.

4. I.W. Yu and G.P. Sendeckyj, *Journal of Applied Mechanics* 41 (1974) 215–221.

5. K. Suyehiro, "The distribution of stress in a tension strap having a circular hole filled with a plug," *The Society of Mechanical Engineers*, Japan (1914).

6. E.G. Coker and L.N.G. Filon, in *A Treatise on Photoelasticity*, Cambridge (1931) 497–500.

7. L.M. Keer, J. Dundurs and K. Kiattikomol, *International Journal of Engineering Science* 11 (1973) 1221–1233.

8. D.B. Bogy, *Journal of Applied Mechanics Transactions* ASME 35 (1968) 460–466.

9. V.L. Hein and F. Erdogan, *International Journal of Fracture Mechanics* (1971) 317–330.

10. E.S. Folias, *International Journal of Fracture* 39 (1989).

11. E.S. Folias, *Journal of Applied Mechanics* 42 (1975) 663–674.

12. E.S. Folias and J.J. Wang, "On the Three-dimensional Stress Field Around a Circular Hole in a Plate of Arbitrary Thickness", *Proceedings of the 19th, Midwestern Mechanics Conference*, Columbus, Ohio, September 1985. Also *University of Utah Technical Report*, May 1986.

13. L.V. Kantorovich and V.I. Krylov, *Approximate Methods of Higher Analysis*, Noordhoff, Holland (1964).

14. F.E. Penado, and E.S. Folias, "The Three-Dimensional Stress Field Around a Cylindrical Inclusion in a Plate of an Arbitrary Thickness," Interim Report, College of Engineering, University of Utah, September 1987.

Résumé. On résoud par voie analytique les équations de Navier à trois dimensions relatives au cas d'une inclusion cylindrique de rayon "a" noyée dans une tôle d'épaisseur arbitraire 2*h*. On suppose que les matériaux constituant l'inclusion et la tôle sont homogènes et isotropes, et qu'ils ont des propriétés mécaniques différentes. Une liaison parfaite de leur interface est également supposée. La mise en charge est réalisée par une tension uniforme appliquée dans le plan de la tôle, en des point suffisamment distants de l'inclusion.

L'analyse montre que toutes les contraintes comprises dans l'octoèdre des tensions de cisaillement sont influencées par le rapport *a/h* et par les propriétés des matériaux. A la limite, lorsque le rapport des modules de cisaillement de l'inclusion et de la tôle tend vers zéro, ou vers un, on retrouve respectivement les résultats relatifs à un trou circulaire et à une plaque continue. De même, on retrouve la solution d'état plan de contrainte lorsque *a/h* → ∞, ce qui serait le cas d'une tôle très mince. En outre, lorsque le rapport des modules est supérieur à un, les résultats numériques confirment la présence d'une singularité de contrainte près du point d'intersection de l'inclusion et de la surface libre de la tôle.

International Journal of Fracture 39: 147–161 (1989)
© Kluwer Academic Publishers, Dordrecht

The effect of damping on the spring-mass dynamic fracture model

J.G. WILLIAMS and M.N.M. BADI
Department of Mechanical Engineering, Imperial College of Science & Technology, London, UK

Received 20 September 1987; accepted 1 April 1988

Abstract. A recent paper has analysed high speed crack growth using a lumped mass-spring model which includes the effect of the contact stiffness. It was shown that the energy release rate G could be computed for this dynamic situation by using the kinetic energy of the mass. The model has been particularly successful in describing dynamic effects in the Charpy impact test. This paper extends the model by including damping in parallel with the specimen stiffness which is a reasonable representation of a visco-elastic material. It is shown that G can be greatly influenced by even a small amount of damping if short time fracture is involved. The method is based on the energy rate balance approach pioneered by MLW.

1. Introduction

An earlier publication [1] discussed the use of a lumped-spring model to describe dynamic fracture. It was noted that the inclusion of a contact stiffness was important in determining the motion of the mass and hence the kinetic energy of the system. The utility of the model is that a complete energy balance can be made since the motion of every part can be determined and from this the energy release rate may be computed. For the complete energy balance of this dynamic case this must, of course, be equal to the crack resistance.

Comparisons of these results have been made with experiment [2] and with finite element computations [3] for three point bend test geometry and good agreement has been found. This illustrates that, for this geometry at least, the lumped mass assumption is a good approximation in that most of the dynamic effects arise from the specimen motion and that stress waves, which are not included here, are of secondary importance. A limitation of this, and indeed, of most dynamic analysis is that a balance is made of external work, elastic and kinetic energies and it is assumed that all the necessary dissipation rate to effect this balance arises from the crack growth and is embodied in R. In many cases, however, this is not true since there is energy dissipation in the bulk of the body and remote from the crack tip. This may be incorporated into the analysis by some form of visco-elastic or visco-plastic modelling of the body [4, 5] but such computations are lengthy and difficult. Given that the lumped-mass model gave helpful insights into the dynamic behaviour it was felt that some useful results could be found by including a damping factor with the specimen stiffness. It is true that some damping in practice comes from friction at the load point and supports and might better be included in the contact stiffness. However, in the interest of reasonable simplicity all the damping is placed in parallel with the specimen stiffness and $G(= R)$ is computed for such a system.

2. Formulation

The model used is shown in Fig. 1, where the contact stiffness k_1, the mass m, and the damping μ are constant but the specimen stiffness k_2, is a function of crack length which may be computed from the usual static results. More complex cases in which $m(a)$ is discussed are in [1] but here we will confine our attention to the three point bend test for which m is constant. We will also consider only the case of constant velocity impact so that the load point displacement is Vt while that of the specimen is u. The equation of motion for the system without crack growth is:

$$m \frac{\partial^2 u}{\partial t^2} + \mu \frac{\partial u}{\partial t} + (k_1 + k_2)u = k_1 Vt \tag{1}$$

and the solution has the form

$$u = e^{-t/\tau}(A \sin \omega t + B \cos \omega t) + \frac{k_1}{k_1 + k_2} \cdot \frac{V}{\omega}\left(\omega t - \frac{2\omega\tau}{\omega^2\tau^2 + 1}\right), \tag{2}$$

where

$$\omega^2 = \frac{k_1 + k_2}{m} - \frac{1}{\tau^2} \quad \text{and} \quad \tau = \frac{2m}{\mu}$$

giving damped oscillatory motion for

$$\frac{k_1 + k_2}{m} > \frac{1}{\tau^2}.$$

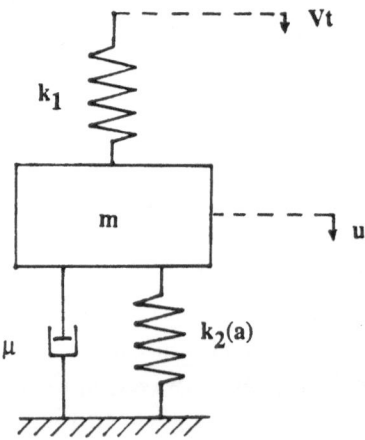

Fig. 1. Lumped-mass spring model with damping.

For the impact condition the boundary conditions are $u = \partial u/\partial t = 0$ at $t = 0$ giving a final result of

$$u = \left(\frac{k_1}{k_1 + k_2}\right)\frac{V}{\omega} \cdot \left\{e^{-\xi/\bar{\xi}}\left[\left(\frac{1 - \bar{\xi}^2}{1 + \bar{\xi}^2}\right)\sin\xi + \left(\frac{2\bar{\xi}}{1 + \bar{\xi}^2}\right)\cos\xi\right] + \left[\xi - \frac{2\bar{\xi}}{1 + \bar{\xi}^2}\right]\right\},$$

(3)

where $\xi = \omega t$ and $\bar{\xi} = \omega\tau$.

A limiting case of some interest is that for quasi-static loading where the inertial and viscous terms are ignored (equivalent to m and $\mu \to 0$) giving the result

$$u = \frac{k_1 Vt}{k_1 + k_2} = \left(\frac{k_1}{k_1 + k_2}\right) \cdot \frac{V\xi}{\omega}.$$

(4)

We may now write the energy balance for a perturbation when the crack grows;

$$\dot{U}_e = \dot{U}_s + \dot{U}_k + \dot{U}_d + BG\dot{a},$$

where U_e is the external work, U_s is the strain energy, U_k the kinetic energy and U_d is the dissipated energy. Since this is a total energy sum $G = R$ the crack resistance, (B is the thickness), we may now write

$$BG = (\dot{U}_e - \dot{U}_s - \dot{U}_k - \dot{U}_d)/\dot{a}.$$

(5)

From the model

$$\dot{U}_e = k_1(Vt - u)V$$

$$\dot{U}_s = k_1(Vt - u)(V - \dot{u}) + k_2 u\dot{u} + \frac{u^2}{2}\frac{dk_2}{da} \cdot \dot{a}$$

$$\dot{U}_k = m \cdot \dot{u} \cdot \ddot{u}$$

(6)

and

$$\dot{U}_d = \mu \cdot \dot{u} \cdot \dot{u},$$

where the time derivatives for $u\{t, k_2[a(t)]\}$ are

$$\dot{u} = \frac{\partial u}{\partial t} + \dot{a} \cdot \frac{\partial u}{\partial k_2} \cdot k_2'$$

$$\ddot{u} = \frac{\partial^2 u}{\partial t^2} + 2\dot{a} \cdot \frac{\partial^2 u}{\partial k_2 \partial t} \cdot k_2' + \dot{a}^2 \left(\frac{\partial u}{\partial k_2} \cdot k_2'' + \frac{\partial^2 u}{\partial k_2^2} \cdot k_2'^2 \right) + \ddot{a} \frac{\partial u}{\partial k_2} \cdot k_2' \tag{7}$$

where $k_2' = dk_2/da$ etc.

Substituting (6) and (7) into (5) and noting Eqn. (1) for the absolute derivatives in time we have

$$\frac{G}{G_s} = \left(\frac{u}{Vt} \right)^2 + \frac{4m}{(Vt)^2} \left(\frac{\partial u}{\partial t} + \dot{a} \frac{\partial u}{\partial k_2} \cdot k_2' \right) \left[\frac{\partial^2 u}{\partial k_2 \partial t} + \frac{1}{\tau} \frac{\partial u}{\partial k_2} \right.$$

$$\left. + \frac{\dot{a}}{2} \left(\frac{\partial u}{\partial k_2'} \cdot \frac{k_2''}{k_2'} + \frac{\partial^2 u}{\partial k_2^2} \cdot k_2' \right) + \frac{\ddot{a}}{2\dot{a}} \frac{\partial u}{\partial k_2} \right] \tag{8}$$

where $BG_s = -(k_2'/2)(Vt)^2$ the static value.

3. The quasi-static case

To illustrate the effect of damping let us consider the quasi-elastic result for the slow initiation case for which $\dot{a} \rightarrow 0$ and $\ddot{a} = 0$. Equation (8) becomes

$$\frac{G}{G_s} = \left(\frac{u}{Vt} \right)^2 + \frac{4m}{(Vt)^2} \cdot \frac{\partial u}{\partial t} \left(\frac{\partial^2 u}{\partial k_2 \partial t} + \frac{1}{\tau} \frac{\partial u}{\partial k_2} \right) \tag{9}$$

and since the kinetic effects here are all due to the specimen motion and not $\dot{a}$ the various derivatives of k_2 are not involved. From (4) we have

$$\frac{\partial u}{\partial t} = \frac{k_1 V}{k_1 + k_2}, \quad \frac{\partial u}{\partial k_2} = -\frac{k_1 Vt}{(k_1 + k_2)^2} \quad \text{and} \quad \frac{\partial^2 u}{\partial k_2 \partial t} = -\frac{k_1 V}{(k_1 + k_2)^2}.$$

On substituting into (9) we have

$$\frac{G}{G_s} = \left[1 - \frac{4}{\xi^2} \cdot \frac{\bar{\xi}^2}{1 + \bar{\xi}^2} \cdot \left(1 + \frac{\xi}{\bar{\xi}} \right) \right]. \tag{10}$$

It should be noted that for zero damping ($\mu = 0$), $\bar{\xi} \rightarrow \infty$ and $G/G_s = (1 - 4/\xi^2)$, while for the heavily damped case $\bar{\xi} \rightarrow 0$ and

$$\frac{G}{G_s} = \left[1 - 4 \left(\frac{\bar{\xi}^2}{\xi^2} + \frac{\bar{\xi}}{\xi} \right) \right]. \tag{11}$$

A further case of interest here is the heavily damped solution of (1) for which;

$$\frac{1}{\tau^2} \gg \frac{k_1 + k_2}{m}$$

and the displacement becomes

$$u = \left(\frac{k_1 V}{k_1 + k_2}\right)\left[t - \frac{\tau}{2}(1 - e^{-2t/\tau})\right]$$

from which $(\partial u/\partial t)$, $(\partial u/\partial k_2)$ and $(\partial^2 u/\partial k_2 \partial t)$ may be found, thus giving

$$\frac{G}{G_s} = \left\{\left[1 - \frac{\bar{\xi}}{2\xi}(1 - e^{-2\xi/\bar{\xi}})\right]^2 - 2\cdot\frac{\bar{\xi}^2}{\xi^2}(1 - e^{-2\xi/\bar{\xi}})\left(1 + \frac{2\xi}{\bar{\xi}} - e^{-2\xi/\bar{\xi}}\right)\right\}. \qquad (12)$$

This is the most accurate representation of the heavily damped case and is compared with (11) in Fig. 2. The solution is of interest since it might be possible to conduct a highly damped test for which τ is low so that the correction factor tends to unity for reasonably short times. The factor is 0.95 at $(t/\tau) = 100$ and 0.99 at $(t/\tau) = 500$. It is important to note that such heavily damped solutions are useful if G_s is determined from striker displacement since the damping removes the oscillations and makes the motion of the specimen coincide with the quasi-static case after short times. Solutions based on load or energy are likely to be inaccurate since large values are observed and only a small proportion goes into fracture.

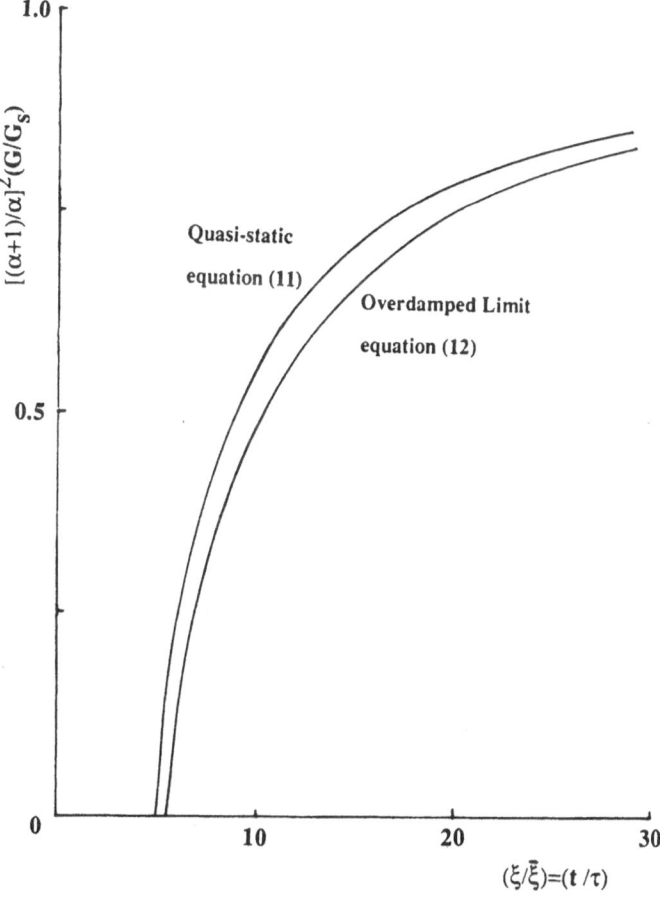

Fig. 2. The heavily damped solution for $\dot{a} \rightarrow 0$.

4. Slow crack ($\dot{a} \to 0$) initiation with damping

This case may be evaluated by taking the general equation of motion for subcritical damping, (3), and deriving the various derivatives in (9). The final solution is

$$\frac{G}{G_s} = \left[F_1^2 + 4 \frac{F_2}{\bar{\xi}^2} F_6 \right],$$
(13)

where

$$F_1 = \left[1 - \left(\frac{2\bar{\xi}}{1 + \bar{\xi}^2} \right) \frac{1}{\xi} - \frac{e^{-\xi/\bar{\xi}}}{\xi} \left(\frac{\bar{\xi}^2 - 1}{\bar{\xi}^2 + 1} \cdot \sin \xi - \frac{2\bar{\xi}}{\bar{\xi}^2 + 1} \cos \xi \right) \right]$$

$$F_2 = \left[1 - e^{-\xi/\bar{\xi}} \left(\cos \xi + \frac{1}{\bar{\xi}} \sin \xi \right) \right]$$

and

$$F_6{}^* = \frac{\bar{\xi}^2}{1 + \bar{\xi}^2} \cdot \left\{ -\left(1 + \frac{\xi}{\bar{\xi}} - \frac{4}{1 + \bar{\xi}^2} \right) + e^{-\xi/\bar{\xi}} \left[\frac{\xi}{2} \left(\frac{\bar{\xi}^2 - 1}{\bar{\xi}^2} \right) + \frac{(3\bar{\xi}^2 - 1)}{\bar{\xi}(\bar{\xi}^2 + 1)} \right] \sin \xi \right.$$

$$\left. + e^{-\xi/\bar{\xi}} \left[\left(\frac{\bar{\xi}^2 - 3}{\bar{\xi}^2 + 1} \right) - \frac{\xi}{\bar{\xi}} \right] \cos \xi \right\}.$$
(14)

For light damping we retrieve the solution given in [1]; i.e., for $\bar{\xi} \to \infty$

$$F_1 \to \left(1 - \frac{\sin \xi}{\xi} \right)$$

$$F_2 \to \left(\frac{1 - \cos \xi}{\xi} \right) \quad \text{and} \quad F_6 \to \tfrac{1}{2} \sin \xi - F_2.$$

This, and the lightly damped cases, give $G/G_s < 0$ for $0 < \xi < \sim 4$. For critical damping $\omega \to 0$ so both ξ and $\bar{\xi} \to 0$. For small $\bar{\xi}$ we have

$$\left. \begin{aligned} F_1 &\to \left[\left(1 - \frac{2\bar{\xi}}{\xi} \right) + \left(1 + \frac{2\bar{\xi}}{\xi} \right) e^{-\xi/\bar{\xi}} \right] \\ F_2 F_6 &\to \left(\frac{\bar{\xi}}{\xi} \right)^2 \left[1 - \left(1 + \frac{\xi}{\bar{\xi}} \right) e^{-\xi/\bar{\xi}} \right] \cdot \left[\left(3 - \frac{\xi}{\bar{\xi}} \right) - \left(3 + 2\frac{\xi}{\bar{\xi}} + \frac{1}{2} \left(\frac{\xi}{\bar{\xi}} \right)^2 \right) e^{-\xi/\bar{\xi}} \right]. \end{aligned} \right\}$$
(15)

---

* F_6 is also given from Appendix 1 as $(\xi^2/(1 + \bar{\xi}^2)) ((\xi/\bar{\xi})F_3 + (\bar{\xi}^2/\xi^2)F_5)$, see Section 5.

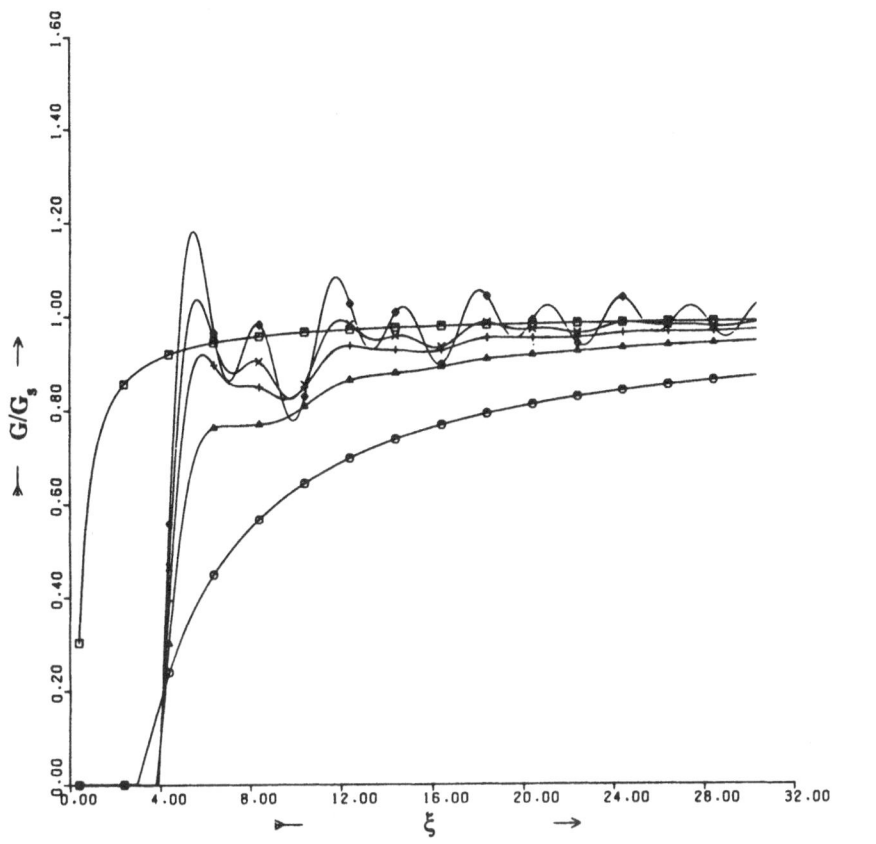

Fig. 3. Slow crack initiation with varying $\bar{\xi}$, $[(a/D) = 0.1$ and $(\dot{a}/C) \to 0]$.

The full solution, (13) and (14), is given in Fig. 3 showing that there is only a slight effect on the undamped solution for $\bar{\xi} > 20$ but for $\bar{\xi} < 5$ the oscillations fade and the curve is depressed. As $\bar{\xi}$ tends to zero, the critical condition, G/G_s rises again as given by (15). The basic mechanism here is that for moderate damping G/G_s decreases, since some energy is dissipated, but as the damping increases the motion is suppressed thus removing some of the kinetic energy effects and causing G/G_s to rise but to remain less than unity because of dissipation.

5. Full dynamic solution

The full solution involving the crack speed effects is obtained from the various differentials derived from (3) which are then substituted into (8). In these cases the stiffness variations with crack length a are needed and these may be conveniently expressed in terms of

$$k_2, \frac{dk_2/da}{k_2} \quad \text{and} \quad \frac{d^2k_2/da^2}{dk_2/da}.$$

Numerical solutions may be derived from the usual stress intensity factor calibration factors, Y^2, and are

$$
\left.
\begin{aligned}
k_2 &= \frac{4EBD^3}{S^3} \cdot \psi_1 \\[2ex]
\frac{1}{k_2} \frac{dk_2}{da} &= -\frac{18\pi a}{SD} \cdot \psi_2 \\[1ex]
\text{and} & \\[1ex]
\frac{d^2 k_2/da^2}{dk_2/da} &= \frac{1}{a} \cdot \psi_3
\end{aligned}
\right\}
\tag{16}
$$

where

$$
\left.
\begin{aligned}
\psi_1 &= \frac{1}{1 + \dfrac{18D}{S} \displaystyle\int^x Y^2 \cdot x \cdot dx} \\[3ex]
\psi_2 &= \frac{Y^2}{\pi} \cdot \psi_1 \\[2ex]
\psi_3 &= 1 - 2x \left(\psi_2 \cdot \frac{18\pi D x}{S} - \frac{dY/dx}{Y} \right) \\[2ex]
\text{and} & \\[1ex]
x &= \frac{a}{D}
\end{aligned}
\right\}
\tag{17}
$$

The functions $\psi(a/D)$ may be computed from $Y^2(x)$, [6], and are shown in Table I, for $(S/D) = 4$ and 8, and as $x \to 0$, $\psi_1 = \psi_2 = \psi_3 = 1$. Quite accurate approximations may be computed (which avoids the use of numerical values) by assuming the infinite plate solution $Y^2 = \pi$.

The results become

$$
\left.
\begin{aligned}
\psi_1 &= \psi_2 = \frac{1}{1 + 9\pi(D/S)x^2} \qquad &\text{-a} \\[2ex]
\text{and} & \\[1ex]
\psi_3 &= 1 - 4 \cdot \frac{9\pi D/S x^2}{\left(1 + 9\pi \dfrac{D}{S} x^2 \right)} \qquad &\text{-b}
\end{aligned}
\right\}
\tag{18}
$$

These results are compared with the numerical values from Table I, in Fig. 4 and show quite close agreement up to $(a/D) = 0.3$.

Table 1. Calibration functions

$x = a/D$	$S/D = 4$			$S/D = 8$		
	ψ_1	ψ_2	ψ_3	ψ_1	ψ_2	ψ_3
0.1	0.931	0.904	0.651	0.962	0.992	0.789
0.2	0.783	0.753	0.275	0.872	0.905	0.630
0.3	0.613	0.665	−0.157	0.745	0.875	0.414
0.4	0.447	0.616	−0.593	0.600	0.880	0.149
0.5	0.303	0.604	−1.063	0.448	0.948	−0.209
0.6	0.189	0.624	−1.883	0.303	1.061	−0.956

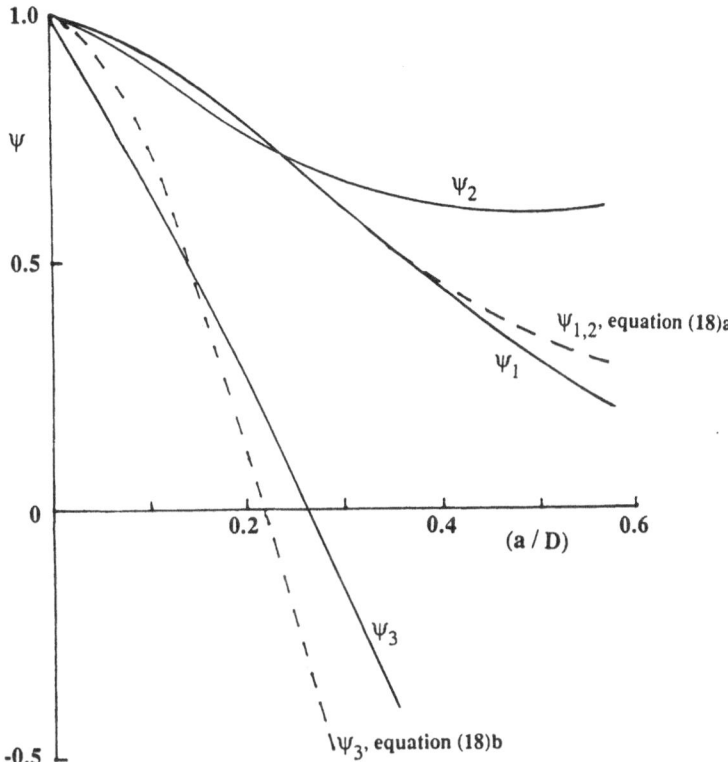

Fig. 4. Comparison of calibration functions for $(s/D) = 4$, dotted line indicating $Y^2 = \pi$.

The final results may be deduced from (8) with $\ddot{a} = 0$ by rearranging it in the following form:

$$
\frac{G}{G_s} = F_1^2 + 2\left[F_2 + \dot{a}\left(\sqrt{\frac{m}{k_2}} \cdot \frac{k_2'}{k_2}\right)\frac{F_3}{(1 + \alpha)^{3/2}}\frac{(1 + \bar{\xi}^2)^{1/2}}{\bar{\xi}}\right]
$$

$$
\times \left\{\left[1 + \frac{\dot{a}}{2}\left(\sqrt{\frac{m}{k_2}} \cdot \frac{k_2''}{k_2'}\right)\frac{(1 + \bar{\xi}^2)^{1/2}}{(1 + \alpha)^{1/2}}\right]\frac{2\bar{\xi}}{(1 + \bar{\xi}^2)\xi} \cdot F_3 \right. \tag{19}
$$

$$
\left. + \dot{a}\left(\sqrt{\frac{m}{k_2}} \cdot \frac{k_2'}{k_2}\right) \cdot \frac{F_4}{(1 + \alpha)^{3/2}} \cdot \frac{\bar{\xi}}{\xi(1 + \bar{\xi}^2)^{1/2}} + \frac{2\bar{\xi}^2}{(1 + \bar{\xi}^2)} \cdot \frac{F_5}{\xi^2}\right\},
$$

where

$$F_1 = \left(\frac{u}{V_0 t}\right), \quad F_2 = \left(\frac{1}{V_0}\frac{\partial u}{\partial t}\right), \quad F_3 = \frac{(1 + \alpha)k_2}{V_0 t}\frac{\partial u}{\partial k_2}$$

$$F_4 = \frac{(1 + \alpha)^2 k_2^2}{V_0 t}\cdot\frac{\partial^2 u}{\partial k_2^2}, \quad F_5 = \frac{(1 + \alpha)k_2}{V_0}\frac{\partial^2 u}{\partial k_2 \partial t} \quad \text{and} \quad V_0 = V\frac{\alpha}{(1 + \alpha)},$$

The functions $F(\xi, \bar{\xi})$ may all be deduced from (3) and are listed in Appendix 1. The various stiffness and mass parameters, may be deduced [1] from the usual beam theory and (16) giving

$$\sqrt{\frac{m}{k_2}}\cdot\frac{k_2}{k_2} = -\frac{18\pi Sa}{CD^2}\sqrt{\frac{17}{140}}\cdot\frac{\psi_2}{\psi_1^{1/2}}$$

$$\sqrt{\frac{m}{k_2}}\cdot\frac{k_2''}{k_2} = \frac{S^2}{C.aD}\cdot\sqrt{\frac{17}{140}}\cdot\frac{\psi_3}{\psi_1^{1/2}}$$

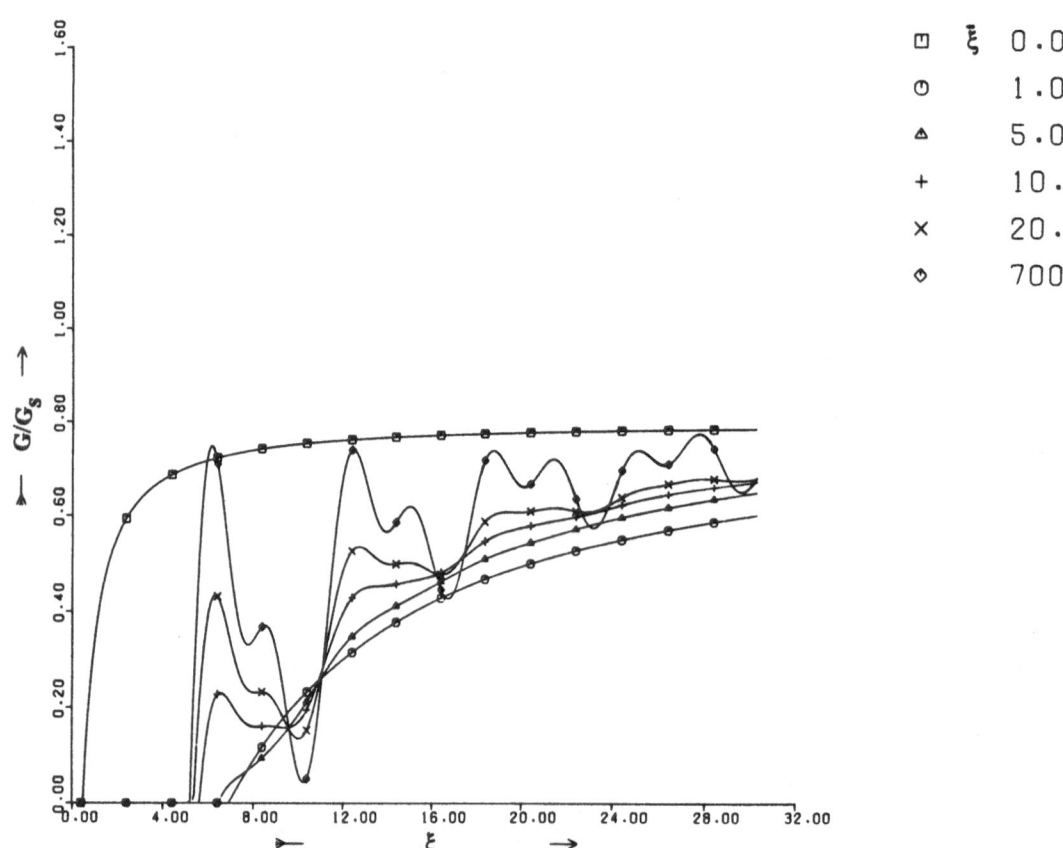

Fig. 5. Effect of increasing crack speed, $(\dot{a}/C)$, on (G/G_s) plots in Fig. 3, $[(a/D) = 0.1$ and $(\dot{a}/C) = 0.05]$.

$$\alpha = \frac{k_1 S^3}{4EBD^3} \cdot \frac{1}{\psi_1} \quad \text{where} \quad C = \sqrt{\frac{E}{\varrho}}.$$

Figure 5 shows the effect of increasing the crack speed $(\dot{a}/C) = 0.05$ for the same values of ξ as in Fig. 3 and $\alpha = 10$. There is an overall reduction in the level of G/G_s as in the undamped solution [1], with a pattern of behaviour with damping similar to the $\dot{a} \to 0$ in Fig. 3. Figure 6 shows the variation of the $\xi = 20$ case with the increase in crack length (a/D). Here the level of G/G_s is decreased with decreasing (a/D), as k_2 decreases.

The force at the load point P_1 is shown in Figs. 7 and 8 non-dimensionalised by $P_0 = k_1 V/\omega$:

$$P_1/P_0 = \xi\{1 - [\alpha/(1 + \alpha)]F_1\}.$$

Figure 7 shows the (P_1/P_0) plot for $(a/D) = 0.1$ and decreasing damping resulting, as expected, with increasing oscillations around a ramp function of $P_1/P_0 = \xi/(1 + \alpha)$. Large damping gives a positive displacement of the average load line. Figure 8 shows (P_1/P_0) for $\xi = 20$ and increasing (a/D). This indicates that, as expected, there is a drop in the force magnitude as the crack is increased in size but there are no major changes in dynamic behaviour by varying (a/D).

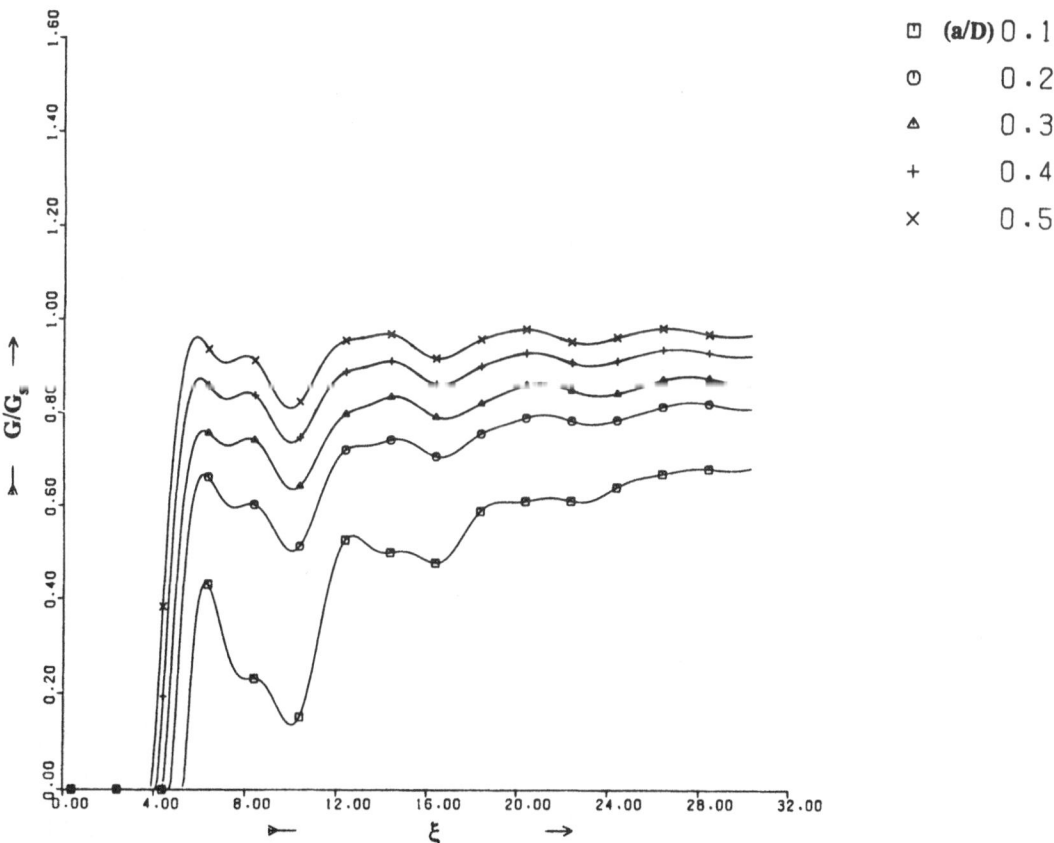

Fig. 6. Effect of increasing crack length, (a/D), for typical damping parameter and crack speed, $[(\dot{a}/C) = 0.05$ and $\xi = 20]$.

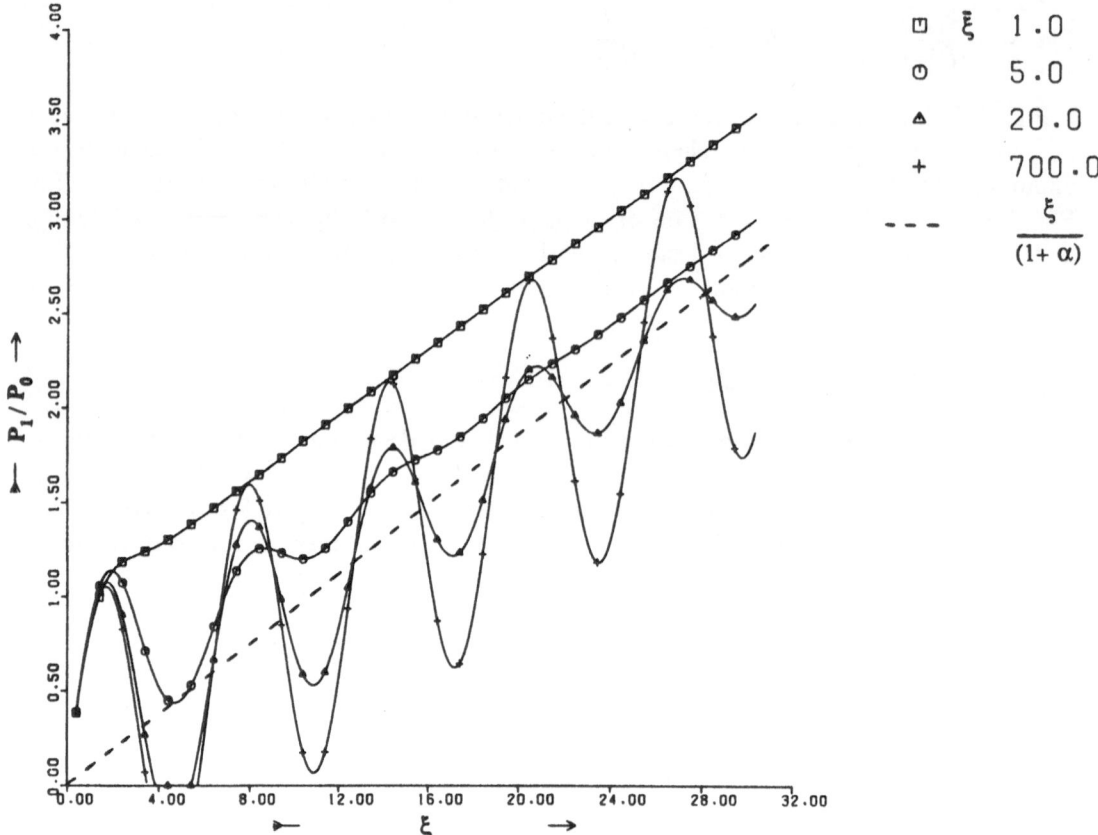

Fig. 7. Force, P_1, in contact stiffness, k_1, for $(a/D) = 0.1$, $(P_0 = k_1 V/\omega)$.

Figure 9 shows a typical experimental load curve taken from [3] and a theoretical curve. The theoretical curve is obtained from the model used in conjunction with the specimen parameters (given in Appendix 2). The following equations are utilised in order to dimensionalise the force plots

$$P_0 = \frac{2V\alpha_0 Cm}{D \cdot (S/D)^2} \cdot \sqrt{\frac{35(1 + 1/\bar{\xi}^2)}{17(\psi_1 + \alpha_0)}}$$

$$\omega = \frac{2C}{D \cdot (S/D)^2} \cdot \sqrt{\frac{35(\psi_1 + \alpha_0)}{17(1 + 1/\bar{\xi}^2)}}$$

and

$$\alpha_0 = \frac{k_1 S^3}{4EBD^3}.$$

The parameters $\bar{\xi}$ and α_0 are adjusted, by trial and error, until the best agreement between experiment and theory is obtained and the values of $\bar{\xi} = 35$ and $\alpha_0 = 5.8$ provide the best

---

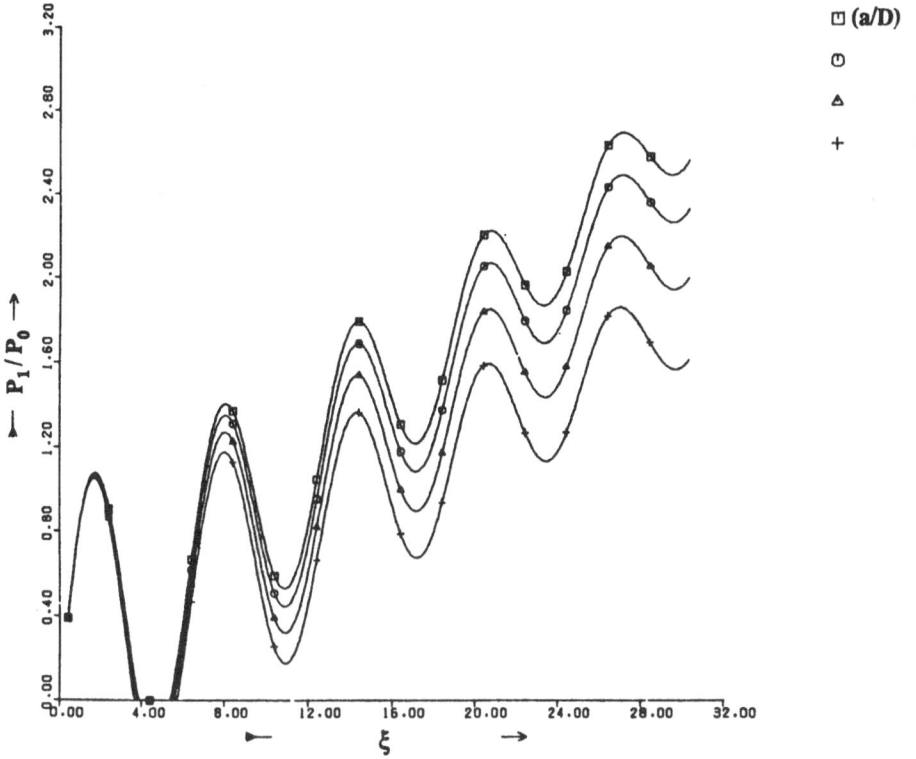

Fig. 8. Variation of force P_1, with (a/D), $(\bar{\xi} = 20)$.

fit. It should be noted that $\alpha_0 = 5.8$ is the same as the value given in the undamped model [3], which is expected since damping does not alter the short time behaviour significantly, see Fig. 7, and the effect of $\bar{\xi}$ will be seen mostly in the fit to the second peak which is much better here than in [3]. A value of $\bar{\xi} = 35$, however, signifies rather light damping and supports the use of the simple undamped solution.

6. Conclusions

The inclusion of damping in the lumped mass-spring model extends the utility of the model since it enables a further factor to be accounted for. The general trends in G and load are sensible and give some insights into large damping effects. Improved fits to experimental data are possible with an extra parameter but only experience will tell if this is of any real significance.

Appendix 1

$$F_1 = 1 - \frac{2\bar{\xi}}{(1 + \bar{\xi}^2)\xi} - \frac{e^{-\xi/\bar{\xi}}}{\xi}\left[\frac{(\bar{\xi}^2 - 1)}{(\bar{\xi}^2 + 1)}\sin\xi - \frac{2\bar{\xi}\cos\xi}{(\bar{\xi}^2 + 1)}\right]$$

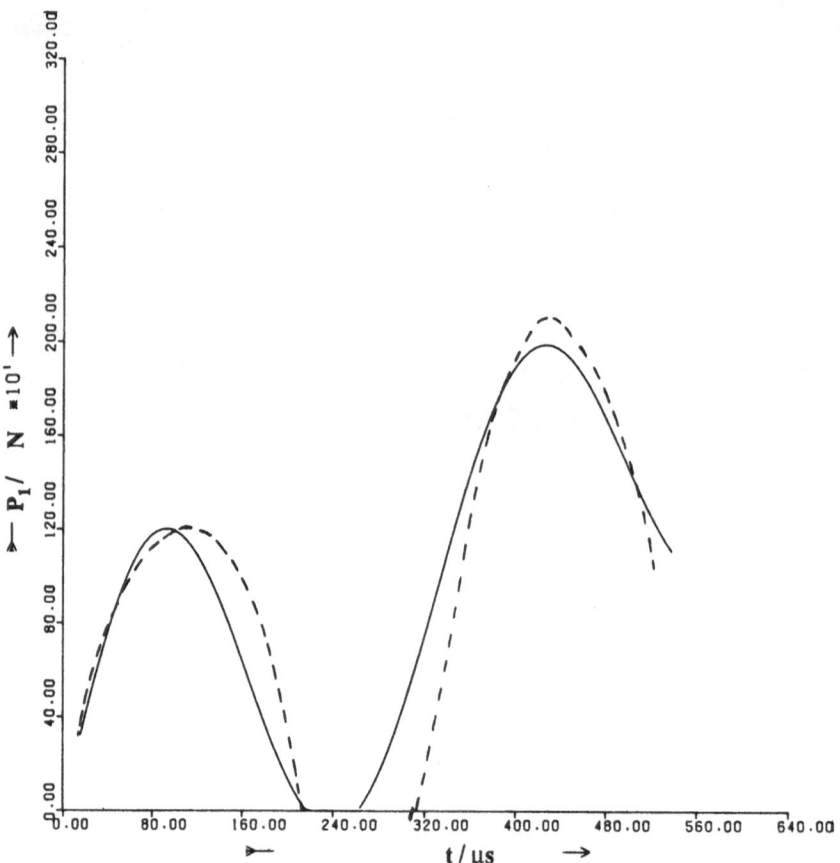

Fig. 9. Comparison of P_1 with experiment (dotted line), $[\bar{\xi} = 35, (a/D) = 0.15,$ and $\alpha_0 = 5.8]$.

$$F_2 = 1 - e^{-\xi/\bar{\xi}}\left(\cos \xi + \frac{\sin \xi}{\bar{\xi}}\right)$$

$$F_3 = -1 + \frac{4\bar{\xi}}{(\bar{\xi}^2 + 1)\xi} + \frac{e^{-\xi/\bar{\xi}}}{2\xi\bar{\xi}^2(1 + \bar{\xi}^2)}$$

$$\times [(3\bar{\xi}^4 - 2\bar{\xi}^3\xi - 6\bar{\xi}^2 - 2\bar{\xi}\xi - 1) \sin \xi - (\bar{\xi}^4\xi + 8\bar{\xi}^3 - \xi) \cos \xi]$$

$$F_4 = 2 - \frac{12\bar{\xi}}{(1 + \bar{\xi}^2)\xi} + \frac{e^{-\xi/\bar{\xi}}}{4\xi\bar{\xi}^4(1 + \bar{\xi})} \{[7\xi\bar{\xi}^6 + (48 - 2\xi^2)\bar{\xi}^5 - 5\xi\bar{\xi}^4 - 4\xi^2\bar{\xi}^3 - 15\xi\bar{\xi}^2$$

$$- 2\xi^2\bar{\xi} - 3\xi] \cos \xi + [(\xi^2 - 15)\bar{\xi}^6 + 18\xi\bar{\xi}^5 + (\xi^2 + 45)\bar{\xi}^4 + 20\xi\bar{\xi}^3$$

$$+ (15 - \xi^2)\bar{\xi}^2 + 2\xi\bar{\xi} + (3 - \xi^2)] \sin \xi\}$$

$$F_5 = -1 + \frac{e^{-\xi/\bar{\xi}}}{2\bar{\xi}^3} [(2\bar{\xi}^3 - \xi\bar{\xi}^2 - \xi) \cos \xi + (\bar{\xi}^3\xi + 3\bar{\xi}^2 + \bar{\xi}\xi + 1) \sin \xi]$$

Appendix 2

Effective dimensions of Charpy specimen (Araldite), and test parameters from [3];

$D = 0.05\,\text{m}\ S/D = 4\ a/D = 0.15$

$m = 0.035\,\text{kg}\ V = 1.9\,\text{m/s}$

$\alpha_0 = 5.8,\ \bar{\xi} = 35$

and $C = 2040\,\text{m/s}$.

Specimen geometry

References

1. J.G. Williams, *International Journal of Fracture* 33 (1987) 47–59.
2. J.G. Williams and G.C. Adams, *International Journal of Fracture* 33 (1987) 209–222.
3. B.A. Crouch and J.G. Williams, *Journal of the Mechanics and Physics of Solids*, in press.
4. R.J. Dexter, S.J. Hudak, Jr., K.W. Reed, E.Z. Polch, and M.F. Kanninen, in *Proceedings 4th International Conference on Numerical Methods in Fracture Mechanics*, Pineridge Press (1987) 173–190.
5. J. Ahmad, in *Proceeding 4th International Conference on Numerical Methods in Fracture Mechanics*, Pineridge Press (1987) 837–849.
6. W.F. Brown and J.E. Srawley, *ASTM STP 410* (1966).

Résumé. Dans un mémoire récent, on a analysé la croissance à grande vitesse d'une fissure en utilisant un modèle masse tombante-ressort incluant l'effet de la raideur à l'endroit du contact. On a montré que la vitesse de relaxation de l'énergie *G* pouvait être calculée dans une telle situation dynamique en utilisant l'énergie cinétique de la masse. Ce modèle s'est révélé particulièrement satisfaisant pour décrire les effets dynamiques dans l'essai de résilience Charpy. La présente étude élargit le modèle en introduisant un amortissement en parallèle à la raideur du matériau, ce qui constitue une représentation raisonnable d'un matériau viscoélastique. On montre que *G* peut être considérablement influencé même par un petit peu d'amortissement si une rupture à brève échéance est en cause. La méthode est basée sur l'équilibre des vitesses des énergies en présence, une approche dont Max L. Williams a été le pionnier.

International Journal of Fracture 39: 163-189 (1989)
© Kluwer Academic Publishers, Dordrecht

On the mechanics of crack closing and bonding in linear viscoelastic media

R.A. SCHAPERY

Civil Engineering Department, Texas A&M University, College Station, Texas 77843, USA

Received 12 July 1988; accepted 29 July 1988

Abstract. The mechanics of quasi-static crack closing and bonding of surfaces of the same or different linear viscoelastic materials is described. Included is a study of time-dependent joining of initially curved surfaces under the action of surface forces of attraction and external loading. Emphasis is on the use of continuum mechanics to develop equations for predicting crack length or contact size as a function of time for relatively general geometries; atomic and molecular processes associated with the healing or bonding process are taken into account using a crack tip idealization which is similar to that used in the Barenblatt method for fracture. Starting with a previously developed correspondence principle, an expression is derived for the rate of movement of the edge of the bonded area. The effects of material time-dependence and the stress intensity factor are quite different from those for crack growth. A comparison of intrinsic and apparent energies of fracture and bonding is made, and criteria are given for determining whether or not bonding can occur. Examples are given to illustrate use of the basic theory for predicting healing of cracks and growth of contact area of initially curved surfaces. Finally, the effect of bonding time on joint strength is estimated from the examples on contact area growth.

1. Introduction

In 1966 Williams [1] observed that ". . . adhesive failure is not normally analyzed as a fracture problem in continuum mechanics". He then demonstrated that cohesive and adhesive fracture may be analyzed by similar methods through consideration of the stress singularity at crack tips and an energy balance. Following this pioneering paper the understanding of adhesive failure has advanced greatly, clearly aided by the use of fracture mechanics theory as urged by Williams.

Quasi-static analysis of adhesive and cohesive fracture of linear viscoelastic media has also reached a relatively mature state through application of fracture mechanics principles [e.g., 2–7]. There are, however, still important unsolved problems of cohesive and adhesive fracture of nonhomogeneous media; in many cases the stresses on the crack plane are functions of the local viscoelastic properties, which significantly complicate the analysis [5, 8]. Much progress has been made as well on dynamic fracture of linear homogeneous viscoelastic materials by Walton [9], among others; at sufficiently high crack speeds the stresses depend on viscoelastic properties.

The problem which is the opposite of fracture, the bonding of surfaces, has received far less attention in the context of fracture-like mechanics. For elastic media under quasi-static conditions, the analysis of bonding is essentially the same as that of fracture. Johnson et al. [10] and Roberts and Thomas [11] studied by experiment and linear theory the bonding and unbonding of flat and curved surfaces of rubber with other materials. They used energy methods analogous to those employed in elastic fracture mechanics, recognizing that contact mechanics and fracture mechanics have to be combined. Although some viscoelastic effects were observed, essentially elasticity theory was used. Later, Greenwood and Johnson [12] made a viscoelastic fracture analysis of the separation of a sphere from a plane, allowing for dissimilar materials, but assuming there are no shear stresses along the interface. The force of attraction acting across the interface was based on a theoretical model of the force between planes of atoms as a function of separation distance.

In a series of six papers, Anand and coworkers [13–18] studied the so-called problem of autohesion, which is the bonding of surfaces of the same material. Contact area growth between irregular (locally curved) surfaces was analyzed using linear viscoelastic contact mechanics. The influence of the force of attraction between surfaces was implicitly neglected compared to that due to a constant externally applied compression when predicting contact area growth. The increase of tensile joint strength with contact time was assumed to be due to the area increase. Other studies of polymer–polymer bonding have assumed complete contact is achieved immediately, and the time-dependence of strength is due to interdiffusion of molecular segments across the interface and the related formation of entanglements; many references may be found in the recent paper on crack healing by Kausch et al. [19]. Jud et al. [20] and Stacer and Schreuder-Stacer [21] discussed the time-dependence of the wetting or contact process in comparison to that for interdiffusion; some explicit estimates of contact time-dependence are in [20].

In this paper we address the problem of bonding between linear, isotropic viscoelastic media, accounting for the effect of interfacial forces of attraction (or surface energy) *and* external loading on the rate of growth of bonded area. Processes are assumed to occur slowly enough that inertia effects are negligible and, also for simplicity, we consider only the mode I problem (in which only normal stresses act across the bond surface). The theory shows how crack healing and contact area growth may occur under either external tensile or compressive forces, and a relatively simple method is developed for predicting this behavior.

It is believed the results will be useful for the development of basic models of the healing of damage in composites, insofar as this damage consists of micro- or macrocracks. Significant amounts of healing have been observed experimentally for asphalt [22] and solid propellant [23], which are particulate composites. More generally, the theory may be used to account for viscoelasticity and surface force or energy effects in adhesion and autohesion processes when externally applied compression is not large enough to produce complete, immediate surface-to-surface contact. Although the surface energy may be small, over a long period of time it can produce considerable deformation and contact area growth in soft materials. With a minor modification the theory in this paper would provide a means for predicting the effect of a concentrated shear stress at the trailing edge of so-called "waves of detachment" observed in the sliding of rubber on smooth surfaces [24]; addition of an interfacial shear stress may be done in the same way as described in [5] for fracture.

In order to illustrate the type of behavior studied here, consider the problem in Fig. 1. Figure 1a shows the effect of (positive) surface energy on the contact between a deformable spherical surface and a rigid flat surface. Without surface energy, Fig. 1b, the spherical surface is tangent to the flat surface at the contact edge, and the load necessary to produce contact to the radius a_c is given by the Hertz theory [25]. Less compressive force is needed in Fig. 1a for the same contact radius; indeed, with surface energy a tensile force can be supported. The dashed lines in Fig. 1 indicate the positions of undeformed spherical surfaces after experiencing the same amount of movement at the remote loading point as in the deformed body. When the body is viscoelastic, the contact radius varies with loading rate

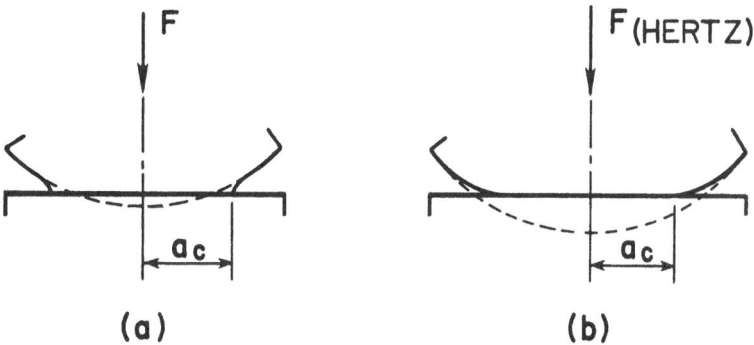

Fig. 1. Contact configurations (a) with surface energy and (b) without surface energy. After [12].

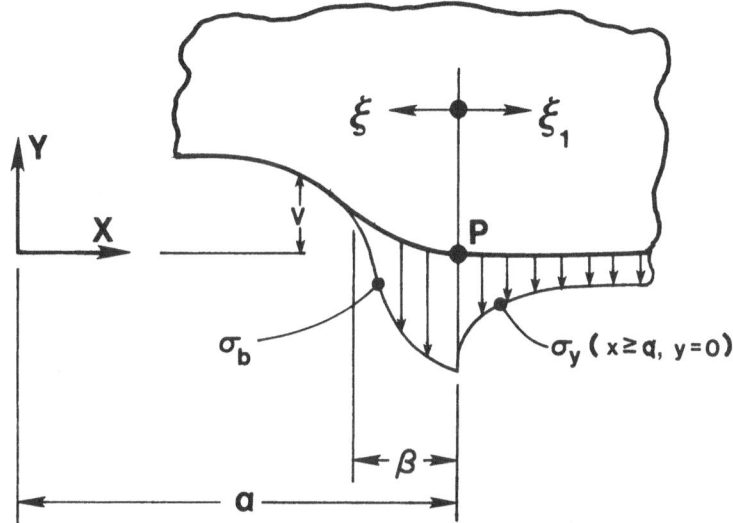

Fig. 2. Normal stress and displacement along the bond surface in the neighborhood of the bond edge.

and, at the same load, may be considerably different for positive bonding speed ($\dot{a}_c > 0$) than for debonding or crack growth ($\dot{a}_c < 0$). With a small tensile or zero external force, the interfacial force of attraction acting across the cusp-shaped separation at the contact edge, Fig. 2, causes the contact radius to grow; if the long-time modulus of the sphere is zero (e.g., noncrosslinked rubber) the contact will grow indefinitely, just as for a liquid. The force required to separate totally the two bodies may be large if considerable viscoelastic dissipation accompanies the separation process. The separation process was analyzed by Greenwood and Johnson [12] using the author's theory of crack growth [4]. Here, a theory is developed for predicting the viscoelastic bonding process for this and other geometries and material combinations, as well as for healing of internal cracks.

Section 2 introduces the notation and geometry used in describing the edge of the bonded area. This edge region, Fig. 2, is assumed to be very small compared to the contact size; for example, it exists only at the contact edge in Fig. 1a where the parabolic surface profile touches the lower flat plate. Without the local tangential separation shown in Fig. 2, the stresses at the contact edge would be infinite, according to linear viscoelasticity theory. With edge movement, these infinite stresses produce infinite strain rates and lead to physically unacceptable predictions of bonding or crack speed. In the latter case, it is well-known that the singular stresses lead to a physically unacceptable prediction of crack growth. Difficulties arising from use of the classical singular stresses were recognized by Williams [2], who circumvented the problem by replacing the local geometry of a crack by a spherical flaw. Later, Knauss [3] and the author [4] used the Barenblatt method (or what is often called the Dugdale-Barenblatt or Dugdale method) to remove the singularity. The latter approach is used in this paper; it amounts to adjusting β and/or σ_b so that the stresses on the bond surface (and in the continua) are finite.

Section 3 indicates how the stresses and displacements in viscoelastic media may be obtained from elastic solutions. This approach is used in Section 4, along with Barenblatt's method for elastic media, to obtain the local viscoelastic interfacial stress and displacement. In contrast to the crack growth problem, the stresses in elastic and viscoelastic bodies are different. This difference complicates the bonding problem, and leads to viscoelastic effects in the neighborhood of the bond edge which are considerably different than predicted with crack growth. Nevertheless, in certain cases it is possible to derive relatively simple equations for predicting bonding speed and the *bonding-zone length* β, Fig. 2, in terms of the stress intensity factor. In Section 5 the interface *bonding stress* σ_b, Fig. 2, is assumed constant, and exact expressions are obtained for a material characterized by a generalized power law creep compliance; approximate expressions are then developed for a more general form of creep compliance. In the Appendix a simple power law compliance is used with a bonding stress σ_b that varies with distance between the surfaces to derive an exact equation for bonding speed; the dependence of bonding speed on stress intensity factor is found to be the same as for constant σ_b. Section 6 is concerned with the energy needed for crack growth as compared to that for bonding of surfaces, accounting for viscoelastic dissipation effects. Examples of crack healing and the bonding of initially curved surfaces are given in Section 7, using the theory developed in Section 5. The theory can be readily extended to certain types of orthotropic linear and nonlinear media, as discussed in the concluding remarks in Section 8.

2. Description of the contact edge

Referring to Fig. 2, the (x, y) coordinate system is stationary, with the y-axis located at any convenient position in the continuum; $y = 0$ defines the surface along which bonding occurs. We assume this surface is *locally* flat, but do not restrict its shape far from the contact edge (relative to β). The bodies above and below $y = 0$ are assumed to be in contact to the right of the point designated as P. In practice, surface irregularities may prevent complete contact of the adjacent areas; however, for our purposes, it will be sufficient to assume only that the scale of this local roughness is small compared to the length of the bonding zone β which is defined below.

To the left of the position $x = a - \beta$ the two surfaces are assumed to be far enough apart that the intermolecular force of attraction is negligible. Over the length β the force of attraction, which on the scale of the continuum is represented by the tensile bonding force per unit area, σ_b, tends to draw the adjacent surfaces together until complete contact exists (in the sense indicated earlier).

The edge of complete contact (which can also be thought of as a crack tip), whose intersection with the plane of the page is the point P, is in general a curved line in space which may be closed or else open and ending on the edges of the continuum. It is necessary to assume the radius of curvature of this line in the neighbourhood of the point P is large relative to β in order to be able to use the equations of plane strain in a local bonding analysis; this neighborhood is defined by the size β.

The materials above and below the bond surface are assumed to be linearly viscoelastic, homogeneous and isotropic, with the possible exception of a thin surface layer of damaged material; this layer, which may be the result of the surface being formed in the first place by means of crack growth or any other process, will have no practical affect on our analysis if its thickness is small compared to β. As noted previously, the scale of local surface roughness is similarly assumed to be small. These surface features will, of course, affect the particular values of σ_b which are to be used in the theory. Roughness on a scale much larger than β is permitted.

3. Viscoelastic stresses and displacements

By following arguments based on the Laplace transform similar to those given in [4] for the case of the opening mode of crack growth, we can express viscoelastic stresses and displacements in terms of elastic solutions for crack shortening; this approach amounts to the use of the extended correspondence principle for contact or indentation problems [6].

With crack growth or no growth, i.e., $\dot{a} \geqslant 0$ in Fig. 2, the normal stress acting along the plane of crack prolongation is the same for both elastic and viscoelastic bodies (which are assumed to be identical with respect to geometry and applied loads) if this stress for the elastic body is independent of elastic constants. This result in turn leads to viscoelastic displacements which can be written explicitly in terms of the history of the elastic values.

In contrast, it may be shown by using the Laplace transform that with crack shortening (or, equivalently, in problems with growing contact area) the normal displacements and stresses take on opposite roles with respect to elastic solutions if the elastic normal surface displacement is independent of the elastic constants. This simplicity in the solutions does not necessarily exist for the type of boundary value problems of interest here unless the Poisson's ratio is constant and all traction boundary conditions are converted to specified displacement boundary conditions. When the Poisson's ratio is constant (or, at least, the effect of its time-dependence is small enough that it may be assumed constant) the more straightforward method given in [26] for constructing viscoelastic solutions from elastic solutions may be used. The method is not based on Laplace transform theory although it uses a so-called correspondence principle, designated as CP-III in [26]. This correspondence principle may be used with certain types of nonlinearity (including distributed damage), anisotropy, and aging. However, for simplicity we consider here only linear, isotropic, nonaging media.

Let us now state the relevant version of CP-III and then define the terms:

If $\partial n_i / \partial t = 0$ on the surface S_T and if $\mathrm{d}S_T/\mathrm{d}t \leqslant 0$, the viscoelastic solution is

$$\sigma_{ij} = \{E \, \mathrm{d}\sigma_{ij}^R\} \tag{1}$$

$$u_i = u_i^R, \tag{2}$$

where σ_{ij}^R and u_i^R satisfy the field equations and boundary conditions of the so-called *reference elastic problem* (or, more briefly, the *elastic problem*) defined below.

The viscoelastic and elastic bodies are assumed to be geometrically identical in all respects, and to have tractions specified over the portion of the (external and internal) surfaces designated by S_T and displacements specified over the remaining surfaces S_U; the complete boundary is designed by S. Mixed conditions (e.g., normal displacement and shear traction) along a boundary may be specified; but for notational simplicity this situation is not explicitly stated in CP-III. All stresses σ_{ij} and displacements u_i in the viscoelastic problem, and σ_{ij}^R and u_i^R in the elastic problem (where $i, j = 1, 2, 3$) are referred to an orthogonal set of Cartesian coordinates x_i of the undeformed bodies. The n_i are components of the outer unit normal at any point on the undeformed surface.

The braces { } denote a convolution integral,

$$\{E\,\mathrm{d}f\} \equiv E_R^{-1} \int_{-\infty}^{t} E(t - \tau) \frac{\partial f}{\partial \tau}\, \mathrm{d}\tau, \tag{3}$$

where f is a function of time such as σ_{ij} or σ_{ij}^R, E_R is an arbitrarily selected constant (usually with units of modulus) and $E(t)$ is the uniaxial, linear viscoelastic relaxation modulus; for an elastic material, E is the Young's modulus. In many cases the quantity $\{D\,\mathrm{d}f\}$ will be used; it is defined using the creep compliance $D(t)$,

$$\{D\,\mathrm{d}f\} \equiv E_R \int_{-\infty}^{t} D(t - \tau) \frac{\partial f}{\partial \tau}\,\mathrm{d}\tau, \tag{4}$$

where $D(t)$ and $E(t)$ are interrelated according to

$$\int_0^t E(t - \tau) \frac{\mathrm{d}D}{\mathrm{d}\tau}\, \mathrm{d}\tau = \int_0^t D(t - \tau) \frac{\mathrm{d}E}{\mathrm{d}\tau}\,\mathrm{d}\tau = 1 \tag{5}$$

for $t > 0$; also $E = D = 0$ for $t < 0$. The lower limit is to be interpreted as 0^- in order to include the singularity in the derivatives at $\tau = 0$.

Additional relationships which will be needed later, and which follow directly from (3)–(5), are concerned with inverses of the linear operators denoted by the braces. Namely, if

$$g = \{D\,\mathrm{d}f\} \tag{6}$$

then

$$f = \{E\,\mathrm{d}g\} \tag{7}$$

or, equivalently,

$$f = \{E\,\mathrm{d}\{D\,\mathrm{d}f\}\} \tag{8}$$

and

$$f = \{D\,\mathrm{d}\{E\,\mathrm{d}f\}\}. \tag{9}$$

Equations (8) and (9) may be easily verified by Laplace transforming them and then using the Laplace transform of (5); the two-sided Laplace transform can be used in the event that f and g do not vanish at negative times.

In the earlier work [26] we assumed that all stresses and strains vanish for $t < 0$, while here this assumption is not made. One can, of course, always use a lower limit of zero in (3) and (4) by shifting the time axis so that the body is unstressed and unstrained for $t < 0$. However, when using CP-III in problems for which $dS_T/dt > 0$ (such as crack growth) before $dS_T/dt < 0$ (such as crack healing) it may be useful to select $t = 0$ to follow the $dS_T/dt > 0$ process; with this convention stresses and strains may exist at $t = 0$ without invalidating CP-III if they can be calculated without regard to the $dS_T/dt > 0$ process. Such a situation arises if a crack first grows and then is arrested for a long enough time that the mechanical state of the continuum at $t = 0$ is the same as if the crack's geometry had not changed when $t < 0$. Consequently, one can solve the crack shortening problem for $t \geqslant 0$ by assuming there is no change in crack dimensions for $t < 0$.

Returning to CP-III, the elastic problem is defined by the standard equilibrium equations with a body force F_i^R,

$$\partial \sigma_{ij}^R / \partial x_j + F_i^R = 0, \tag{10}$$

strain-displacement equations,

$$\varepsilon_{ij}^R = \tfrac{1}{2}(\partial u_i^R/\partial x_j + \partial u_j^R/\partial x_i) \tag{11}$$

and stress–strain equations,

$$\varepsilon_{ij}^R = [(1 + v)\sigma_{ij}^R - v\delta_{ij}\sigma_{kk}^R]/E_R, \tag{12}$$

where $k = 1, 2, 3$. The usual summation convention is used in that summation over the range of the repeated index is implied. Also, E_R and v are the Young's modulus and Poisson's ratio, respectively, for the elastic problem, and δ_{ij} is the Kronecker delta (i.e., $\delta_{ij} = 1$ if $i = j$ and $\delta_{ij} = 0$ if $i \neq j$). The boundary conditions are

$$\sigma_{ij}^R n_j = T_i^R \quad \text{on} \quad S_T \tag{13}$$

$$u_i^R = U_i^R \quad \text{on} \quad S_u. \tag{14}$$

The quantities T_i^R and F_i^R specified in the elastic problem are *not* the same as the traction T_i and body force F_i specified in the viscoelastic problem of interest. Rather, they are to be found using the convolution integrals,

$$F_i^R \equiv \{D\,dF_i\} \tag{15}$$

$$T_i^R \equiv \{D\,dT_i\}. \tag{16}$$

However, the specified displacements are the same for both problems,

$$U_i^R = U_i. \tag{17}$$

The *viscoelastic boundary value problem* is defined by the field equations

$$\partial\sigma_{ij}/\partial x_j + F_i = 0, \quad \varepsilon_{ij} = \tfrac{1}{2}(\partial u_i/\partial x_j + \partial u_j/\partial x_i) \tag{18}$$

$$\varepsilon_{ij} = [(1 + v)\{D\,d\sigma_{ij}\} - v\delta_{ij}\{D\,d\sigma_{kk}\}]/E_R \tag{19}$$

and boundary conditions,

$$\sigma_{ij}n_j = T_i \quad \text{on} \quad S_T \tag{20}$$

$$u_i = U_i \quad \text{on} \quad S_u. \tag{21}$$

Observe that (19) is the constitutive equation for a viscoelastic material with a constant Poisson's ratio. The factor E_R^{-1} is introduced to cancel the factor E_R in the definition of $\{\ \}$ in (4).

That the viscoelastic solution obtained using CP-III, (1) and (2), satisfies the governing equations (18)–(21) may be shown by substituting it into these equations, and then employing (6)–(9) and the fact that the elastic solution satisfies (10)–(17) and that $\partial n_i/\partial t = 0$ on S_T; the last condition does not appear to be a practical restriction. It should be noted that the traction T_i^R, (16), in the elastic problem is in general different from that in the viscoelastic problem. However, the condition $dS_T/dt \leqslant 0$ stated in CP-III implies the latter is known at all past times and thus T_i^R may be explicitly calculated.

When considering the problem of bonding of two different materials, a complication arises because the interfacial stresses acting across the bonding surface are not necessarily equal in the elastic problem. With reference to Fig. 2, let us use superscripts $(+)$ and $(-)$ to denote quantities above and below the surface $y = 0$, respectively. For the viscoelastic problem $T_i^+ + T_i^- = 0$ along this surface, whereas from (16) the elastic traction sum is

$$T_i^{R+} + T_i^{R-} = \{D^+\,dT_i^+\} + \{D^-\,dT_i^-\}$$

$$= \{(D^+ - D^-)\,dT_i^+\}. \tag{22}$$

If

$$E_R^+ D^+ = E_R^- D^- \tag{23}$$

then $T_i^{R+} + T_i^{R-} = 0$ or, equivalently, the interfacial elastic stresses above the bonding surface equal those below it. If D^- is proportional to D^+ we can select E_R^-/E_R^+ so that the traction sum in (22) vanishes. When the materials are not the same and the Poisson's ratios are not one-half, surface tractions include material-dependent shear tractions (unless the friction coefficient vanishes) and thus a pure mode I state does not exist at the tip [27].

In order to avoid mathematical complexities that would make it difficult to understand the basic physics of the viscoelastic bonding problem, we shall assume that if bonding is not between bodies of the same material then, at least, their Poisson's ratios are one-half and their creep compliances are proportional, (23). If one body is much stiffer than the other, so that its deformation can be neglected, clearly we may use the results from an analysis based on these conditions even if the relatively rigid body does not satisfy them.

4. Results from crack-tip analysis

The opening displacement v above the bond surface in the neighborhood of the crack tip, Fig. 2, may be obtained from CP-III and the elastic solution in [4-I],

$$v = C_R \int_0^\beta \sigma_b^R(\xi') F(\xi'/\xi) \, d\xi', \tag{24}$$

where

$$F(\xi'/\xi) \equiv \frac{2}{\pi} \left[2(\xi/\xi')^{1/2} - \ln \left| \frac{1 + (\xi'/\xi)^{1/2}}{1 - (\xi'/\xi)^{1/2}} \right| \right] \tag{25}$$

and

$$C_R \equiv (1 - v^2)/E_R. \tag{26}$$

The surface normal stress σ_b may be treated as a specified quantity, and therefore its counterpart σ_b^R in the elastic problem is obtained from (16), which yields

$$\sigma_b^R = \{D \, d\sigma_b\} = E_R \int_{-\infty}^t D(t - \tau) \frac{\partial \sigma_b}{\partial \tau} \, d\tau. \tag{27}$$

The superscript $(+)$ is omitted until needed later to distinguish between quantities above and below the bond surface.

The stress singularity at the crack tip or bond edge is removed by using the Barenblatt result for elastic media [4],

$$K_I^R = \left(\frac{2}{\pi}\right)^{1/2} \int_0^\beta \sigma_b^R(\xi) \, \xi^{-1/2} \, d\xi, \tag{28}$$

where K_I^R is the stress intensity factor for an elastic material with Young's modulus E_R and Poisson's ratio v. This stress intensity factor is found from an elastic analysis for which $\sigma_b^R \equiv 0$, and is considered to be a known function of time in the present analysis. Equation (28) implies that β and/or σ_b^R must change with K_I^R if the stress at the crack tip is to be finite.

5. Prediction of quasi steady-state bonding speed when σ_b is constant

Equations (24)–(28) will be used along with an energy criterion for bonding to predict β and bonding speed $\dot{a}_b \, (= -\dot{a})$ as functions of K_I^R. It is assumed that σ_b is independent of time and is spacially uniform over $0 < \xi < \beta$. Also, $\dot{a}_b$ is assumed to be essentially independent

172   R.A. Schapery

of time during the time it takes for the tip P to move the distance β; $\dot{a}_b$ may vary appreciably with time over propagation distances which are large compared to β, and thus the terminology *quasi steady-state bonding* is used. In the Appendix a formulation involving a spacially varying σ_b is analyzed, as mentioned previously.

Consider now the process of bonding, $\dot{a} < 0$, and identify in Fig. 2 a material particle ($x = $ constant, $y = 0$) which is currently on the unloaded surface, $\xi > \beta$. Suppose at time $t = t_1$ the left end of the bonding zone, $\xi = \beta$, arrives. This particle is acted on by the constant σ_b until $t = t_2$ when the right end of the zone, $\xi = 0$, arrives. The work per unit surface area done on the particle by σ_b is $\sigma_b v_b$, where v_b is v at $\xi = \beta$, assuming that $v = 0$ at $\xi = 0$. The lower material may not be the same as the upper material and therefore the displacements may be different. This difference will be identified later; but, for now, we shall consider only the displacement above the bond plane.

The problem at hand is to use (24)–(28) to obtain v_b and thus $\sigma_b v_b$. By adding this work to that for the lower surface and equating the result to the so-called *bond energy* denoted by $2\Gamma_b$ (per unit area) an equation for predicting $\dot{a}_b$ is obtained. First we need σ_b^R, which may be found from (27) by noting that $\sigma_b = 0$ for $t < t_1$ and σ_b is constant for $t_1 < t < t_2$; thus, for a particle at fixed x and with $t_1 < t < t_2$,

$$\sigma_b^R = E_R \sigma_b D(t - t_1). \tag{29}$$

The relationship between t and ξ follows from the quasi steady-state assumption,

$$\xi = (t_2 - t)\dot{a}_b \quad \text{and} \quad \beta = (t_2 - t_1)\dot{a}_b \tag{30}$$

from which

$$t - t_1 = (\beta - \xi)/\dot{a}_b. \tag{31}$$

Generalized power law creep compliance

The creep compliance of many materials can be well-represented over several decades of time by the so-called generalized power law,

$$D(t) = D_0 + D_1 t^m, \tag{32}$$

where D_0, D_1 and m are positive constants and $0 < m \leqslant 1$. We shall use this form to obtain exact results and then, guided by these findings, develop an approximate formula for more general forms of creep compliance. From (29), (31), and (32),

$$\sigma_b^R = E_R \sigma_b [D_0 + D_1(1 - \eta)^m (\beta/\dot{a}_b)^m], \tag{33}$$

where

$$\eta \equiv \xi/\beta. \tag{34}$$

Substitute (33) into (28) and find

$$K_I^R = (8/\pi)^{1/2} E_R \sigma_b \beta^{1/2} [D_0 + D_1 \gamma_m (\beta/\dot{a}_b)^m], \tag{35}$$

where

$$\gamma_m \equiv (\pi/4)^{1/2} \Gamma(1 + m)/\Gamma(1.5 + m) \tag{36}$$

and $\Gamma(\cdot)$ is the gamma function. Observe that the relationship between β and K_I^R depends on bonding speed, whereas the analogous relationship for crack growth (which relates "failure zone" or "fracture process zone" length α to K_I) does not depend explicitly on speed [4-I]. Next, use (33) in (24) to find the displacement at $\xi = \beta$, and thus the work $\sigma_b v_b$. It is helpful to integrate-by-parts the term involving the logarithm and obtain, finally,

$$W_b \equiv \sigma_b v_b = \frac{4}{\pi} (1 - v^2) \sigma_b^2 \beta [D_0 + D_1 c_m \gamma_m (\beta/\dot{a}_b)^m], \tag{37}$$

where

$$c_m \equiv (2m + 1)/(m + 1). \tag{38}$$

Previously we introduced superscripts $(+)$ and $(-)$ to denote quantities above and below the bond plane. In view of (23),

$$E_R^+ D_0^+ = E_R^- D_0^- \quad \text{and} \quad E_R^+ D_1^+ = E_R^- D_1^-. \tag{39}$$

Recall that the Poisson's ratios are assumed equal and constant. The superscripts can be omitted in (35) as only these products appear. However, the work above the bond plane, which will now be denoted by W_b^+, may be different from that below, W_b^-, since E_R does not appear as a factor. Indeed, from (37),

$$W_b^-/W_b^+ = D_0^-/D_0^+ = D_1^-/D_1^+. \tag{40}$$

The criterion for bonding may be written as

$$2\Gamma_b = W_b^+ + W_b^- = (1 + D_0^-/D_0^+) W_b^+ \tag{41}$$

or, equivalently,

$$2\Gamma_b' = W_b^+ \tag{42}$$

where

$$\Gamma_b' \equiv \Gamma_b/(1 + D_0^-/D_0^+). \tag{43}$$

For identical materials (42) reduces to $\Gamma_b = W_b^+$, whereas if the lower material is rigid, $2\Gamma_b = W_b^+$. We shall use the *effective bond energy* Γ_b' in all subsequent work so that the notation in the results does not have to be changed when considering the bonding of materials with the same or different stiffnesses; the effect of the lower material will appear only in (43) until we apply the theory in Section 7.

Equations (35) and (42), where (37) is used for W_b^+, may be solved to obtain β and $\dot{a}_b$ in terms of K_I^R, given σ_b and Γ_b'. We assume these latter two quantities are constants and obtain, finally,

$$\beta/\beta_m = Z(1 + \lambda)^{-2}c_m^2, \tag{44}$$

where

$$\beta_m \equiv \pi\Gamma_b'/2\sigma_b^2(1 - \nu^2) D_0^+ c_m^2 \tag{45}$$

is a constant with the dimension of length. Also,

$$\dot{a}_b/\dot{a}_m = Z(1 + \lambda)^{-2}\lambda^{-(1/m)}c_m^{(2+1/m)}, \tag{46}$$

where

$$\dot{a}_m \equiv \beta_m(D_1\gamma_m/D_0 c_m)^{1/m} \tag{47}$$

is a constant with dimensions of velocity. The quantity λ is a function of Z, as found from the solution to a quadratic equation,

$$\lambda = 0.5[c_m Z + (c_m Z)^{1/2}(c_m Z + 4/c_m - 4)^{1/2}] - 1. \tag{48}$$

In turn,

$$Z \equiv (K_I^R/K_{I0}^R)^2, \tag{49}$$

where

$$(K_{I0}^R)^2 \equiv 4\Gamma_b'D_0^+ (E_R^+)^2/(1 - \nu^2). \tag{50}$$

is a constant with dimensions of stress intensity factor squared.

For the range $1 < Z < \infty$, we find both β and $\dot{a}_b$ are monotone decreasing functions of Z and that

$$\lambda \to 0 \text{ as } Z \to 1 \quad \text{and} \quad \lambda \simeq c_m Z \text{ for } Z \gg 1. \tag{51}$$

This behavior yields from (44),

$$\beta/\beta_m \to c_m^2 \text{ as } Z \to 1 \quad \text{and} \quad \beta/\beta_m \simeq Z^{-1} \text{ for } Z \gg 1 \tag{52}$$

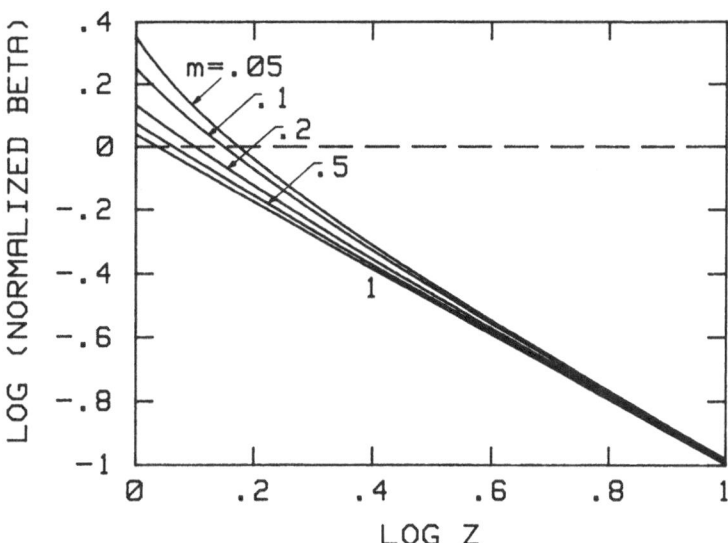

Fig. 3. Variation of normalized bonding-zone length with normalized stress intensity factor squared, Eqns. (44) and (49).

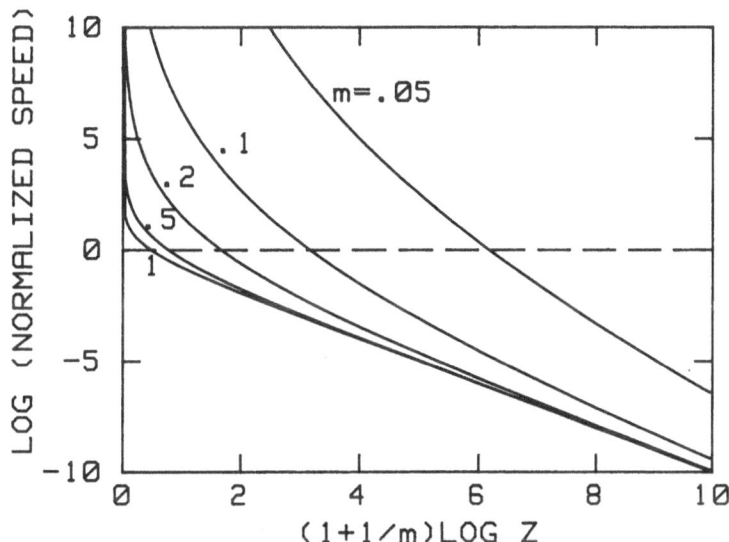

Fig. 4. Variation of normalized bonding speed with normalized stress intensity factor squared, Eqns. (46) and (49).

and from (46),

$$\dot{a}_b/\dot{a}_m \to \infty \text{ as } Z \to 1 \quad \text{and} \quad \dot{a}_b/\dot{a}_m \simeq Z^{-(1+1/m)} \text{ for } Z \gg 1. \tag{53}$$

Figures 3 and 4 show the normalized bonding-zone length (44) and bonding speed (46) as functions of Z on logarithmic scales (log $\equiv$ log$_{10}$). The range of values of m and Z for which the behavior is essentially that for the high stress intensity factor range in (52) and (53) may be seen. In this range the results are the same as the exact solution over $0 < K_I^R < \infty$ for pure power-law material, (32) with $D_0 = 0$.

In order to obtain explicit analytical results for $Z \gg 1$, β_m and $\dot{a}_m$ may be substituted into (44) and (46) in order to eliminate D_0. The results for $Z \gg 1$ are

$$\beta = 2\pi[\Gamma_b' E_R^+ /(1 - v^2)\sigma_b c_m K_I^R]^2 \tag{54}$$

$$\dot{a}_b = \pi[4\Gamma_b']^{(2+1/m)}[(1 - v^2)D_1^+ \gamma_m/c_m]^{1/m}[E_R^+/(1 - v^2)K_I^R]^{2(1+1/m)}/8\sigma_b^2 c_m^2. \tag{55}$$

These forms, i.e.,

$$\beta = k_1(K_I^R/E_R^+)^{-2}, \quad \dot{a}_b = k(K_I^R/E_R^+)^{-q}, \quad q \equiv 2(1 + 1/m) \tag{56}$$

(where k_1 and k are positive constants) are like those for the crack growth problem; but the exponents on stress intensity factor in the fracture problem are positive. Here, both the bonding-zone length and speed increase with decreasing K_I^R. Immediate bonding occurs when $K_I^R = K_{I0}^R$ since $\dot{a}_b = \infty$ at this limit; the corresponding value for β is $\beta_m c_m^2$, (52).

For $K_I^R < K_{I0}^R$, (48) yields the result $\lambda < 0$. A physical interpretation of this low K_I^R behavior may be made by noting that for $K_I^R > K_{I0}^R$, $\dot{a}_b$ increasess with increasing effective bond energy Γ_b', for K_I^R fixed (cf. (45)–(50)). When $K_I^R < K_{I0}^R$ there is more work available than required to draw the surfaces together, neglecting the effects of material inertia. Hence, for $K_I^R \leqslant K_{I0}^R$ a dynamic analysis would be needed to predict finite, high speed bonding.

General creep compliance

The foregoing results based on the generalized power law compliance (32) are helpful in extending the theory to a more general creep compliance, which we write in the form

$$D(t) = D_0 + \Delta D(t), \tag{57}$$

where D_0, as before, is the initial compliance; thus, $\Delta D(0^+) = 0$. This compliance and (29)–(31) yield

$$\sigma_b^R = \sigma_b^R(\xi) = E_R\sigma_b\{D_0 + \Delta D[(\beta - \xi)/\dot{a}_b]\}. \tag{58}$$

This function is to be used in (28), which will be evaluated approximately using a method like that used in crack growth analyses [4-II]. First, rewrite (28) by introducing a logarithmic transformation to change the integration variable to L,

$$L \equiv \log(1 - \eta), \tag{59}$$

where, as before, $\log \equiv \log_{10}$ and $\eta = \xi/\beta$. Equation (28) becomes

$$K_I^R = \left(\frac{2}{\pi}\right)^{1/2} (\ln 10) \beta^{1/2} \int_{-\infty}^0 \sigma_b^R w_1 \, dL, \tag{60}$$

where

$$w_1 \equiv (1 - 10^L)^{-1/2} 10^L \tag{61}$$

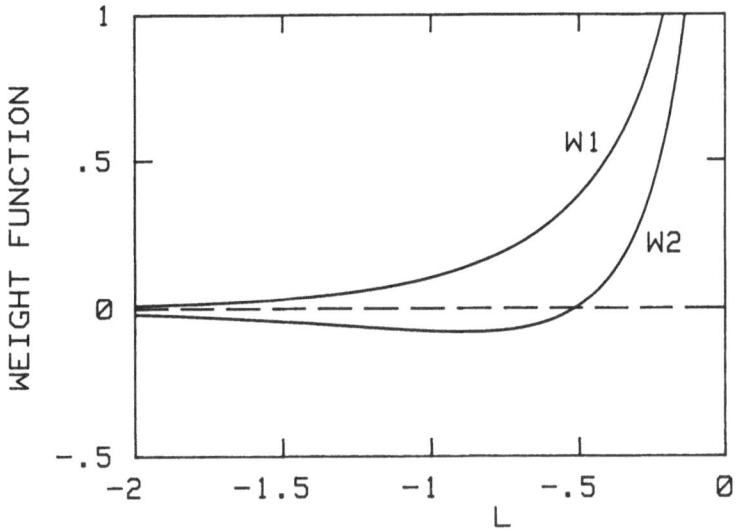

Fig. 5. Weight functions, Eqns. (61) and (66).

is a weight function; it is drawn in Fig. 5. Because this function is very small except for L close to zero (corresponding to $\xi = 0$), and ΔD in (58) decreases with decreasing L (or increasing ξ), only the behavior of $\Delta D(t)$ for $t \simeq \beta/\dot{a}_b$ needs to be considered (cf. (58)). Let us assume that ΔD may be approximated by a power law $\Delta D = D_1 t^m$ over at least a one-decade time range (say), i.e., $\beta/10\dot{a}_b \leqslant t \leqslant \beta/\dot{a}_b$ (which corresponds to $-1 \leqslant L \leqslant 0$) and that behavior for smaller times is unimportant because w_1 is so small when $L < -1$. Then (35) may be used, in which

$$m \equiv \frac{\mathrm{d} \log \Delta D}{\mathrm{d} \log t}, \tag{62}$$

where $t = \beta/\dot{a}_b$. A more compact way of writing (35) is

$$K_I^R = \left(\frac{8}{\pi}\right)^{1/2} E_R \sigma_b \beta^{1/2} D(t'), \tag{63}$$

where

$$t' \equiv \gamma_m^{1/m} \beta/\dot{a}_b. \tag{64}$$

It is found that $\gamma_m^{1/m}$ is practically constant; this factor is a monotone increasing function of m, in which $0.541 < \gamma_m^{1/m} \leqslant 2/3$ for $0 \leqslant m \leqslant 1$.

The displacement (24) is needed to obtain the work $W_b = \sigma_b v_b$, where $v_b \equiv v$ at $\xi = \beta$. Using (59) we find

$$v_b = \frac{4}{\pi} C_R \beta(\ln 10) \int_{-\beta}^{0} \sigma_b^R w_2 \, \mathrm{d}L, \tag{65}$$

where

$$w_2 \equiv 0.5 \left[2\eta^{-1/2} - \ln \left| \frac{1 + \eta^{1/2}}{1 - \eta^{1/2}} \right| \right] 10^L \tag{66}$$

and $\eta = 1 - 10^L$. The graph of the weight function w_2 in Fig. 5 shows that, just as for w_1, the behavior of σ_b^R (and thus ΔD) is relatively unimportant for $L < -1$. Just as (35) has been replaced by (63), we may replace (37) by

$$W_b = \frac{4}{\pi} (1 - v^2) \sigma_b^2 \beta D(t''), \tag{67}$$

where

$$t'' \equiv \gamma_m'^{1/m} \beta / \dot{a}_b \tag{68}$$

and

$$\gamma_m' \equiv c_m \gamma_m. \tag{69}$$

The factor $\gamma_m'^{1/m}$ is a monotone decreasing function of m, in which $1 \leqslant \gamma_m'^{1/m} < 1.47$ for $1 \geqslant m \geqslant 0$.

The bonding criterion (42) may now be written as

$$2\Gamma_b' = \frac{4}{\pi} (1 - v^2) \sigma_b^2 \beta D^+ (t''), \tag{70}$$

which together with (63) (recalling that $E_R D = E_R^+ D^+$) provides the pair of equations for finding β and $\dot{a}_b$. The previous results, (44) and (46), could be used as first approximations in an iterative solution approach. Here, we shall discuss only the low and high speed limiting conditions for bonding.

Previously, it was found that β and $\dot{a}_b$ are monotone decreasing functions of K_I^R, for $K_I^R > K_{I0}^R$. However, $\dot{a}_b$ may vanish at a finite value of K_I^R, as will be shown by using (70) together with (54) for β. Although the latter equation is exact only for a pure power law compliance, it is expected to be a good approximation whenever the compliance can be approximated by a power law over one-decade time intervals. (Observe that β depends on m through c_m, (38), but not on D_1; m is now given by (62).) Substitute (54) for β into (70) and obtain

$$(K_I^R)^2 = 4\Gamma_b' D^+ (t'')(E_R^+)^2 / (1 - v^2) c_m^2, \tag{71}$$

where t'' is in (68). With increasing K_I^R, $D^+ (t'')$ must increase according to (71). However, if the creep compliance has a finite upper limiting value D_∞^+, i.e., $D_\infty^+ \equiv D^+ (\infty) < \infty$, then the largest value of K_I^R for which a solution $\dot{a}_b$ exists is $K_{I\infty}^R$, where

$$(K_{I\infty}^R)^2 \equiv 4\Gamma_b' D_\infty^+ (E_R^+)^2 / (1 - v^2). \tag{72}$$

When $t'' \to \infty$, then $D^+ \to D_\infty^+$, and the slope m vanishes; thus $c_m \to 1$ as $t'' \to \infty$.

A further study of (71), in which it is assumed that m in (62) does not increase with t, along with the previous results (44) and (46), leads to the following inequalities and behavior. For $K_{I0}^R < K_I^R < K_{I\infty}^R$, the length β and speed $\dot{a}_b$ are monotone decreasing functions of K_I^R; also,

$$\beta \to \beta_m c_m^2 \quad \text{and} \quad \dot{a}_b \to \infty \quad \text{as} \quad K_I^R \to K_{I0}^R \tag{73}$$

$$\beta \to \beta_m c_m^2 D_0/D_\infty \quad \text{and} \quad \dot{a}_b \to 0 \quad \text{as} \quad K_I^R \to K_{I\infty}^R. \tag{74}$$

Recall that D_0/D_∞ above the bond surface is the same as that below. From (50) and (72),

$$K_{I\infty}^R/K_{I0}^R = (D_\infty/D_0)^{1/2}. \tag{75}$$

6. Comparison of apparent fracture and bond energies

In many situations involving slow loading or unloading of rubber, the global behavior is elastic, with this behavior defined in terms of elastic constants which are essentially equal to the long-time values of the associated viscoelastic properties. On the other hand, at this same time viscoelastic effects may be very pronounced in the neighborhood of a moving crack tip or contact edge because of the high local strain rates.

This situation is easily represented in the theory at hand by taking as the arbitrary reference elastic modulus the reciprocal of the actual long-time compliance; thus $D_\infty^+ = 1/E_R^+$ and $D_\infty^- = 1/E_R^-$. With this global elastic behavior, it is common practice to define an *apparent fracture energy* associated with crack growth by means of an elastic-like equation for fracture; in terms of the notation and theory at hand, in which adhesive and cohesive fracture are considered,

$$\Gamma_{af}' \equiv (1 - v^2) D_\infty^+ K_I^2/4, \tag{76}$$

where K_I is the stress intensity factor which produces a particular crack speed, $\dot{a}$. Similarly, guided by (72), we may define an *apparent bond energy* using the same expression,

$$\Gamma_{ab}' \equiv (1 - v^2) D_\infty^+ (K_I^R)^2/4, \tag{77}$$

where K_I^R produces a bonding rate $\dot{a}_b$; since the actual global behavior is also that of the reference elastic problem, $K_I = K_I^R$.

The apparent energies in (76) and (77) can be related to the (intrinsic) effective fracture energy Γ_f' and bond energy Γ_b'. From [4-II], after allowing for adhesive fracture through an equation analogous to (43),

$$\Gamma_f' = (1 - v^2) D^+ (\alpha/3\dot{a}) K_I^2/4 \tag{78}$$

and by eliminating K_I^2 between (76) and (78) we find

$$\Gamma_{af}' = \Gamma_f' D_\infty/D(\alpha/3\dot{a}). \tag{79}$$

The compliance ratio is the same above and below the bond surface, as assumed previously. Similarly, from (71) and (77),

$$\Gamma'_{ab} = \Gamma'_b D(t'')/D_\infty c_m^2. \tag{80}$$

The ratio $D_\infty/D(i)$ increases monotonically (from unity) as time t decreases from infinity to zero, and for rubber D_∞/D_0 may exceed one hundred [28]. Hence, the ratios Γ'_{af}/Γ'_f and Γ'_b/Γ'_{ab} may themselves exceed one hundred. Moreover, if the effective energies are equal, $\Gamma'_f = \Gamma'_b$, the ratio of apparent energies is bounded according to

$$1 \leqslant \frac{\Gamma'_{af}}{\Gamma'_{ab}} \leqslant (c_m D_\infty/D_0)^2. \tag{81}$$

In experiments on the time-dependent adhesion of rubber to glass, Roberts and Thomas [11] found that the ratio $\Gamma'_{af}/\Gamma'_{ab}$ is indeed very large over a wide range of bonding and debonding speeds. This ratio was reported to exceed one thousand at the shortest testing time, which was approximately ten seconds.

Finally, it is of interest to observe for this case of global elastic behavior that (72) reduces to

$$K_{I\infty}^2 = 4\Gamma'_b/D_\infty^+(1 - v^2) \tag{82}$$

and that $\dot{a}_b > 0$ if $K_I < K_{I\infty}$. For the fracture problem $\dot{a} > 0$ if $K_I > K_{I\infty}^f$, where

$$(K_{I\infty}^f)^2 \equiv 4\Gamma'_f/D_\infty^+(1 - v^2). \tag{83}$$

Consequently, if the effective bond and fracture energies are equal, (82) or (83) define a value of the stress intensity factor below which there is bonding and above which there is crack growth.

7. Examples

Cohesive crack healing

Consider the Griffith type of problem in which there is an isolated circular crack with radius a, or a through-the-thickness crack of length $2a$. A remote, uniform tensile stress $\sigma(t)$ normal to the crack plane is given. In the reference elastic body the stress intensity factor is [29],

$$K_I^R = ca^{1/2}\sigma^R, \tag{84}$$

where c is a constant; $c = 2/\pi^{1/2}$ for a circular crack and $c = \pi^{1/2}$ for a through-crack. Equation (16) or the inverse of (1) provides σ^R,

$$\sigma^R = E_R \int_{-\infty}^t D(t - \tau)\frac{d\sigma}{d\tau}d\tau, \tag{85}$$

which must be non-negative for (84) to be valid. Let us use the power law in (56) for $\dot{a}_b$; then with (84) for K_I^R and with $\dot{a}_b = -\dot{a}$, we find

$$a/a_0 = (1 - I_t)^{1/p}, \tag{86}$$

where a_0 is the size at $t = 0$. Also,

$$p \equiv q/2 + 1 = (1 + 2m)/m \tag{87}$$

$$I_t \equiv k_2 p a_0^{-p} \int_0^t (c\sigma^R/E_R)^{-q} \, \mathrm{d}t. \tag{88}$$

The *healing time* t_h is defined as the time at which $a = 0$; namely, $I_t(t_h) = 1$. An explict solution for t_h may be easily found if σ is a constant which is applied long before $t = 0$, so that (85) yields $\sigma^R = E_R D_\infty \sigma$, where $D_\infty \equiv D(\infty)$ is the equilibrium or long-time creep compliance. Equation (88) reduces to

$$I_t = k_2 p a_0^{-p} (cD_\infty \sigma)^{-q} t \tag{89}$$

and thus

$$t_h = a_0^p (cD_\infty \sigma)^q / k_2 p. \tag{90}$$

Observe that, as expected, the higher the tensile stress σ, the longer the healing time. Also, notice that the arbitrary constant E_R does not appear in the prediction of a and t_h because σ^R/E_R is independent of E_R (cf. (85)).

It should be recalled that whether or not healing or crack growth occurs depends on the vaue of K_I^R relative to $K_{I\infty}^R$ in (72). For the cohesive healing problem, $\Gamma_b' = \Gamma_b/2$ and $D_\infty^+ = D_\infty$. Using also (84) and (85) the criterion for bond growth is found as

$$ca^{1/2} \int_{-\infty}^t D(t - \tau) \frac{\mathrm{d}\sigma}{\mathrm{d}\tau} \mathrm{d}\tau < [2\Gamma_b D_\infty/(1 - v^2)]^{1/2}. \tag{91}$$

For a pure power law compliance (which includes that for a Newtonian body, $m = 1$) $D_\infty = \infty$, and (91) is always satisfied if $(\sigma, t) < \infty$. But crack growth analysis shows that the criterion for growth also is always satisfied [4-II]. This quasi-steady state analysis is thus not sufficient to determine the sign of $\dot{a}$ for a power law material.

Growth of contact area between initially curved surfaces

Figure 6 shows the contact region for two bodies which are pressed together by the load F and are also drawn together under bonding stress σ_b. The length scale β over which σ_b acts is assumed to be very small compared to the contact size. Just as in the fracture problem (i.e., contact area reduction) the opening displacement is parabolic near the contact edge except very close to the edge where the cusp shape of Fig. 2 exists, shown in Fig. 6 by the blackened edge area. In order to predict $\dot{a}_b$ ($= \dot{a}_c$) the stress intensity factor for the reference elastic

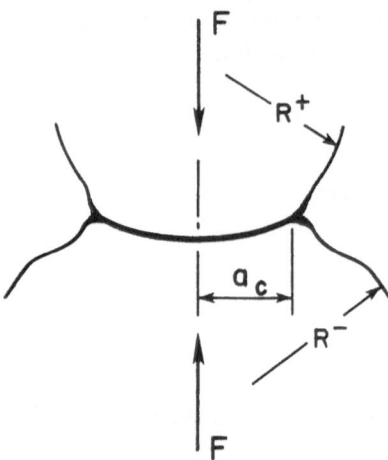

Fig. 6. Bonded contact between initially curved surfaces.

problem K_I^R is needed. Here, we shall consider two cases: (i) contact between initially spherical surfaces with radii R^+ and R^- and (ii) the same type of problem, except the surfaces are cylindrical so that the deformation is two-dimensional. An approach to finding K_I^R will be used which is similar to that in [12] for contact between a plane and a spherical surface.

First, in the absence σ_b, Hertz's solutions for contact size may be used for the spherical surfaces,

$$a_c^3 = 3F_H^R/8BE_R' \tag{92}$$

and the cylindrical surfaces,

$$a_c^2 = 2F_H^R/\pi BE_R', \tag{93}$$

where F_H^R in (92) is the total force, while in (93) it is force per unit length along the cylinder axis. Also,

$$B \equiv (1/R^+ + 1/R^-)/2, \quad \frac{1}{E_R'} \equiv \left(\frac{1}{E_R^+} + \frac{1}{E_R^-}\right)(1 - v^2). \tag{94}$$

Now, let us reduce the compressive load by an amount ΔF^R; assume the surfaces are bonded so that a_c is fixed at the Hertz value. This change produces an interface tensile stress which is infinite at the contact edges (without the Barenblatt modification), because these edges are really crack tips. The stress intensity factor may be obtained from the solution for the normal stress between a bonded flat-ended rigid punch under tensile load ΔF^R and flat elastic halfspace [25], using the assumption that $a_c \ll R^+$ and $a_c \ll R^-$; the Hertz solutions, (92) and (93), are based on this assumption, which we retain. That this is the same normal stress as produced by ΔF^R in the problem in Fig. 6 may be argued by first noting that it is independent of elastic constants and the initial surface radii, and produces a rigid translation of the contact surface. The same normal stress acts on both upper and lower surfaces, and since the contact surface shape does not change the upper and lower bodies remain in contact. Because the interface conditions are satisfied using this normal stress and the bodies are, in effect, half-spaces (i.e., $a_c \ll R^+$ and $a_c \ll R^-$), it is the one and only solution for interface stress. (However, the conditions for zero interfacial shear stress must be met, as discussed at the end of Section 3.)

For a circular punch with radius a_c, the contact stress is,

$$\sigma_y^R = \frac{\Delta F^R}{2\pi a_c (a_c^2 - r^2)^{1/2}}$$

$$= \frac{\Delta F^R}{2\pi a_c (a_c + r)^{1/2}(a_c - r)^{1/2}}, \tag{95}$$

where r is the radial distance from the contact center to a point on the interface. From the singularity at $r = a_c$ and the definition of stress intensity factor,

$$K_I^R = \lim_{r \to a_c} [2\pi(a_c - r)]^{1/2} \sigma_y^R \tag{96}$$

we obtain

$$K_I^R = \Delta F^R / 2\pi^{1/2} a_c^{3/2}. \tag{97}$$

Similarly, for the plane strain problem,

$$K_I^R = \Delta F^R / (\pi a_c)^{1/2}. \tag{98}$$

Now, $\Delta F^R = F_H^R - F^R$, where F^R is the total load. With this expression and F_H^R from (92) and (93), the stress intensity factor in (97) for spherical surface becomes,

$$K_I^R = 4BE_R' a_c^{3/2} / 3\pi^{1/2} - a_c^{-3/2} F^R / 2\pi^{1/2}. \tag{99}$$

For cylindrical surfaces,

$$K_I^R = \pi^{1/2} BE_R' a_c^{3/2} / 2 - a_c^{-1/2} F^R / \pi^{1/2}. \tag{100}$$

The force F^R in the elastic problem is related to the actual compressive force through $F^R = \{D \, dF\}$, just as the stress in (85). Introduction of K_I^R in terms of $\dot{a}_c$ makes (99) and (100) differential equations for a_c.

As simple examples, the contact size will be found analytically from these differential equations using the power law bonding speed equation (56) for two cases: $F^R = 0$ and $K_I^R \simeq 0$. In the first case, (99) yields the radius

$$a_c = A t_b^s, \tag{101}$$

where t_b is the bonding time, and

$$s \equiv 2/(3q + 2). \tag{102}$$

Also

$$A \equiv [k(3\pi^{1/2}D'/4B)^q/s]^s \tag{103}$$

$$D' \equiv (1 + D_1^-/D_1^+)(1 - v^2). \tag{104}$$

The last equality comes from (39). In deriving (101) we assumed interfacial contact starts at $t = 0$ and neglected the error in the Barenblatt method at short times (when $a_c < \beta$). The size a_c for contact between cylinders has the same time dependence as in (101) because (100) depends on a_c in the same way as (99) when $F^R = 0$. From (56) and (102),

$$s = m/(3 + 4m). \tag{105}$$

If $m = 1$ (viscous material) then $s = 1/7$, which is the largest physically acceptable exponent.

The time dependence of a_c in (101) is in general considerably different from that predicted from the Hertz contact problem. Taking for example a constant load F applied at $t = 0$, the load in the elastic problem is $F^R = E_R^+ D_1^+ t_b^m F$. From (99) and (100), respectively,

$$a_c = (3D'D_1^+ F/8B)^{1/3} t_b^{m/3} \tag{106}$$

$$a_c = (2D'D_1^+ F/\pi B)^{1/2} t_b^{m/2}. \tag{107}$$

(Anand's model [13] predicts this time-dependence for a power law material.)

Now, let us retain K_I^R but assume it is small enough that a_c is close to the Hertz solutions (106) and (107). For the power law of (56), this occurs when $\dot{a}_b$ is sufficiently large. Let a_H be the Hertz solution, which is for $K_I^R = 0$ in (99) and (100). Then $a_c = a_H + \Delta a$, where $\Delta a/a_H \ll 1$ by assumption. Using this approximation in (99) we find

$$\Delta a/a_H \simeq \pi^{1/2} D'(k/\dot{a}_H)^{1/q}/4Ba_H^{3/2}, \tag{108}$$

where a_H is the solution to (99) for $K_I^R \equiv 0$. For example, using (106), $\Delta a/a_H \sim t_b^{-2m^2/3(m+1)}$, and thus the error in the Hertz solution (106) diminishes with time. A similar result, with the same exponents as in (108), is found for the contact between cylinders; using (107), we find again that the error diminishes with time in that $\Delta a/a_H \sim t_b^{-m(1+4m)/4(1+m)}$.

Effect of bonding time on joint strength

As discussed in Section 1, the time-dependent strength of bonds formed across contacting surfaces has received considerable attention. The examples considered in the present section deal only with contact area growth and not the processes following contact, such as polymer–polymer interdiffusion (which may be very significant if the temperature is sufficiently high). Here, we observe only that (99) and (100) may be used to predict strength, if the residual viscoelastic effect of the bonding problem is negligible so that the correspondence principle for crack growth may be used. In this case, the elastic and viscoelastic stresses, loads, and stress intensity factors are the same. Suppose, for example, that following the bonding process the temperature is reduced enough that the critical stress intensity factor K_{IC}

is a constant and that the first term in (99) is negligible at fracture. Then the strength is $F \sim a_c^{3/2} K_{IC}$. Using (101), $F \sim t_b^{3s/2}$, while from (106) $F \sim t_b^{m/2}$; for viscous media $m = 1$, and the strengths are, respectively, $F \sim t_b^{0.21}$ and $F \sim t_b^{0.5}$.

If, following bonding, we use the idealization from [20] in which any nonuniformity in stresses in the bond is ignored, use elastic fracture mechanics, and assume the fracture energy Γ_f' is proportional to bonded area A_b, then $K_{IC} \sim \Gamma_f'^{1/2} \sim A_b^{1/2}$; with a given crack at the interface the joint strength is proportional to $A_b^{1/2}$. From (101), $K_{IC} \sim t_b^{s/2}$ for spheres and $K_{IC} \sim t_b^s$ for cylinders; if $m = 1$, these exponents are 1/14 and 1/7, respectively. On the other hand from (106) and (107), $K_{IC} \sim t_b^{m/3}$ and $K_{IC} \sim t_b^{m/4}$, respectively.

For polymer–polymer bonding of noncrosslinked systems somewhat above the glass transition temperature, the behavior $K_{IC} \sim t_b^{1/4}$ is usually found experimentally [19, 20] (in a range for which $m \simeq 1$ is probably a good assumption); it is also the behavior predicted from interdiffusion theory for fully contacted or wetted surfaces. None of the exponents discussed above yield the 1/4 exponent (for the physically acceptable range of viscoelastic exponents $0 < m \leqslant 1$) except for the last one of $m/4$, if $m \simeq 1$, for contacting cylindrical surfaces. Anand [17] reported that when nearly flat surfaces are bonded, the initially irregular contacting areas become in time contacting cylinders, and that the conversion to cylindrical contact surfaces occurs early in the bonding process. For this case, one could not use an experimentally determined exponent of 1/4 to conclude that the increase in joint strength is due to interdiffusion as opposed to contact growth. Of course, if the externally applied contact pressure is high enough during bonding so that essentially full contact is quickly established, contact area growth will not be an important source of time-dependence of the joint strength.

8. Conclusions

A mechanics-based theory of bonding between the same or different linear, isotropic visco-elastic media has been developed and used in some examples. We assumed quasi steady-state bonding and the bonding stress σ_b and energy Γ_b' to be independent of bonding speed $\dot{a}_b$ in order to obtain explicit, closed-form solutions. However, if these material-related quantities vary with speed (which could be due to surface roughness on a scale which is very small compared to the bonding-zone length β) most of the equations do not have to be changed. For example, with speed dependence of these parameters, (55) becomes an implicit equation for $\dot{a}_b$; with power law dependence, the resulting solutions for β and $\dot{a}_b$ obey power laws in K_I^R, as in (56), but the exponents are in general different.

Regardless of how or if σ_b and Γ_b' depend on bonding speed, we may conclude that all dependence of $\dot{a}_b$ and β on external loading and geometry of the continua is through the instantaneous stress intensity factor K_I^R. This simplicity exists as long as β is small compared to local geometric scales such as initial radii of surface asperities or crack length; as mentioned above, very small-scale roughness compared to β is also acceptable.

In contrast to the crack growth problem, K_I^R depends on external loading history, and $\dot{a}_b$ and β decrease with increasing K_I^R. The external loading must be such that $\dot{a}_b$ is nonnegative in order for the correspondence principle used in the analysis to be valid. The correspondence principles used for bond growth and crack growth are different. If crack healing occurs after crack growth, the bonding theory will not be valid unless effects of the prior history of crack growth have essentially faded out. The complexity of the analysis appears to be much greater when the velocity of the boundary of bonded surfaces changes sign and viscoelastic effects from before and after the change are combined; however, methods exist for analyzing this type of problem [30, 31].

Linear theory may not be valid under high stresses, such as when bodies with initially curved surfaces are pressed together under high compressive loads. In such cases, a J integral theory may be applicable. It would be similar to that for crack growth, but based on correspondence principle III (instead of II, which is for crack growth) given in [26].

Although the development here is for isotropic media, through a change in the creep compliance $D(t)$ the theory in Sections 5 and 6 can be used for bonding of surfaces of orthotropic media if the surfaces and bond edge are parallel to principal material planes. This extension is based on the findings in [5] for crack growth, in which the equations for the mechanical state near and at the interface are the same as for isotropic media; in this case $(1 - v^2) D(t)$ is to be replaced by another function of time consisting of a combination of creep compliances of the orthotropic material. (The same generalization can be made in applications involving effectively unbounded media, such as those in Section 7.) However, just as for isotropic materials, property-dependent interfacial shear stresses develop unless the bonding is between the same materials or, if different, between incompressible materials with proportional compliances; one material may be rigid. The present theory has been developed under the condition that there is no shear stress at the interface in order to bring out the essential features of the mechanics of bonding or healing of viscoelastic materials with a minimum of mathematical complexity.

Finally, it should be noted that Poisson's ratio appears in elastic solutions in this paper only in the form $(1 - v^2)$/modulus, as in (26) and (94). For problems in which this is the case, one can show by means of Graham's extended correspondence principle [6] that the theory remains valid when Poisson's ratio is time-dependent, as long as it is the same for the materials above and below the bond surface. In this situation one may use a plane-strain creep compliance $C(t)$ in place of $(1 - v^2) D(t)$ throughout the paper; this replacement is analogous to the generalization made for orthotropic media, as discussed above.

Acknowledgement

Sponsorship of this research by the Office of Naval Research is gratefully acknowledged.

References

1. M.L. Williams, in *Proceedings 5th U.S. National Congress of Applied Mechanics*, American Society of Mechanical Engineers (1966) 451–464.
2. M.L. Williams, *International Journal of Fracture Mechanics* 1 (1965) 292–310.
3. W.G. Knauss, in *Deformation and Fracture of High Polymers*, H. Henning Kausch, John A. Hassell and Robert I. Jaffee (eds.) Plenum Press (1974) 501–541.
4. R.A. Schapery, *International Journal of Fracture* 11 (1975): Part I, 141–159; Part II, 369–388; Part III, 549–562.
5. G.S. Brockway and R.A. Schapery, *Engineering Fracture Mechanics* 10 (1978) 453–468.
6. R.M. Christensen, *Theory of Viscoelasticity, An Introduction*, 2nd edn., Academic Press (1982).

7. J.G. Williams, *Fracture Mechanics of Polymers*, Halstead Press, John Wiley & Sons (1984).
8. R.A. Schapery, *International Journal of Fracture* 14 (1978) 293–309.
9. J.R. Walton, *Journal of Applied Mechanics* 54 (1987) 635–641.
10. K.L. Johnson, K. Kendall and A.D. Roberts, *Proceedings of the Royal Society* (London) A234 (1971) 301–313.
11. A.D. Roberts and A.G. Thomas, *Wear* 33 (1975) 45–64.
12. J.A. Greenwood and K.L. Johnson, *Philosophical Magazine A* 43 (1981) 697–711.
13. J.N. Anand and H.J. Karam, *Journal of Adhesion* 1 (1969) 16–23.
14. J.N. Anand and R.Z. Balwinski, *Journal of Adhesion* 1 (1969) 24–30.
15. J.N. Anand, *Journal of Adhesion* 1 (1969) 31–37.
16. J.N. Anand and L. Dipzinski, *Journal of Adhesion* 2 (1970) 16–22.
17. J.N. Anand, *Journal of Adhesion* 2 (1970) 23–28.
18. J.N. Anand, *Journal of Adhesion* 5 (1973) 265–267.
19. H.H. Kausch, D. Petrovska, R.F. Landel and L. Monnerie, *Polymer Engineering and Science* 27 (1987) 149–154.
20. K. Jud, H.H. Kausch and J.G. Williams, *Journal of Materials Science* 16 (1981) 204–210.
21. R.G. Stacer and H.L. Schreuder-Stacer, *International Journal of Fracture* 39 (1989) 201–216.
22. J.W. Button, D.N. Little, Y. Kim and J. Ahmed, *Proceedings Association of Asphalt Paving Technologists* 5b (1987) 62–90.
23. A.H. Lepie, *Proceedings JANNAF Structures and Mechanical Behavior Working Group* CPIA Publication No. 351 (1981) 233–240.
24. R.A. Schapery, *Tire Science & Technology* 6 (1978) 3–47.
25. S.P. Timoshenko and J.N. Goodier, *Theory of Elasticity*, 3rd edn. McGraw-Hill (1970).
26. R.A. Schapery, *International Journal of Fracture* 25 (1984) 195–223.
27. M.A. Biot, *Quarterly Applied Mathematics* XXX (1972) 379–406.
28. J.D. Ferry, *Viscoelastic Properties of Polymers*, 3rd edn. Wiley (1980).
29. D. Broek, *Elementary Engineering Fracture Mechanics*, 4th edn. Martinus Nijhoff (1986).
30. T.C.T. Ting, *Journal of Applied Mechanics* 33 (1966) 845–854.
31. G.A.C. Graham, *International Journal of Engineering Science* 14 (1976) 1135–1142.

Appendix

The stress σ_b in the bonding zone was assumed constant in Section 5. Here we shall use the form

$$\sigma_b = \sigma_0 f(\bar{v}), \tag{A.1}$$

where $\bar{v} = (v^+ - v^-)/z_0$, in which σ_0 and z_0 are constants with dimensions of stress and length, respectively. The separation between the surfaces is $v^+ - v^-$. The bond energy is

$$2\Gamma_b = z_0\sigma_0 \int_0^\infty f\,\mathrm{d}\bar{v}. \tag{A.2}$$

Greenwood and Johnson [12], in a study of adhesive crack growth, used

$$f = (1 + \bar{v})^{-3} \tag{A.3}$$

based on experimental and theoretical considerations of the force of attraction between smooth surfaces. Figure 7 shows their predicted distributions of f and $\bar{v}$ for elastic materials, where E'_R is defined in (94). Here we will not use (A.3); the only assumption about f we make is that it vanishes rapidly enough with increasing $\bar{v}$ to ensure validity of the Barenblatt

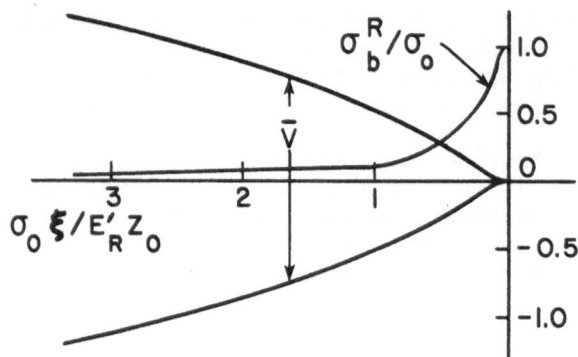

Fig. 7. Distributions of bonding stress and separation between surfaces for elastic materials. After [12].

method (i.e., a small-scale effective bonding-zone length) and convergence of the relevant integrals. In addition, we shall use the creep compliance (32), but with $D_0 = 0$, and assume as before that the compliance of the material below the bonding surface D^- is proportional to that above this surface, D^+; hence,

$$E_R^+ D_1^+ = E_R^- D_1^-. \tag{A.4}$$

The surface stress in the elastic problem, (27), for quasi steady-state bonding, (31), becomes

$$\sigma_b^R = E_R^+ D_1^+ \sigma_0 \dot{a}_b^{-m} \int_\infty^\xi (\xi' - \xi)^m \frac{df}{d\xi'} d\xi' \tag{A.5}$$

with $df/d\xi' = df/d\bar{v} \cdot d\bar{v}/d\xi'$. From (24) and (A.4) with $\beta = \infty$,

$$\bar{v} = (v^+ - v^-)/z_0 = \frac{(1 - v^2)}{E_R^+ z_0} (1 + D_1^-/D_1^+) \int_0^\infty \sigma_b^R(\xi') F(\xi'/\xi) d\xi'. \tag{A.6}$$

Substitute (A.5) into (A.6) and eliminate ξ and ξ' in favor of the dimensionless variable

$$\psi = \psi(\xi) \equiv (D_1^+ + D_1^-) \sigma_0 \xi^{m+1}/\dot{a}_b^m z_0 (m + 1). \tag{A.7}$$

Also, define $\psi' \equiv \psi(\xi')$ and $\psi'' \equiv \psi(\xi'')$; hence,

$$\bar{v} = \int_0^\infty \int_\infty^{\psi'} \left[\left(\frac{\psi''}{\psi'} \right)^p - 1 \right]^m \frac{df}{d\psi''} F[(\psi'/\psi)^p] d\psi'' d\psi'. \tag{A.8}$$

Here, $p \equiv 1/(m + 1)$. Inasmuch as $f = f(\bar{v})$, this is an integral equation for $\bar{v} = \bar{v}(\psi)$. The primary result for our purposes is that $\bar{v}$ is a function of only the dimensionless variable ψ. This observation together with (A.5) permits us to reduce (28) to

$$K_I^R/E_R^+ = k_2 z_0 (D_1^+ \sigma_0/z_0 \dot{a}_b^m)^{0.5/(m+1)}, \tag{A.9}$$

where $k_2 = k_2(m)$ is a dimensionless function of only m. By solving for $\dot{a}_b$ the same exponent q as in (56) is obtained. In order to determine the value of k in the power law (56) it would of course be necessary to find k_2, which requires the solution of (A.8). It should be added that Greenwood and Johnson [12] obtained $\dot{a} \sim K^q$ in the crack growth problem where q is the same as in (56).

We may use arguments similar to those in Section 5 to show that (A.9) is valid for a compliance which does not obey a pure power law, as long as d log D/d log t is a slowly varying function of log t. However, without further analysis it is not possible to indicate just how slow the variation must be to achieve an acceptable degree of accuracy.

Résumé. On décrit le mécanisme de fermeture d'une fissure quasi-statique et de collage des surfaces de matériaux visco-élastiques linéaires indentiques ou différents. On couvre également l'étude de la liaison dans le temps de surfaces à courbure initiale sous l'effet des forces de surface résultant de l'attraction et de charges extérieures. L'accent est placé sur l'utilisation de la mécanique des milieux continus pour l'établissement d'équations permettant de prédire la longueur de fissuration et la dimension des zones et contact en fonction du temps, dans le cas de géométries relativement générales. On prend en compte les processus atmostiques et moléculaires associés aux phénomènes de cicatrisation ou de collage en recourant à une idéalisation de l'extrémité de la fissure comparable à celle utilisée dans la méthode d'analyse des ruptures due à Barenblatt. En partant d'un principe de correspondances établi précédemment, on déduit une expansion décrivant la vitesse de déplacement du bord de la zone collée. On constate que les effets de la dépendance des propriétés du matériau en fonction du temps et du facteur d'intensité de contraintes sont différents de ceux exercés sur la croissance d'une fissure. On procède à une comparaison des énergies intrinsèques et apparentes de rupture et de collage, et on donne des critères pour déterminer si un collage peut ou non avoir lieu. Des exemples illustrent l'emploi de la théorie de base pour décrire la cicatrisation des fissures et la croissance de la surface de contact de surfaces initialement courbes. Enfin, on estime l'effet du temps de collage sur la résistance du joint, à partir des exemples de croissance de la surface de contact.

International Journal of Fracture 39: 191–200 (1989)
© Kluwer Academic Publishers, Dordrecht

Predicting strength of adhesive joints from test results

G. P. ANDERSON[(1)†] and K. L. DeVRIES[(2)]
[(1)]*Morton Thiokol, Inc., Brigham City, Utah 84302, USA;*
[(2)]*Department of Mechanical Engineering, University of Utah, Salt Lake City, Utah 84112, USA*

Received 19 December 1987; accepted in revised form 1 April 1988

Abstract. Using an average-stress criterion can lead to errors of an order of magnitude in predicted load carrying capability of a bonded joint. A fracture mechanics approach is shown to accurately predict failure load in a joint bonded with either a polyurethane or a relatively brittle epoxy when proper consideration is given to loading mode, temperature, and rate. The principal contribution of this paper is in extending fracture mechanics theory to regions where classical singular points do not exist. Analyses are combined with test data to deduce an "inherent" flaw size. It is shown that the combination of this inherent flaw size with the critical energy release rate can be used to predict strength in a variety of bonded joints.

1. Introduction

An engineer is often required to obtain bond strength data from laboratory size samples and infer from these the strength of a given joint. The strength of these bond tests is usually reported as an average failure stress, which is defined as the breaking load per unit of bond area. This stress is then compared with the average stress that exists in the joint being evaluated when its maximum load is applied. However, if the joint geometry, loading time, and other conditions are not identical to that of the laboratory test conditions, such a comparison can lead to unsafe joint designs.

Since the average stress in a bondline is generally not a reliable tool for predicting failure in a bonded joint, an alternative approach is required. If it is hypothesized that failure of a bonded joint occurs when the stress (or some functional of the stresses) reaches a critical value, one might evaluate the stress at each point in a test specimen bondline. The value of the maximum stress(es) at which the bond breaks is termed the bond stress capability. The next step would be to evaluate the stresses at each point in the joint bond when it is subjected to its maximum expected load. The highest resulting stress (or stress functional) at any location in the bond would then be termed the joint requirement. The amount by which the bond capability exceeded the joint requirement would provide the margin of safety of the joint.

This procedure has been very useful in homogeneous materials except when a notch or crack (360 degree notch) is present in the material. In such cases, the stresses are not defined at the notch tip when linear elastic analyses are employed; i.e., the notch tip is a point of stress singularity. One normally relies on a fracture mechanics analysis to predict load carrying capability for such geometries.

Notches in an adhesive and initial debonds are obvious points of bond-line stress singularity. In addition, many bond termination geometries are points of singular stress [1, 2, 3]. Thus, for debonds initiating at bond edges, even in the absence of voids or initial debonds, both the joint requirement and bond capability must be quantified in terms of fracture mechanics parameters. Bond failure does not always initiate at a point of apparent stress singularity. Thus, a failure criterion which is applicable to both singular and non-singular points is highly desirable. Our approach is to determine an "inherent" flaw size from which energy release rates can be calculated and whether failure initiates within a bonded joint or at the bond edge. These inherent flaws may be related to those that exist naturally in all bonds because of air bubbles, local surface discontinuities, etc.

2. Bonds using a linear elastic material

A polyurethane (Solithane 113) to polymethylmethacrylate (PMMA) joint was selected for the initial study. Solithane 113 is a nearly incompressible linear elastic material for temperatures above 70°F and load times longer than 0.005 seconds. The Solithane to PMMA bond strengths is low enough to allow "adhesive" failure.

Three sets of five Solithane/PMMA buttons were tested using the test fixture of Fig. 1. The following average failure loads, P_{cr}, were obtained:

Adhesive thickness, mm (in.)	Initial debond, mm (in.)	P_{cr}, N (lb)
6.38 (0.251)	3.30 (0.130)	286 (64)
6.17 (0.243)	6.45 (0.254)	138 (31)
0.71 (0.028	a_0	1430 (322)

The initial debonds were axisymmetric and located between the Solithane and a PMMA button. The debonds extended from the button outer diameter for the indicated depth.

The average critical energy release rate from the first test set was $23 \, \text{J/m}^2$ ($0.13 \, \text{in.-lb/in.}^2$). With this value of G_c, the failure load for the 6.45 mm (0.254 in.) flaw was predicted to be 146 N (35 lb). This value compares closely to the measured 138 N (31 lb) load.

The inherent flaw size, a_0, is defined as the amount of debond necessary to produce the proper critical energy release rate, $23 \, \text{J/m}^2$ ($0.13 \, \text{in.-lb/in.}^2$), at the measured peak load, 1430 N (322 lb), in specimens with no initial debond. For the Solithane to PMMA bond, the inherent flaw size was determined to be 0.076 mm (0.003 in.).

Once G_c and a_0 are known, the load capability of other bonded joints can be predicted. This is demonstrated by predicting the change in bond strength with adhesive thickness in butt joint tests (Fig. 1). Adhesive thicknesses between 0.15 mm and 25 mm (0.006 in. and 1.0 in.) produced failure loads ranging from 1900 N (430 lb) for thin bonds, to 180 N (40 lb) for thick bonds (Fig. 2). Adhesive failure initiated at the bond edges for joints thicker than 2.5 mm (0.1 in.) and near the centerline for joints thinner than 2.5 mm (0.1 in.). Both the failure loads and debond initiation points were predicted using fracture mechanics theory as illustrated by the solid curve in Fig. 2.

The analytical prediction was made by assuming an inherent flaw of 0.076 mm (0.003 in.) existed at the bond edge and extended around the periphery of the specimen (axisymmetric flaw). With this flaw, an energy release rate was found for each adhesive thickness. A second analysis was then completed by assuming the inherent flaw existed at the specimen center and again evaluating energy release rate as a function of adhesive thickness. The data from the two resulting analyses are plotted as the square root of energy release rate per unit load versus bond thickness in Fig. 3.

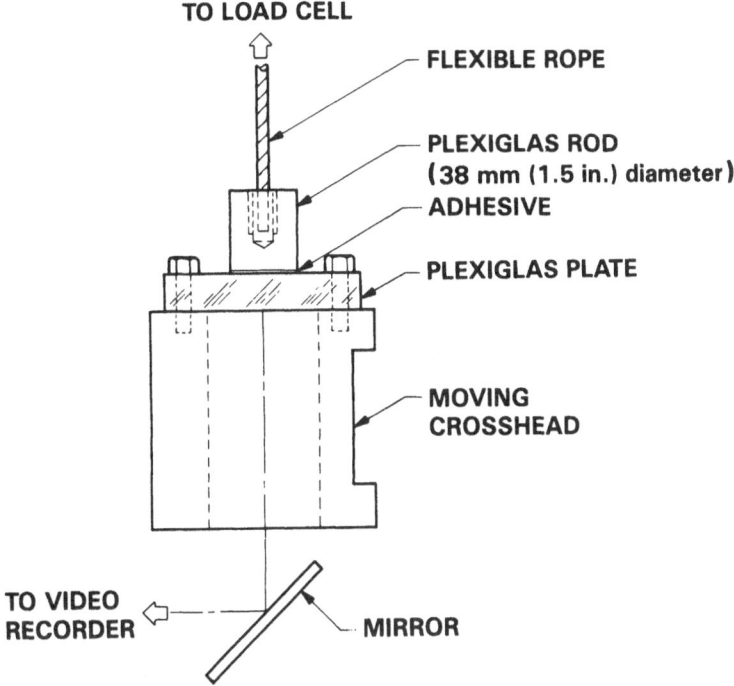

Fig. 1. Transparent tensile test apparatus.

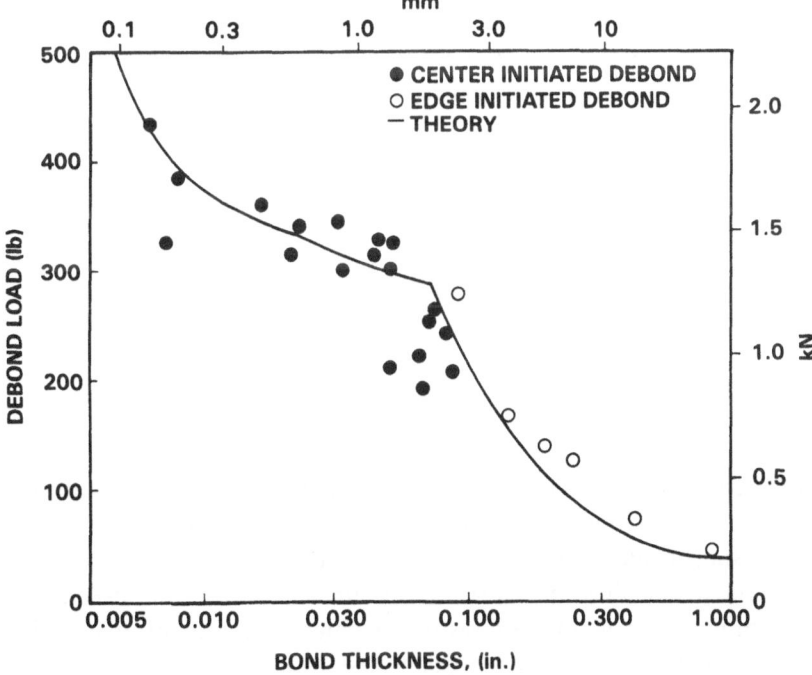

Fig. 2. Effect of adhesive thickness on debond load – polyurethane.

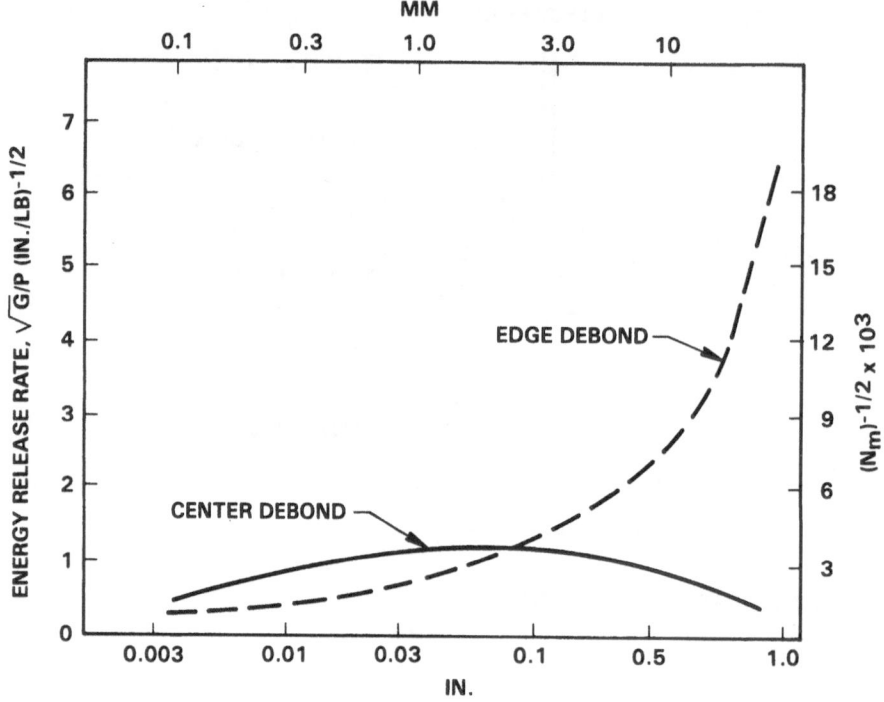

Fig. 3. Energy release rate for 0.076 mm (0.003 in.) debond.

For thin bonds, the energy release rate is greater for a center flaw than for edge flaws, therefore, debond is predicted to initiate at the specimen radial center. The energy release rate increases with bond thickness (strength drops) until the bond thickness is 2.3 mm (0.09 in.). When the adhesive thickness exceeds 2.3 mm (0.09 in.) the energy release rate for edge debonds exceeds that for center debonds. Thus, the transition from center to edge debonds is properly predicted from the analysis.

The use of an inherent flaw size in conjunction with the energy release rate approach to failure was felt to be necessary because of the discontinuity in energy release rate as the debond size approaches zero, and to allow a common failure criterion to be used for edge-initiated debonds (singular point) and internally-initiated debonds.

3. Bonds using a brittle epoxy

The fracture mechanics approach was also evaluated with a nearly linear elastic brittle epoxy, EA 934 (Fig. 4a). Bond tests were completed using tensile button specimens which consisted of two short cylinders bonded together on the cylinder ends in a butt joint. Each cylinder had a groove machined into it so the specimen could be slid into a set of U-shaped grips which are mounted on a universal testing machine. A picture of this specimen and test grips is included as Fig. 5. This specimen was tested in several different configurations. The bondline geometry of the specimen was altered by using spacers to give different thicknesses and by creating intentional debonds by putting Teflon tape on the bonding surface of one

a) EA-934 Epoxy

b) EA-9309.2NA Epoxy

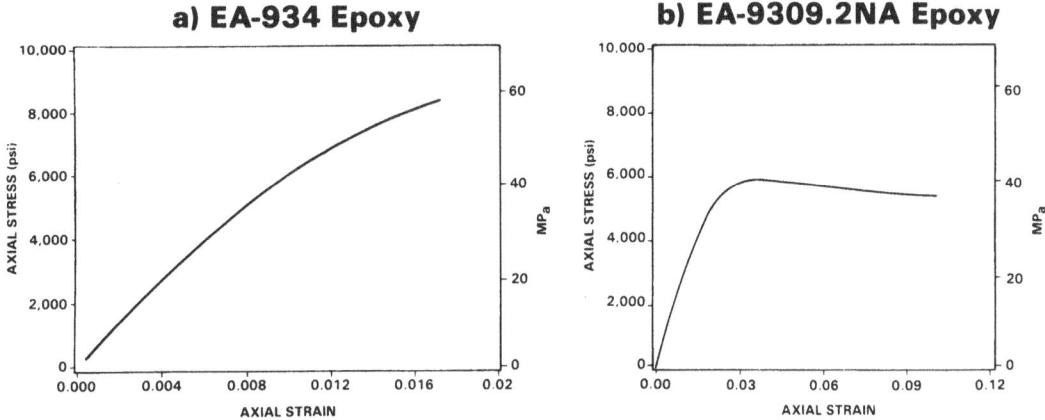

Fig. 4. Stress-strain curves for two epoxy systems.

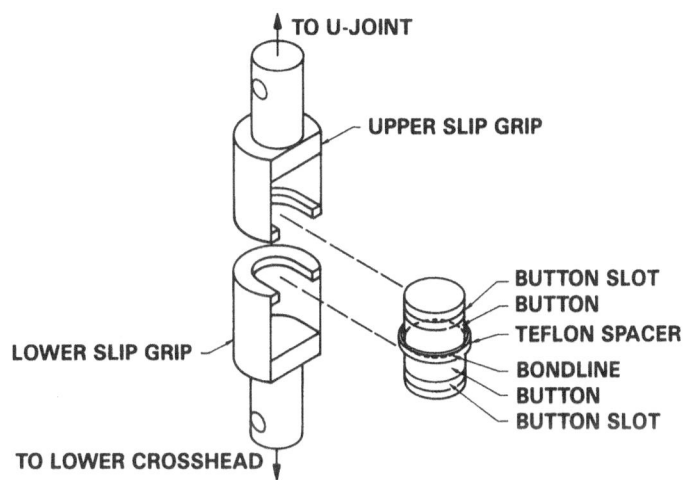

Fig. 5. Tensile adhesion button test setup.

button. This creates an initial flaw of known dimensions. A critical energy release rate of 32 J/m² (0.18 in.–lb/in.²) was obtained for 29 mm (1.13 in.) diameter tensile buttons with an initial debond of 2.5 mm (0.1 in.) and a failure load of 7.1 kN (1600 lb) (average for 10 tests). The critical load of 4.2 kN (954 lb) was then predicted for buttons with 5.1 mm (0.2 in.) initial flaws. This prediction was within one percent of the test results.

A series of 10 specimens with no initial debonds were then tested. These specimens had an adhesive thickness of 1.7 mm (0.68 in.) and failed at an average of 21 kN (4760 lb). These data are summarized in the following table:

Initial debond mm (in.)	Failure load kN (lb)	Result
2.5 (0.1)	7.1 (1600)	$G_c = 32$ J/m² (0.18 in.–lb/in.²)
5.1 (0.2)	4.7 (967)	1 percent prediction error
a_0	21 (4760)	$a_0 = 0.025$ mm (0.001 in.)

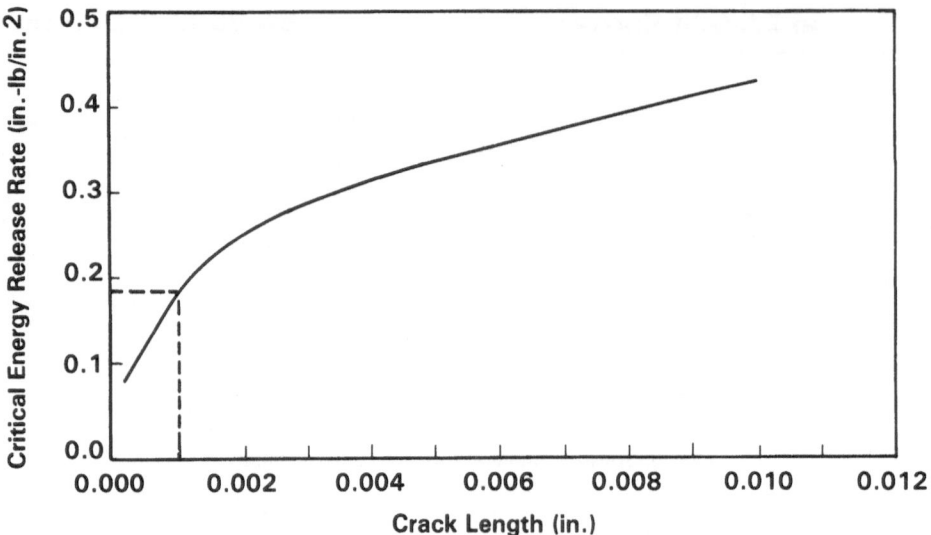

Fig. 6. Critical energy release rate for EA-934 tensile buttons.

To obtain the inherent flaw size, the energy release rate was calculated using finite element techniques (Fig. 6) as a function of debond length for small initial debonds at the specimen outer diameter. A failure load of 21 kN (4760 lb) was used in computing the energy release rate. Since this was the failure load obtained from the test data, the ordinate in Fig. 6 is the critical energy release rate. However, the critical energy release rate was determined to be 32 J/m² (0.18 in.–lb/in.²) from the first set of tests (initial debond 2.5 mm, 0.1 in.). Thus, it can be determined from the Fig. 6 plot, that the bond system has inherent flaws of 0.25 mm (0.001 in.). The two parameters – critical energy release rate and inherent flaw size – are then felt to characterize the bond capability.

A series of tensile button tests was completed where adhesive thickness ranged from 0.127 mm to 13.3 mm (0.005 in. to 0.525 in.). Average values (10 replications) from these tests are presented in Table I. An attempt was made to adjust the load rate to obtain a constant load time to failure. However, the resulting load times were not constant, so additional testing was completed to allow an empirical correction of the first data set (Column 4 of Table I).

These adjusted data are plotted as a function of bond thickness in Fig. 7. Using the critical energy release rate of 32 J/m² (0.18 in.–lb/in.²), the inherent flaw size of 0.025 mm (0.001 in.) and a curve of energy release rate as a function of bond thickness (Fig. 8), the effect of adhesive thickness on bond strength was predicted. The predicted values presented in

Table I. Tensile adhesion tests with EA-934 epoxy adhesive

Adhesive thickness, mm (in.)	Load rate, mm/min (in./min)	Failure load, kN (lb)		
		Measured	Adjusted	Predicted
0.13 (0.005)	1.27 (0.050)	30.1 (6770)	32.3 (7230)	34.8 (7820)
0.13 (0.023)	1.27 (0.050)	24.8 (5570)	25.4 (5720)	26.2 (5900)
1.73 (0.068)	1.27 (0.050)	21.1 (4770)	21.2 (4760)	21.0 (4710)
3.30 (0.130)	1.91 (0.075)	16.2 (3650)	16.1 (3620)	14.6 (3280)
13.30 (0.525)	2.54 (0.100)	10.6 (2380)	10.1 (2280)	9.4 (2120)

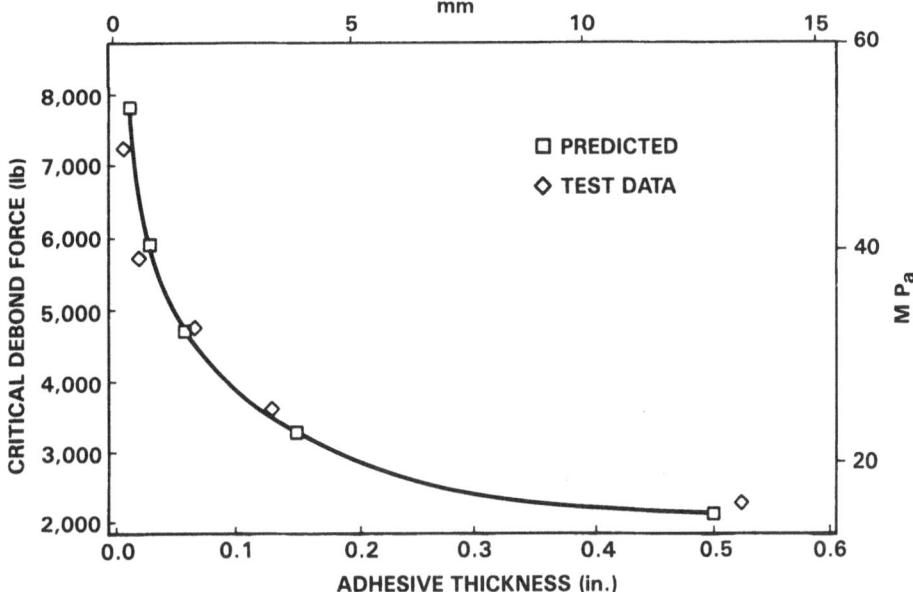

Fig. 7. Effect of adhesive thickness on debond load–EA-934 epoxy.

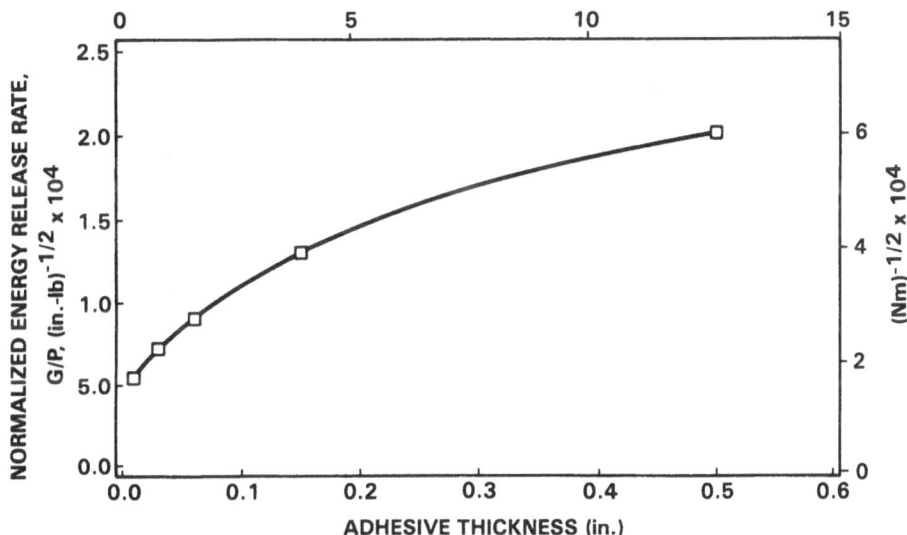

Fig. 8. Energy release rate versus adhesive thickness, Poisson's ratio $= 0.34$, $a_0 = 0.025\,\text{mm}$ (0.001 in.).

Table I and Fig. 7 show very close agreement with measured results. Thus, the reason for adhesive strength changes corresponding to adhesive thickness (at least for polyurethane and brittle epoxies) appears to be completely accounted for by changes in energy release rate.

For all tensile button tests with epoxy, edge-initiated failures are predicted since the energy release rate for an inherent flaw size of 0.025 mm (0.001 in.) is greater for an edge-initiated failure than for failure initiated internally. There was no direct experimental verification of the failure initiation point, since opaque adherends and adhesives are used. The observed failures were primarily cohesive within the adhesive layer. However, in most cases a small

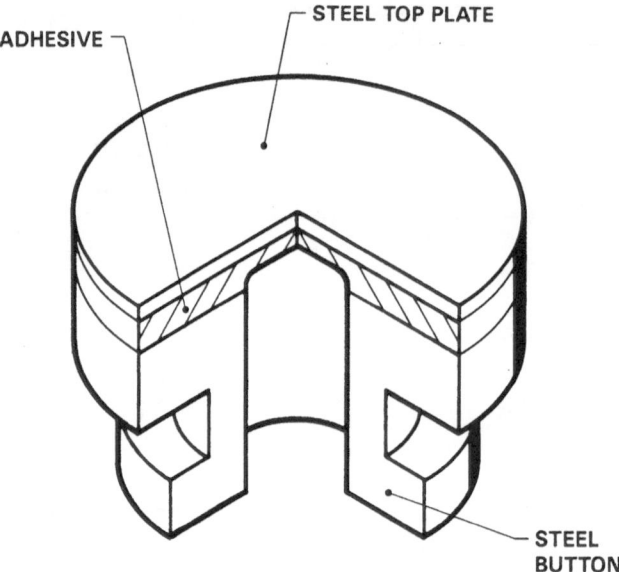

ADHESIVE

STEEL TOP PLATE

STEEL
BUTTON

Fig. 9. Modified blister test specimen.

area of adhesive failure existed near the bond edge. Our hypothesis is that failure initiated at the adhesive/adherend interface adjacent to the bond edge. Failure then propagated into the adhesive to the bond axial center, and continued through the center of the adhesive layer.

The calculated values of critical energy release rate and inherent flaw size were then successfully used to predict failure in the modified blister test illustrated in Fig. 9. Blister test specimens with no initial debonds were tested. The predicted failure pressure as well as the measured values are presented below:

Debond size mm (in.)	Failure pressure		% error
	Predicted kPa (psi)	Measurement kPa (psi)	
a_0	490 (3370)	482 (3320)	2
5.1 (0.2)	125 (860)	132 (910)	6

The close agreement of measured and predicted pressure shows that the fracture mechanics approach was valid for the material and geometries used.

As a second independent verification of this test, a non-tapered double cantilevered beam test was run according to ASTM D-3433. This test involves applying a tensile force normal to the bondline of two parallel plates. A crack is propagated incrementally by loading and unloading the specimen. From this testing a load versus displacement curve is obtained at different crack lengths. From these data the change in compliance of the specimen per change in crack length may be calculated. This allows calculation of G_c according to the following equation:

$$G_c = \frac{p^2}{2b}\frac{dc}{da},$$

where p is the crack propagation load, b is the specimen width, and ds/da is the change in compliance with crack length.

A version of EA934 epoxy which contains no asbestos (EA934NA) was used in the double cantilever beam tests. The resulting critical energy release rate was 91 J/m^2 (0.52 in.–lb/in.^{2}). This is in close agreement with the 82 J/m^2 (0.47 in.–lb/in.^{2}) which was obtained using tensile adhesion tests.

4. Bonds using a nonlinear epoxy

A rubber toughened high strength structural epoxy, EA-9309.2NA, a product of Dexter Hysol, Aerospace Division, was used for the nonlinear mode I testing. This epoxy exhibits a nonlinear stress strain curve as shown in Fig. 4b. As can be seen, this epoxy has an initial linear elastic portion with a larger "plastic" portion.

Tensile adhesion buttons were tested with various bond geometries using this nonlinear epoxy. The testing was performed at one temperature and at one load rate. Buttons were tested which had several different thicknesses and bond termination geometries.

Three failure theories were compared for correctness in analyzing test results. First was the linear fracture mechanics approach which was used with good success for the "linear elastic" material. This approach proved to be inaccurate, which confirmed that a nonlinear approach must be taken to account for plastic deformation. The first nonlinear approach involved the use of the J-integral. The J-integral is a nonlinear energy release rate which reduces to G in the linear case. This approach allowed prediction of loads in the 0.2 inch debonded buttons from results obtained in the 0.1 inch debonded buttons within 5 percent. The critical value of the J-integral, J_c, obtained was 2.0 in.–lb/in.^{2}. Numerical problems arose with the finite element runs to determine the inherent flaw size, a_0. Unreasonably large shear deformation occurred for small flaw sizes near the adhesive/adherend interface. Work is continuing to solve this problem.

5. Conclusions

It is concluded that average stress determined from bond tests is a very poor predictor of failure in bonded joints. Furthermore, other stress failure criteria have serious deficiencies due to stress singularities which exist at notches, debonds, and bond termination points. Fracture mechanics theory has shown very promising results in mode I loading of both a polyurethane and a linear epoxy. Further work is progressing to extend the theory to include nonlinear adhesives.

References

1. M.L. Williams, *Bulletin of the Siesmological Society of America* 49 (1959) 199.
2. V.L. Hein and F. Erdogan, *International Journal of Fracture Mechanics* 7 (1971) 317–330.
3. E.H. Dill, A.L. Deak and W.F. Schmidt, *Handbook for Engineering Structural Analysis of Solid Propellants*, CPIA Publication 214 (1971), Chapter 5.

Résumé. Lorsqu'on veut prédire la capacité de charge portante d'un joint collé, on peut être conduit à des erreurs d'un ordre de grandeur en utilisant le critère de contrainte moyenne. On montre qu'une approche par mécanique de rupture peut permettre de prédire la charge de rupture d'un joint collé réalisé avec un polyuréthane ou une épox relativement fragile, lorsqu'on considère de manière appropriée la mode de mise en charge, la température et la vitesse de sollicitation. La principale contribution de cette étude réside dans une extension de la théorie de la mécanique de la rupture à des domaines où n'existent pas de point singuliers classiques. On joint á l'analyse des données d'essais en vue de déduire une taille de défaut "inhérente". On montre que la combinaison de cette dimension de défaut inhérent et de la vitesse critique de relaxation de l'énergie, peut être utilisée pour prédire la résistance d'une grande variété de joints collés.

International Journal of Fracture 39: 201–216 (1989)
© Kluwer Academic Publishers, Dordrecht

Time-dependent autohesion

R.G. STACER and H.L. SCHREUDER-STACER
Air Force Systems Command, Air Force Astronautics Laboratory, Edwards AFB, CA 93523-5000, USA

Received 22 September 1987; accepted in revised form 16 April 1988

Abstract. A study has been conducted to investigate the relationship between polymeric structure and time-dependent autohesion, measured in terms of autohesive fracture energy, G_a. Using the method of reduced variables, it was found that G_a data as a function of contact time and temperature could be superposed into master curves of temperature-reduced contact times. Autohesion master curves developed in this fashion showed fracture resistance increasing with time along a logarithmic-type curve with monotonically decreasing slope. These data indicate that the generally accepted $\frac{1}{2}$ power law dependency for autohesion only applies over a narrow range of contact times. Modelling of the experimental results was accomplished using a first-order kinetic equation derived to account for contact-area formation. Two diffusion-based models also provided good predictions in specific cases, most notably for the effect of molecular weight on time to equilibrium. However, evidence that diffusion is not the rate controlling process included the pronounced effects of contact pressure on autohesion and the identical time-dependent behavior of nondiffusing crosslinked networks when compared with systems containing mobile polymeric chains.

1. Introduction

Autohesion in polymeric materials is defined as the resistance to separation of two surfaces after they have been joined together for a period of time under a given pressure [1–3]. Research into the autohesion phenomenon provides fundamental insight into the physical processes of adhesive bond formation and failure, as well as addressing practical engineering issues such as crack healing, elastomer tack, polymer fusion and polymer welding. One of the most frequently investigated features of autohesion is its time dependency. A number of investigators [4–11] have experimentally determined that adhesive fracture energy (G_{IC}) is proportional to contact time to the $\frac{1}{2}$ power; $\frac{1}{4}$ power if fracture stress (σ) or the stress intensity factor (K_{IC}) are used. Controversy has arisen, however, in establishing the controlling physical process. Hamed [2] has recently reviewed the two mechanisms believed responsible for the time-dependent nature of autohesion. Number (1) is bond formation via chain entanglements through polymer diffusion across the interface, and number (2) is contact area formation (wetting) through viscous flow. As summarized below, each mechanism has strong literature proponents, and if applied, can lead to widely different theoretical predictions.

Voyutskii [1, 12, 13] appears to be the earliest supporter of mechanism number (1) as the rate controlling process for autohesion. This qualitative argument ascribes the increase in strength with time to diffusion of long chain molecules across the interface. With time, the migrated chain ends entangle with their new neighbors, thus enhancing fracture resistance, and eventually they randomize to fuse the surfaces. More recently, de Gennes [14] has adapted the reptation theory [15] of polymer diffusion to quantify the mass transfer between two contacted surfaces. This "crossing density" model has been expanded by Prager and Tirrell [16–18] to allow predictions of equilibrium strength, as well as the time-dependency, based solely on molecular parameters. Wool and coworkers [7, 10, 19] have incorporated the reptation theory into an alternative "chain pull out" model, which gives somewhat different structure-property relationships. Direct experimental evidence that significant diffusion across polymer-polymer interfaces does occur has been summarized by Wu [3], and is based on optical microscopy, electron microscopy, and luminescence techniques.

Although they do not dispute the claim of mass transfer across the interface, proponents [2, 20–26] of mechanism number (2) argue that this time-dependence of autohesion originates from the kinetics of surface wetting. They point out that immediately upon contact, the rough topography (on a microscale) of the two surfaces resists intimate molecular contact, creating voids at the interface. Under light pressure, the material spreads with time to fill these voids through viscous flow. Thus, bond strength develops by the first-order kinetics of wetting. The controversy over which of the two mechanisms is dominant, if any, has centered on the influence of contact time, temperature and pressure on autohesive strength. Korenev-skaya and coworkers [27], and Wool and O'Connor [7, 10] assert that the time scale of contact area formation is much shorter than the diffusion process. Kelley [20] and Kaelble [22] show much longer time scales for viscous flow. When the effects of temperature on autohesion are plotted in an Arrhenius fashion, the resultant activation energy is appropriate for a diffusion controlled process [1]. Anand [24] argues against this as a deciding factor pointing out that the temperature dependence of viscosity is very similar, if not identical, to that of self-diffusion.

The strongest argument in favor of the wetting mechanism is the effect of pressure on autohesion. In an experiment designed to differentiate between the roles of mechanisms (1) and (2), Hamed [25] systematically varied contact pressure (or load) with time. Autohesion increased with increasing contact pressure at constant time. Since the diffusion mechanisms are independent of pressure, and wetting kinetics are not, he concluded that wetting is the controlling process. It should be noted that diffusion in polymers can not be considered strictly independent of pressure. The influence of temperature on diffusion has been shown to follow free volume arguments, and free volume is dependent on pressure [28]. However, the most comprehensive data available to date on the effects of pressure on free volume [29–31] suggests that several orders of magnitude greater pressure effects would be required to produce the behavior observed by Hamed [25] and others [1, 3].

As has been discussed above, the physical processes that control the rate dependence of autohesion remain a subject of considerable debate. The purpose of this paper is to describe the results of an experimental investigation intended to evaluate the autohesion of well characterized polymer networks and gums over a much broader time scale than those previously reported. These results are then used to determine the applicability of the various autohesion theories and the appropriateness of their assumptions.

2. Experimental

2.1. Materials characterization

Table I lists the seven materials investigated as part of this effort. Included is a series of uncrosslinked polyisobutylene (PIB) samples covering a broad range of molecular weight, with the lower molecular weight materials being nearly monodisperse, as indicated by the

Table I. Material characterization for the seven materials. Viscosity and diffusion data at 25°C

Material	Supplier	M_w g/mole $\times 10^{-5}$	M_w/M_n	η_0 Pa $-$ s $\times 10^{-9}$	D m²/s $\times 10^{21}$
PIB-1	Polysciences	1.9	1.1	0.083	5.6
PIB-2	Polysciences	2.3	1.1	0.11	3.8
PIB-3	Polysciences	1.5	1.1	0.0015	9.1
PIB-4	Aldrich	4.2	1.8	0.90	1.1
PIB-5	Aldrich	21.0	3.6	6.0	0.038
CTPIB	Exxon	0.027	2.0	0.33	–
HTPB	Sartomer	0.030	1.9	3.3	–

polydispersity index, M_w/M_n, being close to unity. Additionally, two crosslinked polymers were employed. These were prepared from the prepolymers carboxy terminated polyisobutylene (CTPIB) and hydroxy terminated polybutadiene (HTPB). The CTPIB material, which is difunctional, was crosslinked using tris 1-(2-methyl)aziridinyl phosphine oxide (MAPO). HTPB, which has a functionality of approximately 2.3, was cured by reaction with isophorone diisocyanate. Both CTPIB and HTPB were crosslinked at an equivalence ratio of 1:1. As can be seen, all the materials are based on the same PIB backbone structure with the exception of HTPB. HTPB was included for comparison with data from earlier studies [7, 8].

Characterization data for each material is also contained in the table. The weight average molecular weights, M_w, were determined by gel permeation chromatography using a polyisoprene calibration curve. Self-diffusion coefficients, D, were obtained from literature values [3] for polyisobutylene at the specific molecular weights.

Steady-state shear viscosities, η_0, were determined from tensile creep-recovery tests using an apparatus described by Meinecke [32]. In this experimental set-up, displacement with time is monitored with a linear variable transformer. Deformations were maintained within the linear range during the 10 ks loading and unloading periods. The longitudinal viscosity, λ_0, was calculated from the ratio of the stress to the rate of accumulation of the irreversible deformation in the manner recommended by Ferry [28]. This value was then converted to the steady-state shear viscosity through the relation

$$\eta_0 = \lambda_0/3. \tag{1}$$

2.2. Autohesion experiment

PIB test specimens were prepared by hot-pressing the chunk rubber between mylar sheets. This approach was taken to eliminate the possibility of molecular weight degradation, as would occur in other forming operations such as open-roll milling. The mylar sheets served to protect the surfaces from contamination until contact. A cloth backing material was embedded into the exterior surface as part of this operation to render the polymer inextensible during subsequent peel testing. CTPIB and HTPB, which are liquids in the experiments described, were cast against a mylar sheet while the cloth backing material was embedded on the opposite face after partial cure. All samples were then cut into 1-cm wide by 8-cm long strips.

Autohesive fracture resistance was monitored versus contact time at five temperatures ranging from -27 to 45°C. Before contact, samples were equilibrated overnight at the given

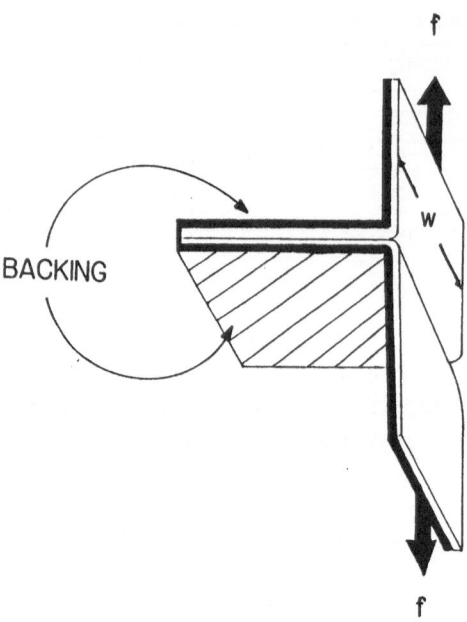

Fig. 1. T-peel test geometry.

temperature. After removal of their protective mylar cover, surfaces were contacted by hand at temperature and placed under a temperature-conditioned pressure plate. A pressure of 3 kPa was used as the reference condition. Samples of PIB-2 were also placed under 0.3 and 30 kPa to evaluate the effect of loading pressure.

Samples were removed from these temperature controlled conditions for periodic evaluation between 17 min and 12 days according to a logarithmic time scale. Evaluation was accomplished using the *T*-peel geometry illustrated in Fig. 1. In an analogous fashion to the work of Rivlin and Thomas [33], Gent [34, 35] and others [36, 37] have shown that the fracture surface energy, G_a, in this geometry for flexible materials is given by

$$G_a = 2f/w, \tag{2}$$

where *f* is the peel force necessary to induce steady crack growth and *w* is the sample width. As with all applications of the Rivlin and Thomas fracture criteria, the two primary assumptions are that crack propagation results from conversion of free energy to fracture energy and that this energy conversion depends on the rate of propagation rather than crack growth history. In this paper, G_a was calculated based upon certain simplifying assumptions. Certainly, viscoelastic dissipation of the rubber layers near the crack tip contributes significantly to a fracture measure, such as the surface energy, G_a. However, this energy contribution is evidenced in the time and temperature dependent feature of the measured and calculated G_a value. All peel tests were conducted in an INSTRON at a temperature of $-35°C$ and crack propagation rate of 1 mm/s. These test conditions were selected for three reasons: (1) no autohesive strength was observed to develop in samples at this temperature until times greatly in excess of the peel test period (including equilibration time) were reached, (2) no stick-slip crack propagation occurred in the vicinity of this temperature/rate condition, and

(3) test specimens were sufficiently above the rubber-to-glass transition region so bending forces were negligible.

3. Results and discussion

3.1. Application of reduced variables

An example of the autohesion data measured for each of the materials is presented in Fig. 2. This figure shows the effect of contact time, t_c, and temperature on the fracture energy of PIB-2 interfaces. Data were collected over three decades of logarithmic time for each of the five temperatures as shown. Fracture resistance increases in a regular fashion with increasing time or temperature, suggesting that time-temperature superposition of the data may be possible. Application of the method of reduced variables to autohesion data for PIB-3 is illustrated in Fig. 3. In a previous paper [38], it was shown that reduced variables could be applied to similar data collected for an adhesively bonded rubber joint, but this represents the first time this approach has been tested extensively or used for autohesion data. The continuous curve shown was obtained by horizontally superposing fracture data at different contact temperatures into a master curve of contact time. The horizontal shift factors, a_T, used to superpose the data are given in the figure inset. A reference temperature, T_s, of 25°C was arbitrarily selected.

Table II presents the WLF (Williams, Landel and Ferry) and free volume constants for the materials investigated. These were determined from the shift factors used to superpose

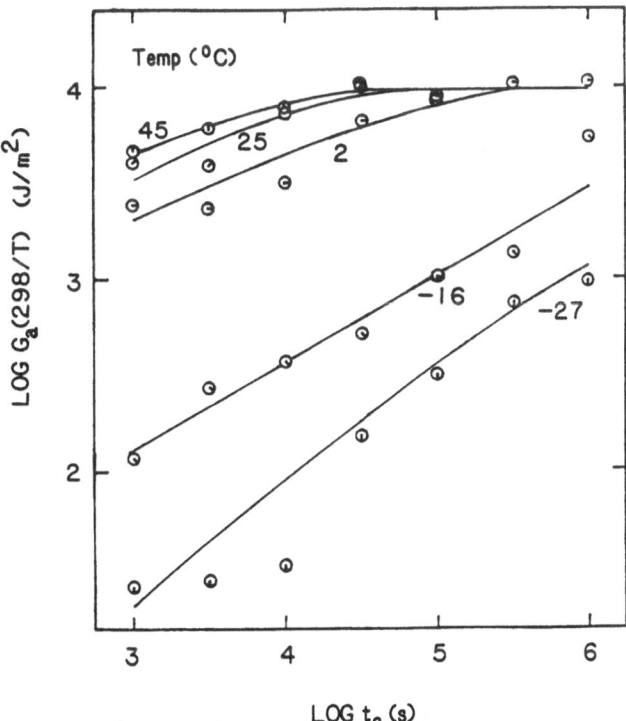

Fig. 2. Effect of contact temperature and time on the fracture energy of PIB-2.

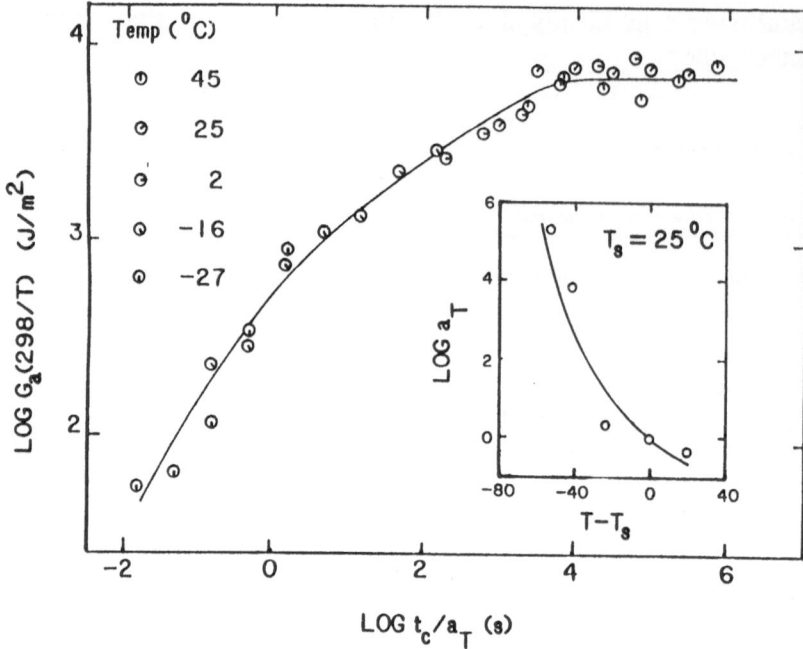

Fig. 3. Autohesion master curve for PIB-3.

Table II. WLF and free volume constants compared with PIB values from [28]. All constants referenced to 25°C. Glass transition temperatures (T_g) obtained from mechanical testing

Material	C_1	C_2 (deg)	T_g (°C)	α (deg^{-1} × 10^{-4})
PIB	8.61	200.4	− 68	2.5
PIB-1	5.17	121.1	− 73	6.9
PIB-2	7.63	154.1	− 73	3.7
PIB-3	6.85	126.3	− 73	5.0
PIB-4	6.65	129.4	− 73	5.0
PIB-5	4.61	123.2	− 73	7.6
CTPIB	8.38	216.9	− 72	2.4
HTPB	6.01	160.4	− 80	4.5

G_a values as illustrated in Fig. 3. In all cases, the fracture data shifted well, with no instances of thermorheologically complex behavior observed. The constants C_1 and C_2 in this table were determined from the WLF equation [39]

$$\text{Log } a_T = -C_1(T - T_s)/C_2 + (T - T_s). \tag{3}$$

Also included in this table is α, the thermal coefficient of expansion of the free volume above T_g, which is obtained from

$$\alpha = B/2.303\, C_1 C_2, \tag{4}$$

where B is a constant in the Dolittle equation assumed equal to unity [28].

Comparing the experimentally determined constants in Table II with literature values obtained from small deformation response testing of PIB [28] leads to several interesting observations. Polyisobutylene is generally regarded as a "sluggish" polymer, meaning its free volume response to thermal changes is less than that of other polymers. This can be seen by comparing its α of $2.5 \times 10^{-4}\,deg^{-1}$ to the larger "universal" value of $4.8 \times 10^{-4}\,deg^{-1}$. Of the six PIB materials tested, only CTPIB displayed the expected low α value. This indicates that the autohesion mechanism in CTPIB is dominated by the same viscoelastic process as its small deformation response, or at least has a very similar temperature dependence.

Although there is some scatter, all of the other PIB materials display significantly greater free volume responses. These large α values suggest the possibility of another viscoelastic process or multiple processes being responsible for time-dependent autohesion. Since diffusion of molecules across the interface is expected for the uncrosslinked PIB's, and not for the crosslinked CTPIB material, this leads to a preliminary conclusion that long range diffusive motion is responsible for the apparent increase in α. Visual observation of autohesion samples after equilibrium conditions had been reached showed fused surfaces with no residual interfaces for the five PIB samples. A distinct interface did, however, remain in CTPIB joints. Although use of the WLF equation for self-diffusion of polymers has not as yet been attempted, data are available for diffusion of small molecules in polymers. As summarized by Ferry [28], the ratio of C_1 values obtained from these diffusion experiments to those determined from mechanical properties measurements range from 0.6 to 1.0, consistent with the data in Table II. Additional experimentation will be required, however, before this can be considered conclusive evidence for a diffusive mechanism of autohesion.

As discussed in the Introduction, a number of investigators have reported a $\frac{1}{2}$ power law for time-dependent autohesion. In all cases, however, data were obtained over a somewhat narrow range of contact times, usually two to three logarithmic decades, leaving in doubt the broader applicability of this relationship. The data in Fig. 3, which spans nearly eight decades in temperature-reduced contact time, provides considerably more information concerning the autohesion phenomena. At long times, equilibrium contact conditions are reached and the fracture energy is independent of contact time. For PIB-3, the time-to-equilibrium, t_∞, is 6 ks, and the equilibrium fracture energy, G_∞, is 6.9 kJ/m^2. At times shorter than t_∞, G_a is dependent on contact time, and appears to asymptomatically approach a slope of $\frac{1}{2}$ at the shortest contact time.

Figure 4 presents all the master curves for the PIB series. Each of these curves shows the same general features as those just discussed for PIB-3. The differences in molecular weight and molecular weight distribution are, however, readily apparent in these five curves. As molecular weight increases, data in Fig. 4 shows both t_∞ and G_∞ also increasing; the latter with the exception of PIB-5. For the three materials with narrow distributions, PIB-1, 2, and 3, increasing molecular weight appears to shift the entire curve to longer times. PIB-4 and 5 do not obey this trend, as curves for these higher molecular weight polymers have lower slopes and cross over those of the other three. It is believed that these lower slopes are the result of broader molecular weight distributions. Quantification of specific relationships between molecular structure and contact time dependency will be discussed in the next section in terms of the various proposed models.

3.2. Effect of molecular weight

As mentioned earlier, Wool and coworkers [7, 10, 19] have proposed a model for time-dependent autohesion which predicts several important molecular structure/physical property

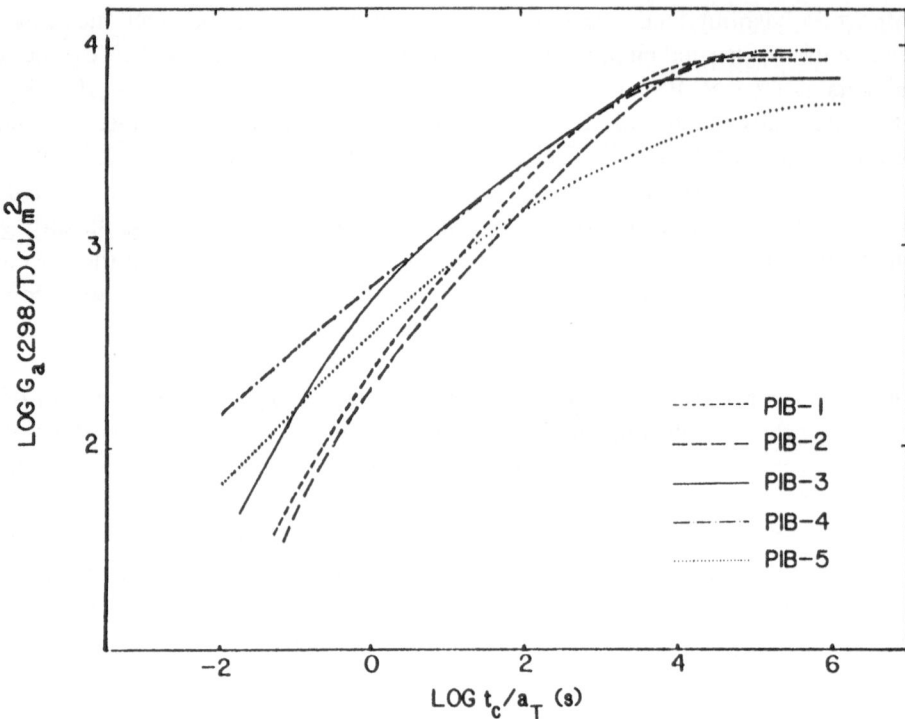

Fig. 4. Autohesion master curves for five PIB materials as indicated.

relationships. Fracture resistance in this model is assumed to be dominated by the molecules which diffuse across the interface with time. Specifically, it is equal to the energy required to pull these molecules out of the tube created by their surrounding physical entanglements and back across the interface. The longer the molecule, or the deeper the interpenetration distance, the greater will be the resistance to fracture. The key scaling relationship of this theory, adapted to the nomenclature and definitions of this discussion, is given as

$$G_a \sim t_c^{1/2} M_w^{-1/2}. \tag{5}$$

This equation predicts the $\frac{1}{2}$ power law, as well as a relationship between fracture resistance and molecular weight. Additionally this model provides a quantitative prediction for the time to equilibrium,

$$t_\infty = \frac{C_\infty b^2}{bAMo} M_w^3 \tag{6}$$

where C_∞ is the characteristic ratio, equal to 6.6 for PIB [40], b is the repeat length of the monomer, A is the constant for $D = A/M_w^2$, and M_0 is the molecular weight of the monomer.

How well (5) and (6) describe the autohesion data collected in this study can be seen in Figs. 5 and 6. Figure 5 shows the autohesion master curve for PIB-3, as well as the predicted $\frac{1}{2}$ slope. In this figure, as well as those previously discussed, time-dependent autohesion manifests itself with increasing time as a logarithmic function with monotonically descreasing

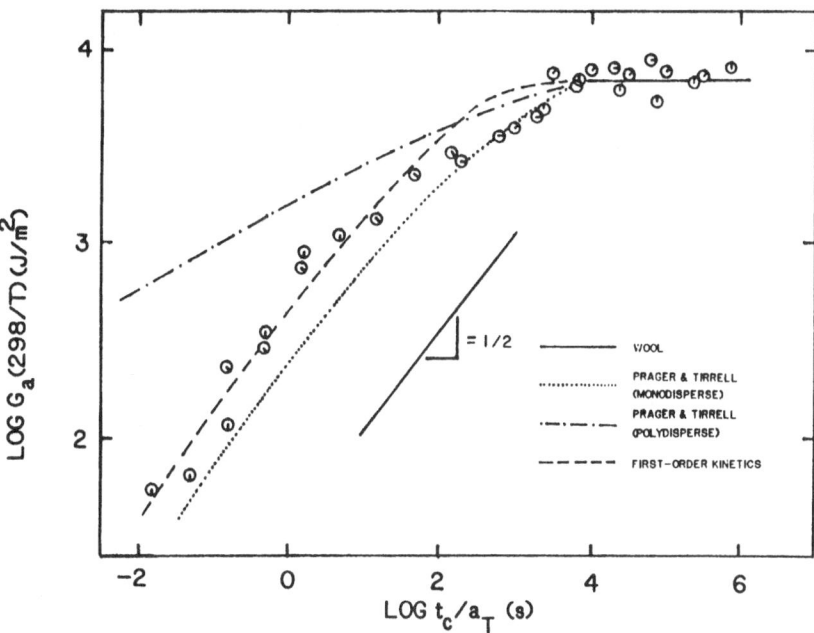

Fig. 5. Comparison of the predictions from several models as indicated with the autohesion curve for PIB-3.

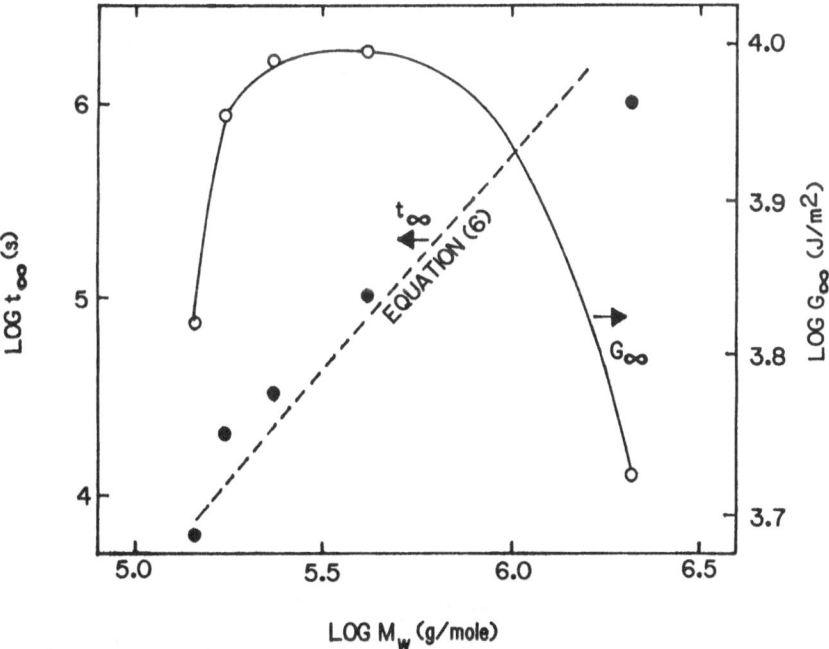

Fig. 6. Effect of molecular weight on time-to equilibrium fracture energy for polyisobutylene.

slope. Although over a limited portion of the curve, the $\frac{1}{2}$ power law may seem appropriate, it does not contain the correct physical characteristics to totally describe time-dependent auto-hesion. A second test of (5) can be seen in Fig. 6. This scaling relationship predicts a constant negative slope of the $\frac{1}{2}$ for G_∞ versus molecular weight when plotted in log–log coordinates. The data, however, show a curve with a broad maximum near $M_w = 300\,000\,\text{g/mole}$. These data are consistent with those of Forbes and McLeod [5], who showed identical behavior for natural rubber. A physical explanation for this behavior has been reviewed by Hamed [2].

In contrast, the predictions of this diffusion model for time-to-equilibrium are excellent. As shown in Fig. 6, Eqn. (6) successfully predicts the molecular weight dependency of t_∞ using only experimentally determined molecular parameters. With respect to the question of whether diffusion or wetting is the dominant physical process, however, it should be noted that t_∞ in (6) is the tube-renewal time in the reptation theory or longest relaxation time [41, 42]. Consequently, the same equation would be appropriate to describe time-to-equilibrium in a viscous flow process, consistent with the wetting mechanism of autohesion.

An alternate approach to incorporate the reptation model of diffusion into a general theory of autohesion has been taken by Prager and Tirrell [16–18]. This model predicts

$$\frac{G_a(t_c)}{G_\infty} = \frac{n\,(t_c, M_w, M_e)}{n\,(t_\infty, M_w, M_e)}, \tag{7}$$

where $n\,(t_c, M_w, M_e)$ is the number of effective interface crossings per unit area; a function of contact time t_c, molecular weight M_w and entanglement molecular weight M_e. For polyisobutylene, M_e is approximately $15\,000\,\text{g/mole}$ [28]. The denominator, $n\,(t_\infty, M_w, M_e)$, is the number of effective crossings per unit area of interface after equilibrium contact conditions have been reached. An effective crossing is defined as a chain which contributes to fracture resistance through either chain scission or viscoelastic dissipation. This model predicts the $\frac{1}{2}$ power dependency as a limiting value at the shortest times, not from zero to equilibrium. Figure 5 shows two curves derived from this model. These curves were obtained from Figs. 6 and 7 of [18] for a M_w/M_e value of 10, appropriate for PIB-3. The lower curve is the solution for a monodisperse polymer, the upper for one that is polydisperse. Agreement is good for the monodisperse prediction in Fig. 5. Additionally, the polydisperse result reasonably describes the behavior of PIB-4 and 5 shown in Fig. 4. In all cases, this model predicts a slightly slower (approximately $\frac{1}{2}$ decade in logarithmic time) increase in G_a with contact time than that experimentally observed.

The above two models both assume diffusion is the dominant process in time-dependent autohesion. We now turn our attention to a model which assumes wetting is the rate controlling mechanism. Korenevskaya and coworkers [27] studied the growth of interfacial contact area between bonded interfaces with time by direct measurement using an optical distortion device of their own design. The first-order kinetics of this process was later [43] found to follow

$$\phi + \ln\,(1 - \phi) = -Pt_c/\eta_0, \tag{8}$$

where ϕ is the fractional interfacial contact area and P is the applied pressure or load. By itself, this equation does not estimate the effect of increased contact area on fracture resistance. However, if we define the fractional fracture energy, G_f, as $G_f = G_a/G_\infty$, and assume this normalized surface energy closely follows ϕ, (8) may be rewritten as

$$G_f + \ln (1 - G_f) = -Pt_c/\eta_0. \tag{9}$$

Application of this later equation to the autohesion master curve for PIB-3 is illustrated in Fig. 5. As can be seen, agreement is excellent except at times just shorter than t_∞. Similar agreement occurred for PIB-1 and 2; however, (9) deviated significantly to lower G_a values at the shortest times for PIB-4 and 5.

In addition to the three models just discussed, four others are known to have been developed to describe the phenomena experimentally observed in this work. Kelley [20] appears to have been the earliest to adapt the first-order kinetics of wetting to autohesion data. His equation is very similar to (9). Vasenin [44] attempted to quantify the diffusion mechanism using Fick's second law in his "momentum exchange" model. The model predicts a $\frac{1}{2}$ power law as does (5) and relies upon a similar physical picture, but does not have nearly the range of applicability and, therefore, testability as the reptation models. A more rigorously developed wetting model was proposed by Anand and coworkers [45, 46]. Using viscoelastic models, this more complex approach describes behavior much the same as (9), with the notable exception that it also predicts an inflection point in the curve. Careful review of the experimental data collected in this study showed no strong evidence for an inflection point; however, PIB-1,2 and HTPB data at the shortest times did suggest a possible inflection point. More experimental data in this region, which is difficult to obtain since the peeling force begins to approach the weight of the sample, would be required to draw a firm conclusion. Finally, Campion [47] has proposed a diffusion-based model using free volume arguments. Rather than predict time-dependent autohesion; however, the purpose of his model is to describe which polymeric structures will display the best auto-hesive strength.

3.3. Effect of contact pressure

As discussed in the Introduction, if autohesion is primarily controlled by a diffusion mechanism, then the process should be essentially independent of pressure. This is not experimentally observed in Fig. 7. Figure 7 presents the autohesion master curves for PIB-2 surfaces contacted under progressively larger applied pressures of 0.3, 3, and 30 kPa. The middle of these three is the reference contact condition discussed previously. All three curves were shifted according to the same WLF coefficients. As pressure or load increases, G_a data appears to shift to shorter times, and t_∞ decreases with no discernible effect on G_∞.

Since all three curves are roughly parallel, this suggests the possibility of transposition along the time axis. Equation (9) predicts just such a transposition which is tested in Fig. 8. In this figure, the time axis has been normalized for applied pressure according to (9) by adding $\log (P/P_0)$, where P_0 equals the reference condition. As can be seen, all three curves reasonably superpose, with the possible exception of the low pressure data at short times. Additional experimentation, using different materials and geometries, would be required before general applicability could be considered; however, these data strongly suggest a contact-controlled mechanism for bond formation.

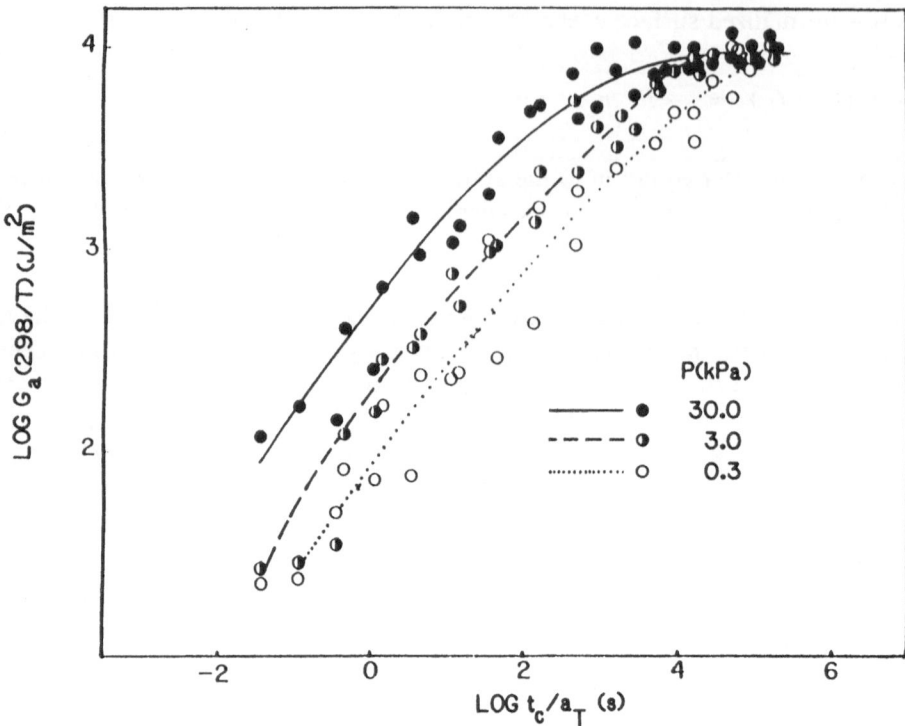

Fig. 7. Effect of contact pressure on the autohesion curve for PIB-2.

3.4. Autohesion in polymer networks

In his summary of factors which affect the autohesive behavior of polymers, Voyutskii [1] stated that crosslinking inhibits bond strength development. Two crosslinked networks were studied as part of this effort. Figure 9 shows autohesion master curves for these two materials, CTPIB and HTPB. Although development of bond strength appears somewhat slower in both cases, specific features appear similar to those previously observed in the PIB series. Of the two, fracture resistance rises earlier in CTPIB. But, instead of displaying an equilibrium at long times, this material shows evidence of a continuing autohesion process. Whether this nonequilibrium is an indication of a long term relaxation process or some other mechanism will require further study.

Application of the diffusion models to the CTPIB and HTPB data is inappropriate since the longest possible chain strand is significantly less than M_e. Consequently, no physical entanglements of long chain molecules can form across the interface. Fusion of the surfaces, or randomization, does not occur. The mechanism of contact area formation is presumed, therefore, to be the physical process of interest. Figure 9 illustrates that the first-order kinetics of this process, as embodied in (9), provide a reasonable prediction. Although the fit of theory to experiment is not as good as those obtained for the uncrosslinked systems, the theoretical curves never deviate, at constant t_c, from actual G_a values by more than 25 percent. One possible explanation for this deviation is the method used to determine η_0. Flow for these crosslinked networks is quite small on the time scale of the elongational viscosity

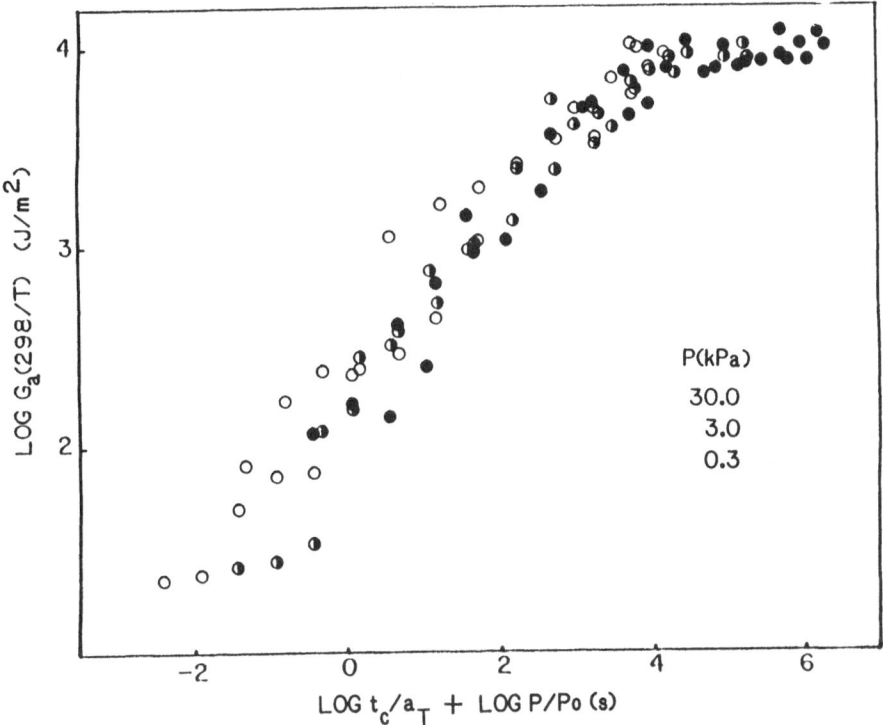

Fig. 8. Data from Fig. 7 replotted by compensating for the different pressures along the contact time axis, according to the wetting model.

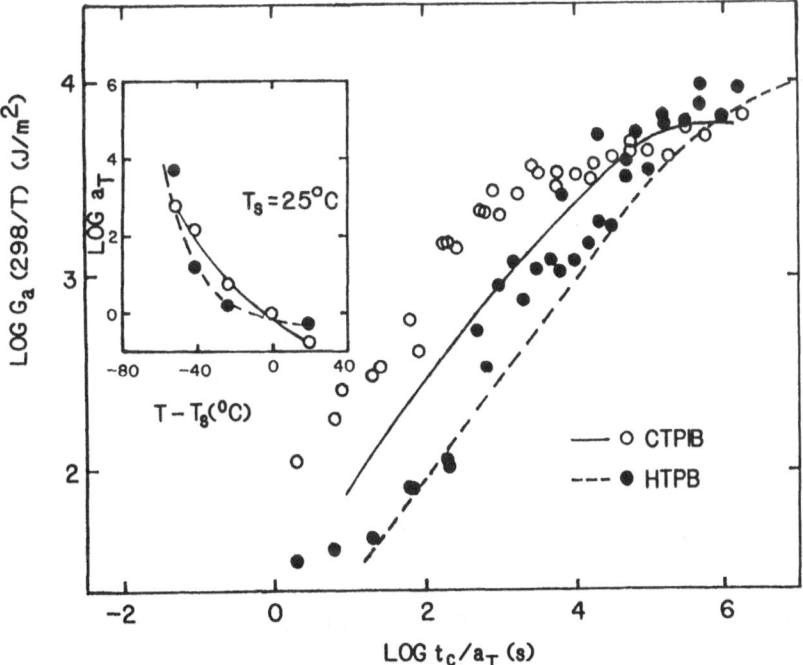

Fig. 9. Autohesion master curves for CTPIB and HTPB in comparison with those predicted from wetting model.

experiment used, and viscosities of 10^9 Pa-s approach the limit of measurability by this method [48].

It is interesting to note that both networks showed the same two orders of magnitude increase in G_a with contact time as did the uncrosslinked gums. Since no mass transfer across the interface is believed to have occurred, and the same degree of autohesion observed, it does not appear necessary to invoke a diffusion controlled mechanism to explain any of the data collected in this study. A possible exception to this could be the aforementioned WLF shift coefficients. Why the two diffusion models discussed in this paper provide such excellent predictions in specific cases should be a topic for further investigation. One possible explanation is the strong relationship between viscosity and self-diffusion [49]. This relationship suggests that (6) and (7) may inadvertently model aspects of viscous flow, in addition to the diffusion process for which they were derived.

4. Conclusions

The following conclusions are obtained:

(a) Autohesive fracture energy versus contact time data collected at various contact temperatures may be superposed into a master curve by the method of reduced variables. This approach allows the autohesion phenomena to be studied over a much broader experimental time scale than that previously achieved. For uncrosslinked polyisobutylenes of different molecular weights, the experimentally determined shift coefficients indicated greater molecular mobility than those in the literature obtained by shifting small deformation response data.

(b) Autohesion master curves developed in this fashion showed fracture resistance increasing with time along a logarithmic-type curve with monotonically decreasing slope. At the shortest times, this curve asymptomatically approached a slope of $\frac{1}{2}$. These data indicate that the $\frac{1}{2}$ power law dependency for autohesion, generally accepted in the literature, only applies over a narrow range of contact times.

(c) Various theories were compared with the experimental results. The Wool diffusion-based model gave an excellent prediction for time-to-equilibrium as a function of molecular weight, but its scaling laws for time-dependency and equilibrium fracture energy gave poor fits to the data. An alternate diffusion-based model of Prager and Tirrell provided good descriptions of time-dependent autohesion data for both monodisperse and polydisperse molecular weight polymers. In general, this model predicted a slightly slower, by approximately $\frac{1}{2}$ decade in logarithmic time, increase in fracture energy with contact time than that experimentally observed. The best description of the data was given by a first-order kinetic equation derived to account for contact-area formation with time.

(d) Joints contacted at different pressures showed an expected shift to shorter times at higher pressures. It was found that this shift could be quantitatively accounted for by first-order wetting kinetics.

(e) Data obtained in this study can be explained in terms of a contact-controlled mechanism for autohesion. Evidence for this mechanism is the pronounced effects of contact pressure on autohesion and the identical time-dependent behavior of nondiffusing crosslinked networks when compared with systems containing mobile polymeric chains. Neither of these features are consistent with a diffusion-based model.

Acknowledgements

The authors would like to acknowledge the efforts of W. Geisler, D. Foxx and R.A. Wurzbach for their contributions to the material characterization efforts. Thanks are also due to T.P. Rudy of United Technologies/Chemical Systems Division for providing the CTPIB and MAPO samples.

References

1. S.S. Voyutskii, *Autohesion and Adhesion of High Polymers*, Interscience, New York (1963).
2. G.R. Hamed, *Rubber Chemistry and Technology* 54 (1981) 576–595.
3. S. Wu, *Polymer Interface and Adhesion*, Dekker, New York (1982).
4. S.S. Voyutskii and B.V. Shtarkh, *Rubber Chemistry and Technology* 30 (1957) 548–553.
5. W.G. Forbes and L.A. McLeod, *Transactions of the Institute of the Rubber Industry* 34 (1958) 154–184.
6. K. Jud, H.H. Kausch and J.G. Williams, *Journal of Material Science* 16 (1981) 204–214.
7. R.P. Wool and K.M. O'Connor, *Journal of Applied Physics* 52 (1981) 5953–5963.
8. R.P. Wool and K.M. O'Connor, *Journal of Polymer Science: Polymer Letters Edition* 20 (1982) 7–16.
9. T.Q. Nguyen, H.H. Kausch, H. Jud and M. Dettenmaier, *Polymer* 23 (1982) 1305–1321.
10. R.P. Wool, *Rubber Chemistry and Technology* 57 (1984) 307–319.
11. H.H. Kausch, D. Petrovska, R.F. Landel and L. Monnerie, *Polymer Science and Engineering* 27 (1987) 149–154.
12. S.S. Voyutskii, *Rubber Chemistry and Technology* 33 (1960) 748–755.
13. S.S. Voyutskii and V.L. Vakula, *Journal of Applied Polymer Science* 7 (1963) 475–491.
14. P.G. de Gennes, *Comptes Readu Academy of Sciences (Paris)*, Series B, 291 (1980) 219–223.
15. P.G. de Gennes, *Journal of Chemical Physics* 55 (1971) 572–579.
16. S. Prager and M. Tirrell, *Journal of Chemical Physics* 75 (1981) 5194–5198.
17. S. Prager, D. Adolf and M. Tirrell, *Journal of Chemical Physics* 78 (1983) 7015–7016.
18. D. Adolf, M. Tirrell and S. Prager, *Journal of Polymer Science: Polymer Physics Edition* 23 (1985) 413–427.
19. Y.H. Kim and R.P. Wool, *Macromolecules* 16 (1983) 1115–1120.
20. F.N. Kelley, PhD dissertation, University of Akron (1961).
21. D.H. Kaelble, in *Treatise on Adhesion and Adhesives*, R.L. Patrick (ed.), Dekker, New York (1967).
22. D.H. Kaelble, *Journal of Adhesion* 1 (1969) 102–123.
23. D.H. Kaeble, *Journal of Macromolecular Science – Reviews in Macromolecular Chemistry* C6 (1971) 85–112.
24. J.N. Anand, *Journal of Adhesion* 5 (1973) 265–275.
25. G.R. Hamed, *Rubber Chemistry and Technology* 54 (1981) 403–414.
26. G.R. Hamed, *Rubber Chemistry and Technology* 55 (1982) 1469–1481.
27. N.S. Korenevskaya, V.V. Laurent'ev, S.M. Yagnyatinskaya, V.G. Rayevskii and S.S. Voyutskii, *Polymer Science USSR* 8 (1966) 1372–1377.
28. J.D. Ferry, *Viscoelastic Properties of Polymers, Third Edition*, Wiley, New York (1980).
29. R.W. Fillers and N.M. Tschoegl, *Transactions of the Society of Rheology* 21 (1977) 51–100.
30. W.K. Moonan and N.W. Tschoegl, *Macromolecules* 16 (1983) 55–59.
31. W.K. Moonan and N.W. Tschoegl, *Journal of Polymer Science: Polymer Physics Edition* 23 (1985) 623–651.
32. E.A. Meinecke, *Rubber Chemistry and Technology* 53 (1980) 1145–1159.
33. R.S. Rivlin and A.G. Thomas, *Journal of Polymer Science* 10 (1953) 291–318.
34. A.N. Gent and A.J. Kinloch, *Journal of Polymer Science: Polymer Physics Edition* 9 (1971) 659–668.
35. A.N. Gent, *Rubber Chemistry and Technology* 47 (1974) 202–212.
36. E.H. Andrews and A.J. Kinloch, *Proceedings of the Royal Society London* A332 (1973) 385–399.
37. E.H. Andrews and A.J. Kinloch, *Proceedings of the Royal Society London* A332 (1973) 401–414.
38. R.G. Stacer, D.M. Husband, and H.L. Stacer, *Rubber Chemistry and Technology* 60 (1987) 227–244.
39. M.L. Williams, R.F. Landel and J.D. Ferry, *Journal of the American Chemical Society* 77 (1955) 3701–3706.
40. P.J. Flory, *Statistical Mechanics of Chain Molecules*, Wiley, New York (1969).
41. M. Tirrell, *Rubber Chemistry and Technology* 57 (1984) 523–556.
42. D.S. Pearson, *Rubber Chemistry and Technology* 60 (1987) 439–496.

43. L.F. Plisko, V.V. Laurentyev, V.L. Vakula, and S.S. Voyutskii, *Polymer Science USSR* 14 (1972) 2501–2506.
44. R.M. Vasenin, *Adhesives Age* 8 (1965) 18–29.
45. J.N. Anand and R.Z. Balwinski, *Journal of Adhesion* 1 (1969) 24–30.
46. J.N. Anand and L. Dipsinski, *Journal of Adhesion* 2 (1970) 16–22.
47. R.P. Campion, *Journal of Adhesion* 7 (1974) 1–23.
48. A.Ya. Malkin, in *Experimental Methods of Polymer Physics*, A.Ya. Malkin (ed.), Prentice-Hall, Englewood Cliffs, New Jersey (1983).
49. F. Bueche, *Physical Properties of Polymer*, Wiley, New York (1962).

Résumé. On a mené une étude sur la relation liant la structure d'un polymère et de l'auto-adhésion dépendant du temps, mesurée en termes de l'énergie de rupture d'auto-adhésion Ga. En utilisant la méthode des réduites, on trouve que les donées relatives à Ga exprimées en fonction de la durée du contact et de la température, peuvent être superposées à des courbes directrices liant la température et les durées de contact réduites. Les courbes directrices d'auto-adhésion développées par cette voie montrent que la résistance à la rupture augmente avec le temps selon loi de type logarithmique, avec une pente à décroissance régulière. Ces données indiquent que la loi de puissance 1/2 qui est généralement acceptée pour l'auto-adhésion ne s'applique que sur une plage de durées de contact relativement étroite. Pour tenir compte de la formation de surfaces de contact, on a accompli une modélisation des données expérimentales en utilisant une équation cinétique du premier ordre. Deux modèles basés sur la diffusion fournissent également de bonne prédictions pour des cas spécifiques, et plus particulièrement pour traiter le problème de l'effet du poids moléculaire sur la durée pour atteindre un équilibre. Toutefois, il est évident que la diffusion n'est pas le processus contrôlant la vitesse. Ceci transparaît par les effets prononcés de la pression de contact sur l'auto-adhésion, et sur le comportement identique par rapport au temps de réseaux à liaison crousées non sujets à la diffusion et de systèmes comportant des chaînes polymères multiples.

International Journal of Fracture 39: 217–234 (1989)
© Kluwer Academic Publishers, Dordrecht

A fracture analysis of cathodic delamination in rubber to metal bonds

K.M. LIECHTI, E.B. BECKER, C. LIN and T.H. MILLER
Department of Aerospace Engineering and Engineering Mechanics, The University of Texas at Austin, Austin, Texas 78712, USA

Received 1 May 1988; accepted in revised form 20 June 1988

Abstract. The purpose of this paper is to provide a method of characterizing the resistance of various adhesives and surface preparations to cathodic delamination using a fracture mechanics approach. The approach is quite similar to that taken in the stress corrosion cracking of metals although no direct accounting is made of the diffusion processes that are involved. The main emphasis is on the large deformation analysis, including the nonlinear material behavior of the rubber, that is required for determining fracture parameters for correlation with debond growth rates.

1. Introduction

There are a number of cases in the shipbuilding and oil industries where rubber is bonded to metal in order to inhibit corrosion or to seal components from seawater intrusion. In some applications the bonds are subjected to mechanical loading which, in a dry environment, is not sufficiently severe to cause debonding. However, the durability of the bond can be a problem in the seawater environment [1] and can be further exacerbated if a cathodic potential exists due to the galvanic action of dissimilar metals. Such a condition arises when zinc is used as a sacrificial anode on mild steel structures. Although the zinc protects the steel from corrosion, the reduction of dissolved oxygen or water in the presence of free electrons can occur at the metal oxide/adhesive interface, thereby increasing the concentration of hydroxide ions which in turn degrades the bond through mechanisms that are not clearly understood at this time. The purpose of this paper is to provide a method of characterizing the resistance of various adhesives and surface preparations to cathodic delamination using a fracture mechanics approach. The approach is quite similar to that taken in the stress corrosion cracking of metals although no direct accounting [2] is made of the diffusion processes that are involved. The main emphasis is on the large deformation analysis, including the nonlinear material behavior of the rubber, that is required for determining fracture parameters for correlation with debond growth rates.

2. Analytical considerations

The finite element calculations of the finite elastic deformations and energy release rates made use of algorithms based on the existence of a strain energy density function, U. These algorithms are quite general, being readily adapted to any particular form of the energy function, and have been implemented in the finite element code TEXPAC-NL, [3]. The following abbreviated description contains the essence of the development.

Let X and x denote the position of a material particle in the undeformed and the deformed configurations respectively. The deformation gradient, F, contains all information about the deformation from undeformed to deformed configurations. The tensor, F, is calculated as

$$F = \frac{\partial x}{\partial X} \tag{1}$$

and the deformation tensor, B, which is used as the argument in elastic constitutive relations is

$$B = FF^T. \tag{2}$$

For isotropic hyperelastic materials, the energy function must be expressible in terms of the three principal invariants of B, namely

$$I_1 = \text{tr } B$$

$$I_2 = \tfrac{1}{2}((\text{tr } B)^2 + \text{tr } B^2) \tag{3}$$

$$I_3 = \det B.$$

These invariants are the coefficients in the characteristic polynomial of B whose roots, the principal values or eigenvalues of B, are the squares of the principal stretches, λ_i.

Various algebraic forms have been used to curve fit the dependence of the strain energy density on the invariants and, alternatively, on the principal stretches. One, two and multiple term polynomials in the I_i have been popularized as Neohookean, Mooney–Rivlin [4], and Rivlin polynomial forms respectively. Valanis and Landel [5] proposed a simplifying assumption which allows the energy function to be fitted using only a single scalar function of one variable, $W(\lambda)$. Ogden [6] has given a simple form of the function W that agrees well with material tests. Peng [7] has shown that values of the important part of W, namely its derivative, W', can be determined directly from test data without the need to assume a functional form. This last alternative is attractive from a computational point of view in that the values of W' can be input directly into the finite element code. As a matter of choice one has either

$$U = U(I_i)$$

or

$$U = U(\lambda_i). \tag{4}$$

For conservative loadings the potential energy of the body is

$$\pi = \int_V U \, dV - \int_V f(x - X) \, dV - \int_S t(x - X) \, dS \tag{5}$$

in which V and S are the undeformed volume and surface of the body and f and t are applied body forces and surface tractions respectively.

The finite element scheme uses isoparametric elements in which we take

$$
\begin{aligned}
X &= \phi(\xi)X \\
x &= \phi(\xi)x.
\end{aligned}
\tag{6}
$$

In (6) ϕ is an array of known shape functions, ξ are the coordinates of the material particle in a reference configuration (parent element); while X and x are arrays of nodal point values which define the shape of the body in the undeformed and deformed states respectively. It follows from (6) and (1) through (5), respectively, that F, B, I_i, (and therefore λ_i), U and π are all functions of the two sets of nodal point values X and x. The functional dependence is shown explicitly as

$$
\pi(X, x) = \int_{V(x)} U(X, x)\,dV - \int_{V(x)} f(x - X)\,dV - \int_{S(x)} t(x - X)\,dS.
\tag{7}
$$

The calculation of the deformed equilibrium state due to the loads f and t is, according to the principle of stationary potential energy (or that of virtual work), find x so that

$$
\frac{\partial \pi}{\partial x}(X, x) = 0.
\tag{8}
$$

The nonlinear system (8) is solved with the use of Newton's method, each iteration of which requires the solution of the linear system

$$
\frac{\partial^2 \pi}{\partial x^2}(X, x)\,dx = -\frac{\partial \pi}{\partial x}(X, x).
\tag{9}
$$

The calculation of the first and second derivatives of potential energy with respect to x is trivial except for the strain energy term. In the algorithms described here, explicit formulae have been developed and coded for the calculation of the derivatives $\partial I_i/\partial x$, $\partial^2 I_i/\partial x^2$ and $\partial \lambda_i/\partial I_j$, $\partial^2 \lambda_i/(\partial I_j \partial I_k)$. Thus, when a particular form of the energy is to be used only the derivatives of (4)

$$
\frac{\partial U}{\partial I_i}, \quad \frac{\partial^2 U}{\partial I_j \partial I_k}
$$

$$
\frac{\partial U}{\partial \lambda_i}, \quad \frac{\partial^2 U}{\partial \lambda_i \partial \lambda_j}
$$

must be determined. This scheme has proven effective in treating a variety of hyperelastic material models.

The energy release rate G is the rate of decrease in the potential energy with respect to increasing crack surface area. In the finite element algorithm, one chooses a pattern of changes to nodal point values, δX, which corresponds to an extension of the crack by an amount δl which causes an increase in surface area $\delta a = 2h\delta l$, for a crack front with width h. Denoting the pattern of nodal point increments by δX we have

$$G = \frac{1}{\delta a}\frac{\partial \pi}{\partial X}(X, x)\delta X. \tag{10}$$

The actual calculation of G employs a scheme similar to that used to calculate derivatives with respect to x. It should be noted that additional terms contribute to $\partial \pi/\partial X$ because of the dependence of the volume and surface (V and S) on the values of X. Although the perturbation of the original geometry will cause a change in the deformed as well as undeformed values, the rate of change of potential energy with respect to x vanishes according to (8).

The evaluation of energy release rate using (10) is a post processing operation in the finite element code. A patch of elements containing the crack tip is selected and a perturbation δX which extends the crack δl is defined. Figure 1 shows such a choice. In principle, the value

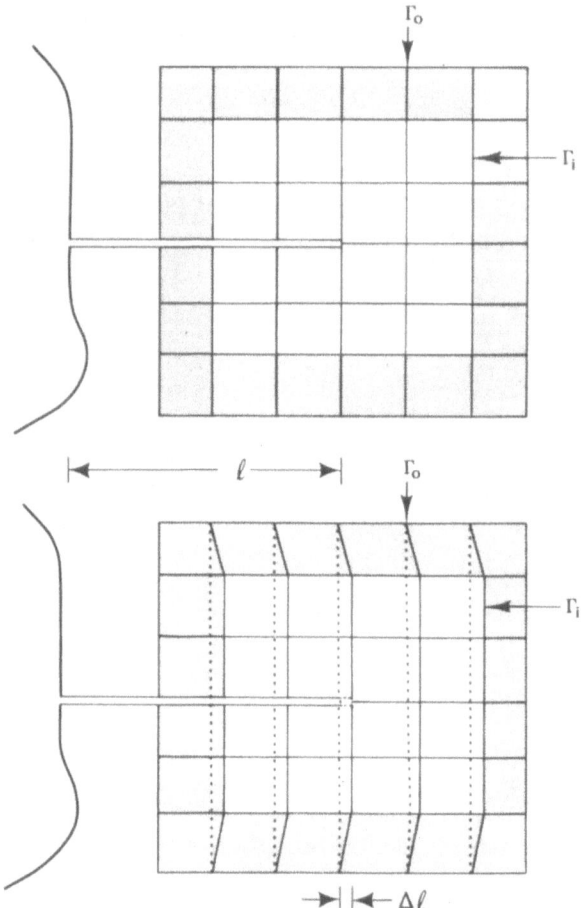

Fig. 1. Patch of elements used for G calculation.

of G that is calculated should be independent of the choice of δX. This "patch independence" is a consequence of the point wise satisfaction of equilibrium and is almost true for reasonably fine grids. It is significant to note that only elements on the boundaries of the path contribute to the calculation and that these can be well removed from the crack tip when large patches are chosen. Finally we note that the method described above is similar to some widely used methods – probably due initially to Parks. The present implementation is distinguished by using derivatives rather than differences so that the magnitude of δl does not affect the results.

3. Experimental procedures

The procedures described in this section were those that were developed for the characterization of the nonlinearly elastic response of Neoprene 5109S rubber and the cathodic delamination tests. The modeling of the rubber response was conducted on the basis of uniaxial and biaxial tests. The cathodic delamination tests were conducted at a fixed temperature and potential using the elastomeric equivalent of the Boeing wedge test – a strip blister test.

3.1. Rubber response

The models of the nonlinearly elastic response of the rubber that were considered here were those due to Mooney [4], Ogden [6] and Peng [7]. For the representations due to Mooney and Ogden (simplest) a uniaxial test is sufficient for the determination of the parameters of the models. Higher order Ogden models and the representation due to Peng require data from at least one biaxial test in addition to that obtained from the uniaxial test. The manner in which the uniaxial and biaxial data was obtained is described before deriving the parameters for each of the models.

The specimen geometry for the uniaxial tension test is shown in Fig. 2. The test was conducted at a constant grip displacement rate of 2.54 mm min^{-1}. The load P was measured with a load cell and recorded as a function of time. A video camera and recorder were used to track two lines that were painted on the specimen to define an initial gauge length of 6.53 mm. Following each test, the stretch was determined as a function of time using a digital image analysis system to automate the process. In one instance, the transverse stretch was also measured and the Neoprene appeared to be incompressible. The uniaxial true stress, σ, vs. stretch, λ, curve shown in Fig. 3 was therefore determined from incompressibility and the longitudinal stretch and synchronizing the load and stretch data.

The biaxial stress state that was used for characterizing the Neoprene was that produced by the strip biaxial geometry shown in Fig. 2. The Neoprene was bonded to the steel rails during vulcanization using the Chemlok 205-220 primer and adhesive system. The load, P, was applied perpendicular to the rails and parallel gauge marks were placed 6.35 mm apart in the center of the specimen. The true stress was taken to be

$$s = \frac{P\lambda}{\omega t},$$

(11)

where w is the initial width and t is the initial thickness.

Uniaxial

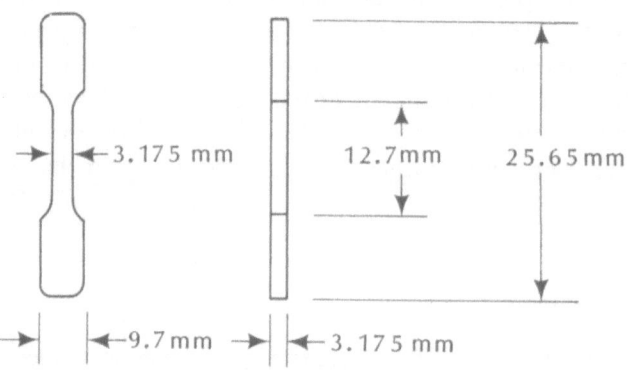

Strip Biaxial

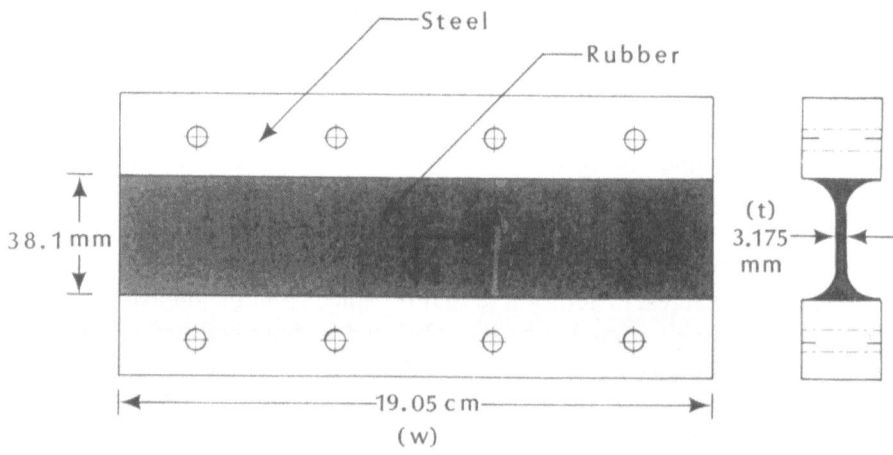

Fig. 2. Specimen geometries for the uniaxial tension and strip biaxial tests.

At first sight this might appear to be an improper true stress since no account is taken of the decrease in w. However, subsequent stress analysis determined that this measure of true stress was the same as the true stress at the specimen center due to the fact that the change in w is negligible and is not monotonic because w first decreases and then, for large stretches, increases. Some measurements also confirmed that the transverse stretch was unity in the central region of the specimen. The strip biaxial true stress stretch data thus determined are also shown in Fig. 3. The full lines in the plot are nine term polynomial fits that were used for interpolation purposes in obtaining the parameters of the models now described.

3.1.1. Mooney–Rivlin material
The strain energy density function, U, for a Mooney–Rivlin material is given by

$$U = C_1(I_1 - 3) + C_2(I_2 - 3), \tag{12}$$

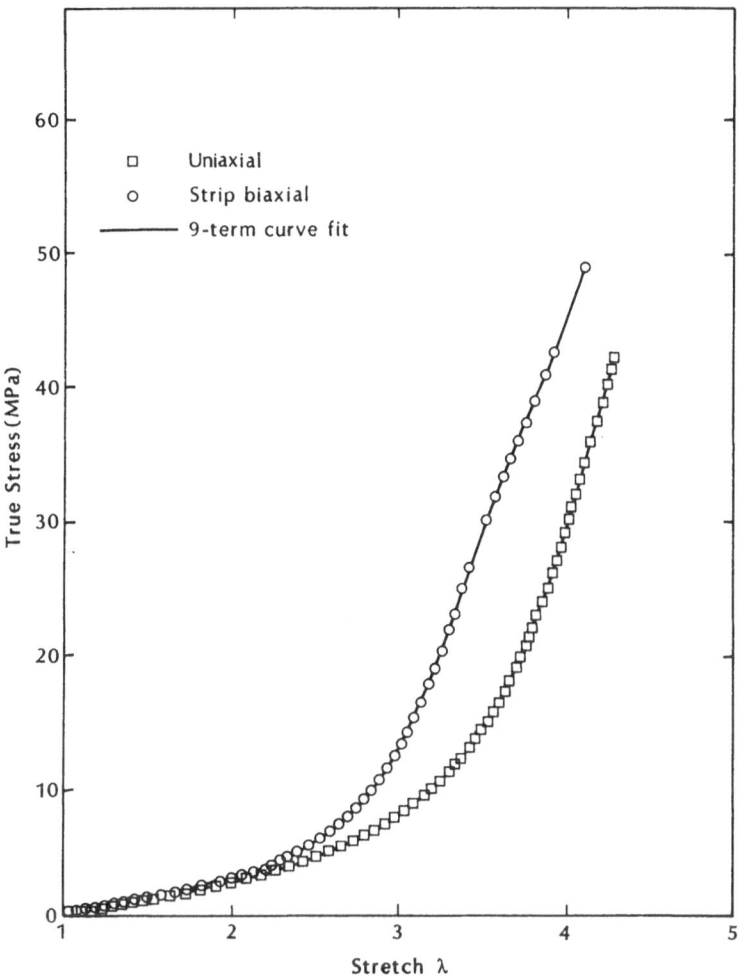

Fig. 3. Uniaxial and strip biaxial true stress stretch behavior of Neoprene 5109S.

where I_1 and I_2 are the first and second principal strain invariants

$$I_1 = \lambda_1^2 + \lambda_2^2 + \lambda_3^2 \tag{13}$$

$$I_2 = \lambda_1^2\lambda_2^2 + \lambda_2^2\lambda_3^2 + \lambda_3^2\lambda_1^2. \tag{14}$$

In a uniaxial tension test the true stress is given by

$$\sigma = 2\lambda^2(\lambda - 1/\lambda^2)(C_1 + C_2/\lambda). \tag{15}$$

The constants C_1 and C_2 were determined from the uniaxial test data plotted as $\bar{\sigma}$ vs. $1/\lambda$ where

$$\bar{\sigma} = \frac{\sigma}{2\lambda^2(\lambda - 1/\lambda^2)} = C_1 + C_2/\lambda. \tag{16}$$

224   *K.M. Liechti, E.B. Becker, C. Lin and T.H. Miller*

The plot was linear for $1 < \lambda < 2.25$ and it was in this range that C_1 and C_2 were found to be 192 and 311 kPa, respectively.

3.1.2. Ogden material
In the representation due to Ogden [6] the strain energy density function, U, depends directly on the principal stretches through

$$U = \mu_r \phi(\alpha_r) \quad \text{(summation on } r.) \quad r = 1, 2, 3, \ldots, \tag{17}$$

where

$$\phi(r) = [\lambda_1^{\alpha_r} + \lambda_2^{\alpha_r} + \lambda_3^{\alpha_r} - 3]/\alpha_r. \tag{18}$$

For simple tension

$$\sigma = \mu_r(\lambda^{\alpha_r} - \lambda^{-\alpha_r/2}). \tag{19}$$

Following the procedures outlined in [6], a two term summation was fit to the data to yield

r	μ_r	α_r
1	1.75 MPa	1.05
2	1.72 kPa	6.85

3.1.3. Peng material
For the Peng material [7], the strain energy density function was written in the form

$$U = \sum_{i=1}^{3} W(\lambda_i). \tag{20}$$

For uniaxial tension

$$\sigma(\lambda) = \lambda W'(\lambda) - \lambda^{-1/2} W'(\lambda^{-1/2}) \tag{21}$$

under strip biaxial tension ($\lambda_1 = \lambda, \lambda_2 = 1, \lambda_3 = 1/\lambda$), the stress, s, in the direction of the applied stretch is

$$s(\lambda) = \lambda W'(\lambda) - \lambda^{-1} W'(\lambda^{-1}). \tag{22}$$

The derivative of the strain energy function, $W'(\lambda)$, can be obtained from the uniaxial and strip biaxial data by developing a recursive relation which yields [7]

$$W'(\lambda) = \frac{1}{\lambda} \sum_{n=1}^{8} [\sigma(\lambda^{(1/2)^n}) - s(\lambda^{(1/2)^{n+1}})]. \tag{23}$$

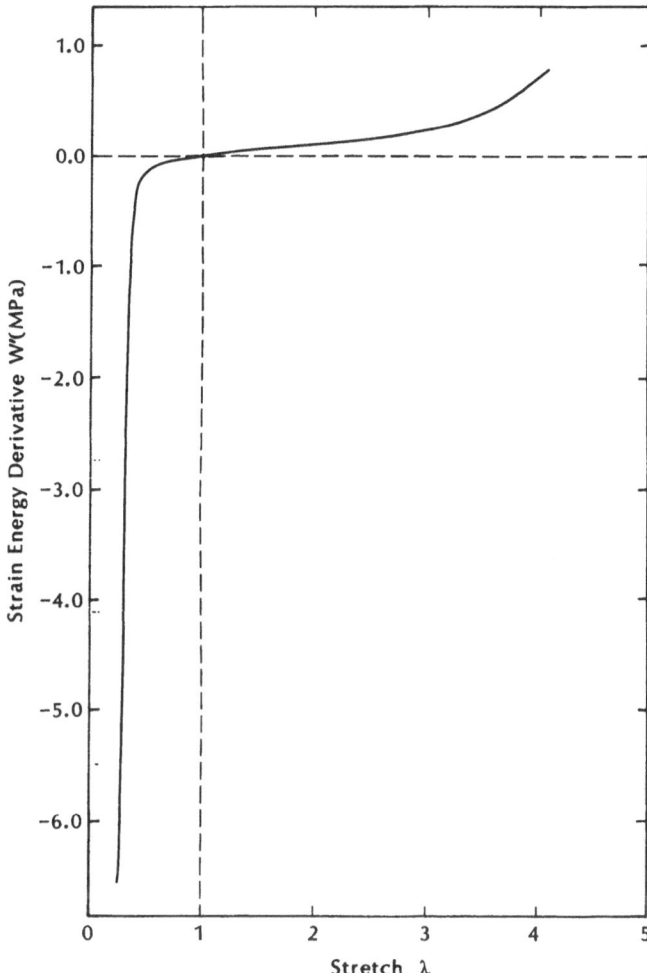

Fig. 4. Strain energy density derivative for the Peng model.

Values of W' for compressive stretches can be obtained from (22) and (23) through

$$W'(\lambda^{-1}) = \lambda[\lambda W'(\lambda) - s(\lambda)]. \qquad (24)$$

The results of the operations (23) and (24) are shown in Fig. 4. The curve is digitized and entered directly to the finite element code.

3.2. Cathodic delamination tests

Since the cathodic delamination tests were to be conducted in a special environment, a fairly compact and self-loading specimen was required. The double strip blister specimen shown in Fig. 5 was arrived at after several iterations [8] and is basically the same as one of the variants of the pressurized blister test discussed by Williams in [9]. For this investigation, Neoprene 5109S was bonded to an ANSI 1026 steel strip using the Chemlok 205-220 primer-adhesive system. Central 25 mm initial delaminations were provided by masking the

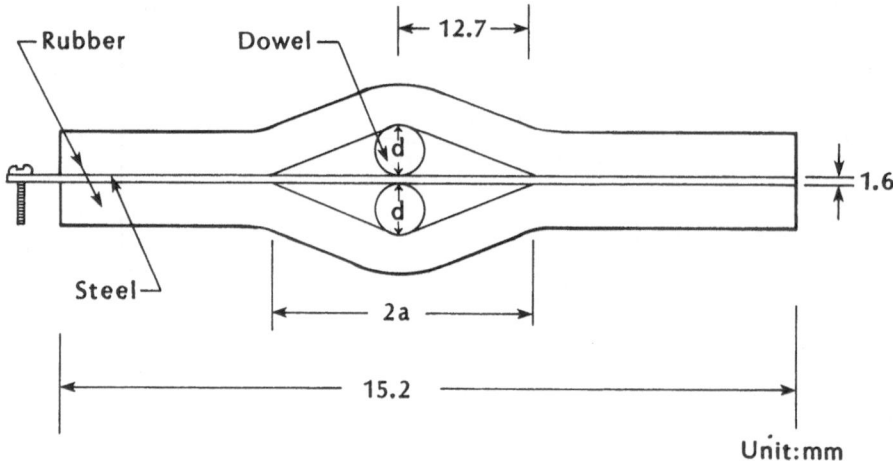

Fig. 5. Double strip blister durability specimen.

steel prior to the adhesive application. Following vulcanization, 9.5 mm dowels were inserted into the center of the initial delaminations and the specimen was then allowed to relax for several days prior to environmental exposure in order to minimize any viscoelastic effects. For the data to be reported here, the particular environment that was considered was a 1 N NaOH solution at 30°C. A potentiostat was used to maintain the steel at −0.9 V relative to a standard calomel electrode. The tank was sparged with air that had been passed through a CO_2 trap in order to prevent calcium carbonate from being produced. Significant levels of calcium carbonate had been found to be detrimental to any activity in the tank. The crack length, *a*, was measured periodically by taking the average of two edge measurements made using calipers. The caliper measurements were made more easily than those made using an optical comparator and revealed no significant deviations. The double strip blister provided four delamination fronts per specimen and eliminated the need to coat steel surfaces as would be the case for a single sided blister specimen. The consistency of the debond growth at the various debond fronts was taken as a check on the surface preparation, bonding procedures, etc.

4. Results

Having described the analytical experiment and procedures, we first examine the adequacy of the models for the nonlinearly elastic response of the Neoprene. Consideration is then given to the stress analysis of the double strip blister specimen which forms the basis for the correlations of debond growth rates with strain energy release rates.

4.1. Comparison of rubber material models

The first step that was taken was essentially a consistency check in that finite element simulations of the uniaxial test were made using the parameters derived for the Mooney–Rivlin, Ogden and Peng models. The results are shown in Fig. 6. The deviation of the Mooney–Rivlin material from the measurements for stretches greater than 2.25 was expected

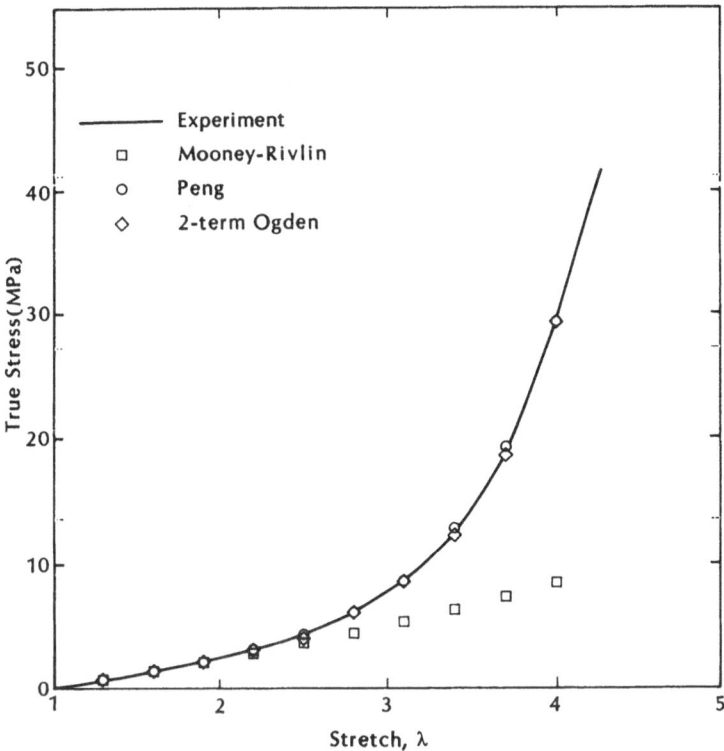

Fig. 6. Simulation of uniaxial test.

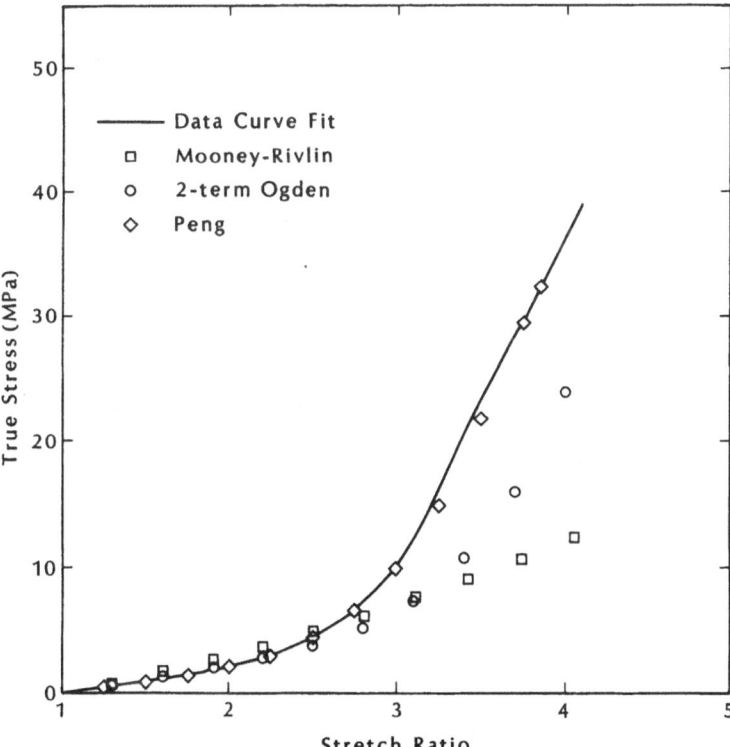

Fig. 7. Simulation of strip biaxial test.

since the parameters were only derived over that range. The two-term Ogden and Peng representations were consistent. The next step was to simulate the strip biaxial test (Fig. 7). The sharpest disagreement between predictions and measurements was again seen for the Mooney–Rivlin representation. The two-term Ogden model provided reasonable agreement up to stretches of 2.75. A third term would be required for better agreement thereafter. The representation due to Peng provided the best agreement with measurements. The agreement was, of course, more of a consistency check since both tests were used to generate $W'(\lambda)$ which was, in turn, used in the simulation.

4.2. Stress analysis of the double strip blister specimen

With some understanding for the characteristics of each material model in hand, the stress analysis of the double strip blister was undertaken. A quarter-symmetric mesh is shown in Fig. 8. The steel substrate was modeled as a rigid surface to which the rubber was directly bonded over the remaining ligament. The direct bonding assumption seemed to be a reasonable first step since the primer and adhesive were very thin (7.6 and 25.4 μm, respectively) and their moduli were at least two orders of magnitude greater than the initial modulus of the rubber. The dowel was modeled as five rigid and linear contact segments, each subtending an 18 degree angle at the dowel center. The 9.53 mm pushout provided in the cathodic delamination tests was applied by moving the (frictionless) contact segments against the rubber; specially reformulated elements were used to handle the incompressibility of the rubber. Fully nonlinear solutions were obtained for Peng and Mooney–Rivlin material representations since they gave the best and worst agreement in the strip biaxial simulation (Fig. 7). A small strain, linearly elastic (consistent initial modulus) solution subject to the same degree of dowel contact as the nonlinear solutions was also obtained. The predicted overall deformed shapes of the rubber crack face are shown in Fig. 9 where the three solutions were compared with measurements that were made along the specimen edge using an optical comparator.

When compared in this way, there are small differences between the solutions which predict larger gaps than the measured one over most of the debond length. The measurements indicate that the degree of sagging from just behind the blunted crack tip to the point of dowel contact should be greater. However, this is most likely a three-dimensional effect, not considered in the plane strain analyses. A more detailed comparison is offered in Fig. 10, where the crack tip region is considered. The two nonlinear solutions differ slightly but capture the blunting very close to the crack front quite well. The linear solution does not predict the blunting so well and predicts a consistently smaller gap than the nonlinear

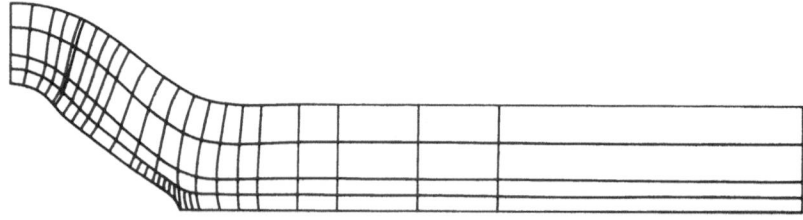

Fig. 8. Finite element mesh for a quarter-symmetric model of the double strip blister specimen.

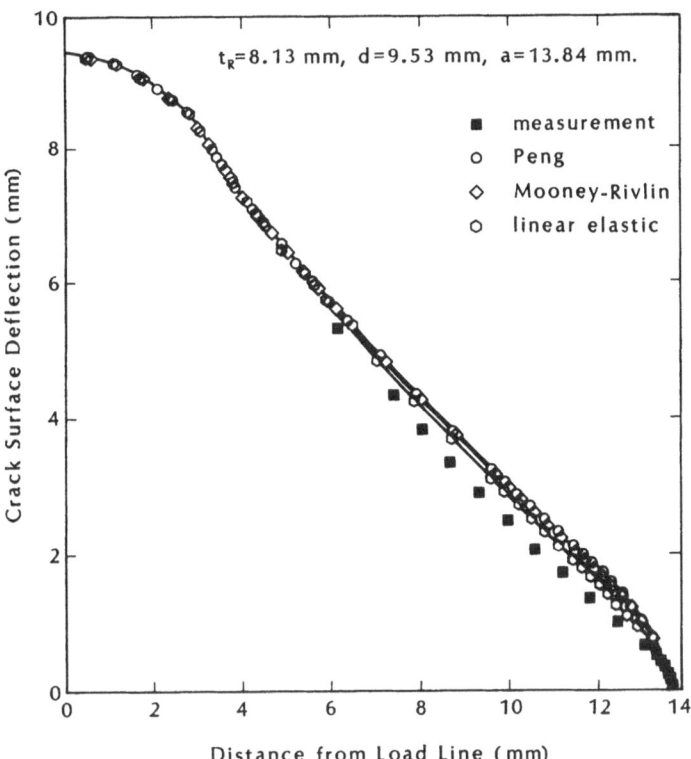

Fig. 9. Overall rubber crack face profile in a double strip blister specimen.

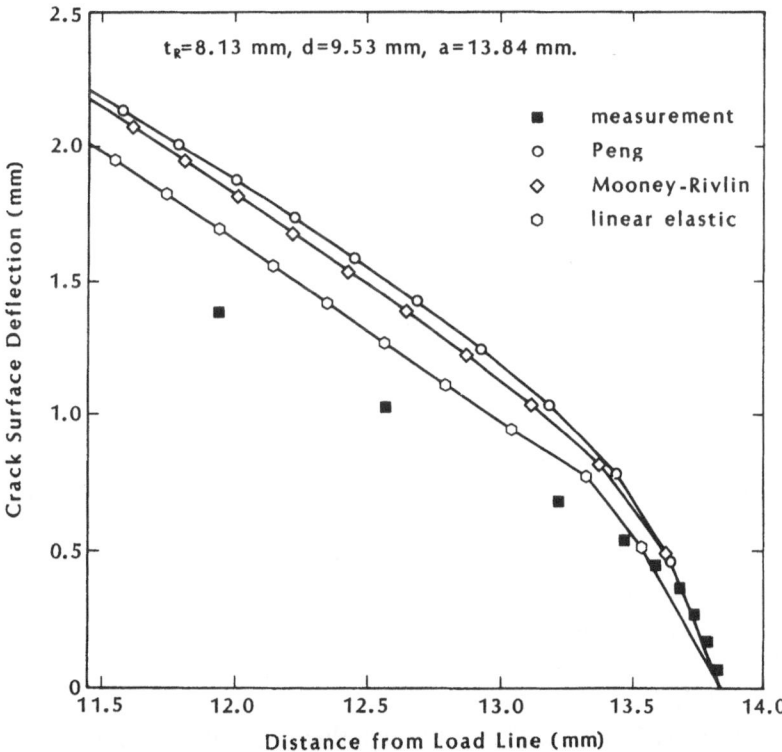

Fig. 10. Crack tip profile in a double strip blister pattern.

230   *K.M. Liechti, E.B. Becker, C. Lin and T.H. Miller*

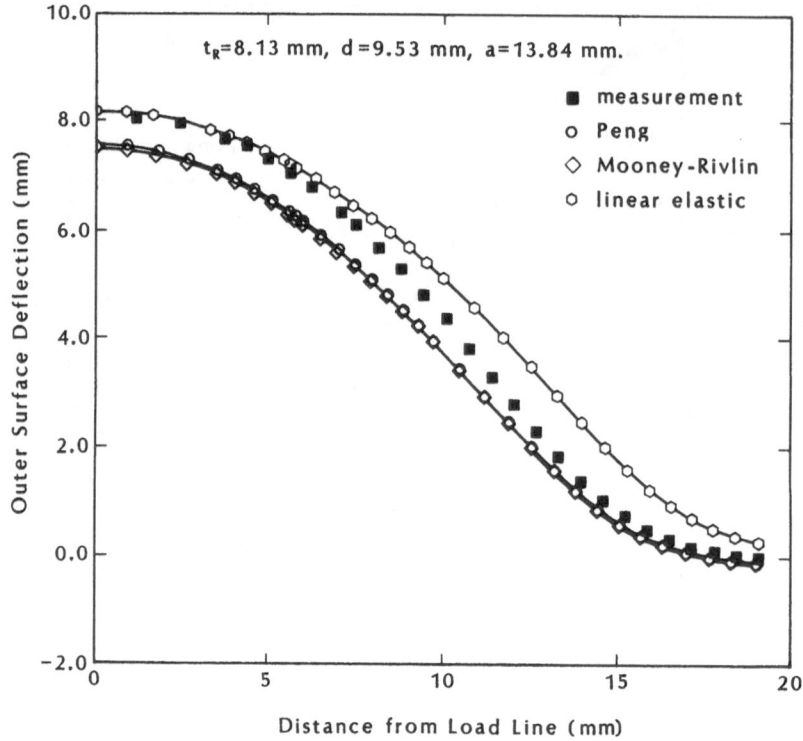

Fig. 11. Deformed shape of the rubber free surface in a double strip blister specimen.

solutions. Similar comparisons were made of the shape of the upper free surface of the rubber (Fig. 11). The two nonlinear solutions are again in very close agreement and are consistently lower than the measurements and the linear solution. The best agreement with the measurements is over the debond tip region. The reason for the disagreement above the dowel is a distinctly observable three-dimensional effect which leads to ridges being formed along the specimen edges over the region of dowel contact. Since the measurements are made along the specimen edge, the deflected shape of the rubber is exaggerated when compared to the plane strain solutions that reflect the behavior along the center of the specimen. The linear solution consistently underpredicts the degree of thinning of the rubber layer. Work is in progress to measure debond opening profiles and rubber free surface shapes across the specimen width using a projection moiré technique in order to assess the strength of three-dimensional effects. It has also been observed that the degree of normality of the debond front to the specimen edge has a noticeable effect on the debond opening profile.

Although the differences between the displacements examined were small, particularly between nonlinear solutions, there was a notable difference in the strain energy release rates that were obtained (Fig. 12) as a function of debond length. The greatest differences occurred for small debonds with the linear solution providing consistently higher *G* values. The values obtained for the Mooney–Rivlin material were always intermediate to those obtained from the linear solution and the solution for the Peng material. The total strain energy release rate derived from the solution due to the Peng material was up to 50 percent lower than that obtained from the linear solution. Thus, although there was little difference in crack opening profiles and rubber free surface displacements due to the constrained nature of the problem

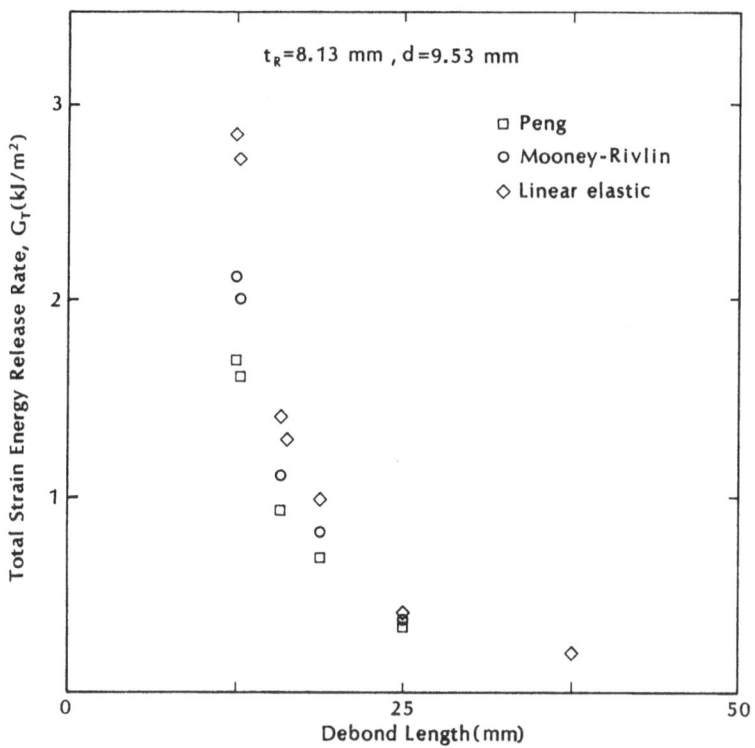

Fig. 12. Total strain energy release rate solutions for the double strip blister specimen.

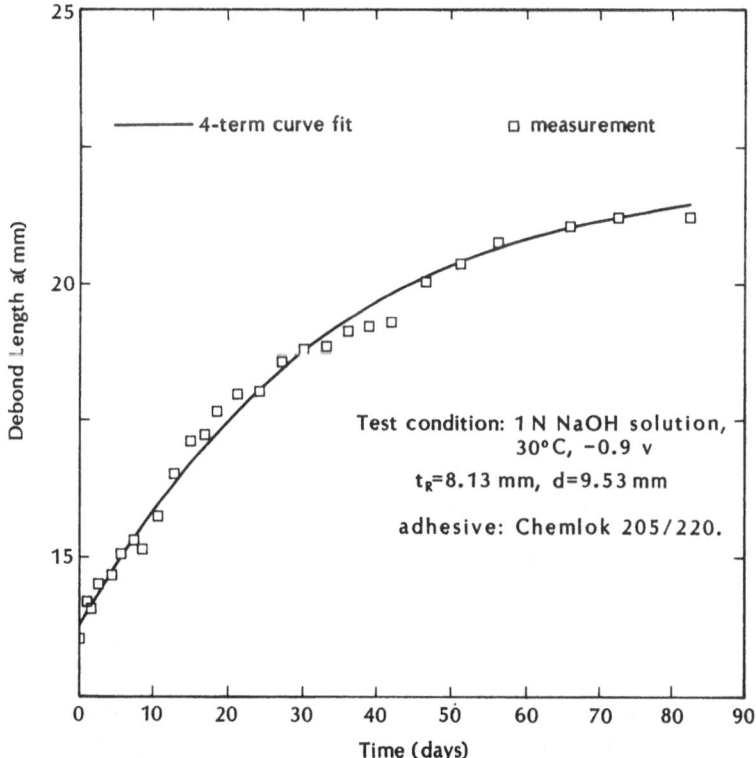

Fig. 13. Debond growth history in a double strip blister specimen.

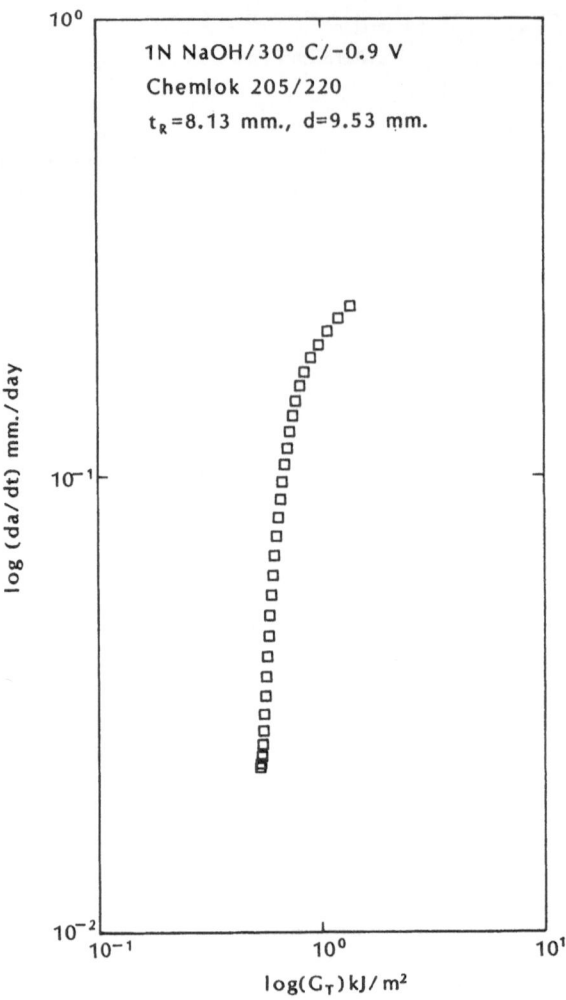

Fig. 14. Debond growth rate correlation.

and the incompressibility of the rubber, the stress fields due to the different material models are quite different.

4.3. Debond growth rate correlations

The debond history for a double strip blister subjected to a 9.53 mm pushout in a 1 N NaOH environment at 30°C with the steel potential at −0.9 V is shown in Fig. 13. The raw data is quite consistent but does display some "stair stepping", indicating periods of essentially zero debond growth followed by relatively rapid growth. In view of the desire to obtain debond growth rates, the data was first smoothed by fitting a four-term polynomial to it. The correlation of debond growth rates obtained from the polynomial and total strain release rates derived from the solution for the Peng material are shown in Fig. 14. There appears to be a threshold value between 0.3 and 0.5 kJ m^{-2} followed by a sharply rising curve where debonding is dominated by diffusion effects. At higher debond growth rates the slope of the correlation curve drops off to approach a region where debond growth rates are nearly independent of the energy release rate. The G values provided by this version of the specimen were not high enough to approach the adhesive fracture toughness of the bond in a dry environment which was approximately $10\,kJ\,m^{-2}$ [8]. The 1 N NaOH environment with applied potential clearly has a strong effect on the durability of the bond.

5. Conclusions

The resistance to cathodic delamination of rubber bonded to steel has been characterized using a fracture mechanics approach. The double strip blister specimen was convenient to use for bond durability tests and provided a suitable range of G values to establish the threshold limit and plateau regions that commonly occur in environmentally assisted cracking problems. Various models of nonlinearly elastic behavior of the rubber were considered in a finite deformation analysis of the double strip blister specimen. The representation due to Peng [6] was the most applicable of the models considered. The G value obtained from solutions due to Peng material were as much as 50 percent lower than fully linearized solutions. The threshold value of energy release rate for the Chemlok 205-220 primer adhesive system under the conditions tested was a factor of 30 less than the fracture toughness in a dry environment.

Acknowledgements

One of the authors (KML) would like to acknowledge the support of the Naval Research Laboratory and the Office of Naval Research through subcontracts let from Texas Research Institute. The respective institutional representatives were Drs R.W. Timme, L.R. Peebles and J.S. Thornton who have been very encouraging. The help of TRI personnel in preparing specimens and W. Adamjee in obtaining the stress/stretch data is also gratefully acknowledged.

References

1. A. Stevenson, *International Journal of Adhesion and Adhesives* 5 (1985) 81–91.
2. E.E. Gdoutos and E.C. Aifantis, *Engineering Fracture Mechanics* 23 (1986) 431–439.
3. E.B. Becker and T.H. Miller, "TEXPAC-NL: A Finite Element Code for Elastomeric Analysis," *Users Manual 1987.*
4. M. Mooney, *Journal of Applied Physics* 11 (1940) 582–592.
5. K.C. Valanis and R.F. Landel, *Journal of Applied Physics* 38 (1967) 2997–3002.
6. R.W. Ogden, *Proceedings Royal Society of London* A326 (1972) 565–584.
7. S.J. Peng, "Nonlinear Multiaxial Finite Deformation Investigation of Solid Propellant," *AFRPL-TR-84-036* (1984).
8. D.A. Dillard, K.M. Liechti, D.R. LeFebvre, C. Lin, J.S. Thornton and H.F. Brinson, *ASTM STP 1981* (1988).
9. M.L. Williams, *Journal of Applied Polymer Science* 13 (1969) 29–40.

Résumé. L'objet de cette étude est de proposer une méthode de caractérisation de la résistance de divers adhésifs et préparation de surface au décollement d'une couche de caoutchouc déposée sur un substrat métallique sous potentiel cathodique. L'approche est très voisine de celle adoptée en corrosion fissurante des métaux, bien que l'on ne tienne pas directement compte (Réf. 2) du processus de diffusion qui y est impliqué. L'accent principal est placé sur l'analyse des grandes déformations, prenant en compte le comportement non linéaire que présente le matériau caoutchouc, qui sont nécessaires à la détermination des paramètres de rupture, en vue d'une correlation avec la vitesse de croissance du décollement.

International Journal of Fracture 39: 235–253 (1989)
© Kluwer Academic Publishers, Dordrecht

A finite elastostatic analysis of bimaterial interface cracks

G.RAVICHANDRAN* and W.G. KNAUSS

*Graduate Aeronautical Laboratories, California Institute of Technology, Pasadena, California 91125, USA
(*now at Department of Applied Mechanics and Engineering Sciences, University of California at San Diego, La Jolla, California 92093, USA)*

Received 24 December 1987; accepted in revised form 1 July 1988

Abstract. A numerical method is developed to study the bimaterial interface problem in Neo-Hookean materials under plane stress conditions. Comparison is made with the analytical predictions for the asymptotic field of the problem. The range of dominance of the asymptotic solution at different load levels is established and the amplitudes of the crack-tip asymptotic field are related to the far field loading. The numerical model is extended to analyze the experiments conducted on specimens with an edge crack at the interface between two dissimilar Solithane plates that are characterized by Mooney–Rivlin material behavior.

1. Introduction

Because of their technological importance in many aspects of structural bonding interfacial cracks between dissimilar materials have received considerable attention. Applications include understanding of debonding between solid rocket propellant grains and the binder, delamination associated with composite structures, applications to adhesive bonding, and bonding between the substrate and coating in advanced materials such as ceramics. The stress analysis of such bimaterial interfaces will enhance the understanding of the relation between the loads on a structure and the prevailing stress field in the vicinity of stress concentrations or singularities. This will help to improve the structural integrity of components and prevent costly failures.

Analyses in the area of interfacial cracks have been mostly confined to the linearized theory of elasticity. Results based on the linear theory of elasticity (Williams [1]) for this problem exhibit oscillatory singularities for the asymptotic solution implying interpenetration of material in the vicinity of the crack tip which is physically inadmissible. Such a solution suggests that results based on the linear theory in order to assess the stress intensity in the vicinity of the crack tip might not yield an accurate picture in predicting failure. England [2] presented the solution for the problem of a uniformly pressurized interface crack of finite length between two dissimilar, linearly elastic semi-infinite slabs. Rice and Sih [3] have analyzed plane problems of cracks in dissimilar media (a center cracked panel) and determined the results for the problem such as the stress intensity factors and obtained the "bond" condition that establishes the relation between the loadings at infinity in order to satisfy the continuity of displacements and tractions along the interface. Knowles and Sternberg [4] analyzed the interface problem for joined half planes of distinct Neo-Hookean sheets and established asymptotic stress and displacement fields at the crack tip. The problem of interpenetration of material is not present when one allows for the large deformations near the tip of an interface crack between dissimilar Neo-Hookean sheets. In this paper we shall present full field solutions for this physically realistic model.

Recently Shih and Asaro [5] have carried out an elastic-plastic analysis of cracks in bimaterial interfaces under conditions of small scale yielding within the framework of infinitesimal deformations. They considered the plane strain problem of a center cracked panel with a material bonded to a rigid substrate. Once again it was found that the oscillatory behavior results in the vicinity of the crack tip under applied remote loads are small in relation to the yield stress of the material.

This paper focuses on the study of a bimaterial interface under *plane stress* conditions for Neo-Hookean solids. The displacement based finite element method is used to analyze the problem of a centrally cracked plate, made of two thin Neo-Hookean sheets, with a crack at the interface. The cracked panel is so large that it may be considered to be of infinite extent. The asymptotic results presented by Knowles and Sternberg [4] are then compared with the numerically obtained results and thus the size of the region of dominance of the asymptotic solution for increasing load level is established. Efforts have been made to understand the relation between the applied far field loading and the near-tip amplitude parameters of the Knowles–Sternberg asymptotic solution.

In the following sections, a brief description of the nonlinear theory of elastostatic plane stress and a summary of the asymptotic results presented by Knowles and Sternberg [4] are reviewed. Then the numerical procedure used to solve the boundary value problems is described. The finite element method is used in the formulation for the current or deformed configuration. In the subsequent section the computational model is presented and particular difficulties associated with the problem are discussed briefly.

The results are presented in the form of a comparison of the numerically obtained solution and the asymptotic predictions of Knowles and Sternberg [4]. The zone of dominance of the asymptotic solution increases with increasing load level. Several different cases are considered including the limiting cases of the homogeneous cracked sheet and the case when one material is rigid. The relation between the far field loading and the near-tip amplitude parameters is explored. The material models considered are the Neo-Hookean and Mooney–Rivlin material models. The results are summarized in the final section and related topics for further investigation are briefly discussed.

The current work is an effort to understand the behavior of material interfaces in connection with an experimental program on a model material (Solithane) substituting for soft materials used in solid rocket propellant designs. Solithane is a mechanically well characterized material [6] and it has been found that it may be reasonably well represented by a Mooney–Rivlin nonlinear elastic material under relaxed conditions (long time, rubbery state) and at suitable temperatures. From the numerical results synthetic caustic patterns are generated and compared with those obtained experimentally. Efforts are underway to establish a failure criterion (crack propagation path) for the bimaterial interface problems and understand the relation between the applied loading and the prevailing near-tip mechanical field.

2. Finite elastostatic plane stress

The elastic potential U for the incompressible Neo-Hookean material is of the form

$$U = \tfrac{1}{2}\mu(I_1 - 3), \tag{2.1}$$

where μ is the Neo-Hookean modulus which equals the shear modulus for infinitesimal deformations. I_1, I_2, I_3 are the principal invariations of the left deformation tensor **G** defined by

$$\mathbf{G} = \mathbf{FF}^\mathrm{T} \tag{2.2}$$

and where **F** is the deformation gradient,

$$\mathbf{F} = \nabla \mathbf{y}, \tag{2.3}$$

with "**y**" as the deformed coordinate in the current configuration and is related to the material coordinate **x** in the undeformed configuration through the relation

$$\mathbf{y} = \mathbf{x} + \mathbf{u}, \tag{2.4}$$

with **u** the displacement vector in the deformation.

The Cauchy stress tensor τ is then given by

$$\tau_{ij} = \mu G_{ij} - p\delta_{ij}, \tag{2.5}$$

where p is an arbitrary pressure arising from the incompressibility condition $I_3 = 1$. If one assumes generalized plane stress, i.e. $\tau_{33} \equiv 0$, then (2.5) can be rewritten as

$$\tau_{ij} = \mu(G_{ij} - \lambda^2 \delta_{ij}), \tag{2.6}$$

where λ is the transverse stretch or the ratio of current thickness of the transverse section to the original thickness.

3. Asymptotic results

Consider a crack of length $2l$ between two homogeneous and isotropic semi-infinite slabs of possibly distinct material properties, as shown in Fig. 1. Let H_1 and H_2 be the half planes $x_2 > 0$ and $x_2 < 0$ as shown in Fig. 1. The solution for the case of two distinct linearly elastic materials forming the half spaces has been presented by England [2] who found that the crack tip singularities were oscillatory; this result implied that overlapping of the crack faces occurred and he therefore rejected this result as physically inadmissible. It can be shown that this kind of interpenetration occurs over a distance δ, with $\delta/l < 4 \times 10^{-8}$ for all physically realistic values of elastic constants [4]. In the limiting case of two distinct incompressible slabs the infinitesimal theory of elasticity leads to singularities that are no longer oscillatory under plane strain conditions but continue to be so under plane stress conditions. Knowles and Sternberg [4] have analyzed asymptotically the problem of a bimaterial interface crack with the two halves of the material being charcterized by a Neo-Hookean material model under plane stress conditions.

For the interface problem the moduli for the two half planes are given by μ_1 and μ_2 and the ratio of the moduli is denoted by $s = \mu_1/\mu_2$. Let r and θ be the material polar coordinates

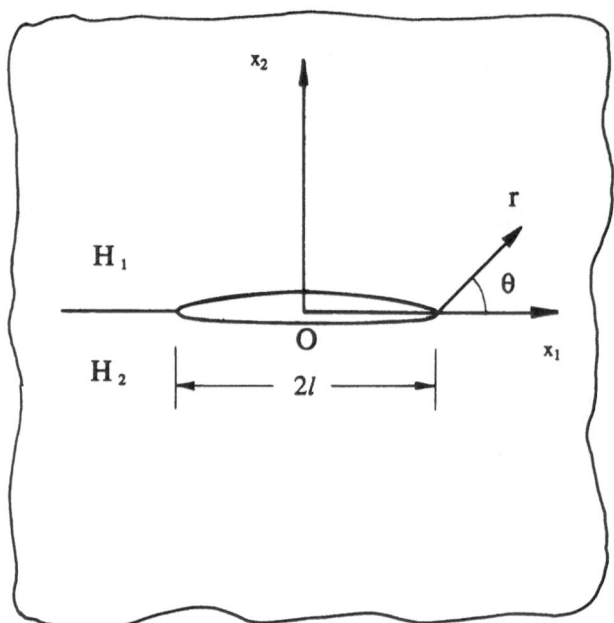

Fig. 1. Interface crack between two semi-infinite dissimilar slabs.

in the undeformed configuration. Also, let y_1 and y_2 be the coordinates in the deformed configuration for a material point x_1, x_2 in the undeformed configuration. The transformation is characterized by the deformation gradient $\mathbf{F}$ (see (2.3)). The two term asymptotic solution presented by Knowles and Sternberg [4] for the deformed coordinate for the interfacial crack problem considered here is,

$$
\begin{aligned}
y_1 &= l + \hat{a}_1 r^{1/2} h(\theta) \sin(\theta/2) + \hat{b}_1 r \cos(\theta) \\
y_2 &= \hat{a}_2 r^{1/2} h(\theta) \sin(\theta/2) + \hat{b}_2 r \cos(\theta),
\end{aligned}
\tag{3.1}
$$

where $h(\theta)$ is the angular function,

$$
h(\theta) = \begin{cases} 1 & 0 \leqslant \theta \leqslant \pi; \\ s & -\pi \leqslant \theta < 0. \end{cases}
\tag{3.2}
$$

$\hat{a}_1$, $\hat{a}_2$, $\hat{b}_1$ and $\hat{b}_2$ are amplitude parameters and elude local asymptotic analysis. These parameters could depend on the stiffness ratio s, on the crack length $2l$ and on the specific loading at infinity. The asymptotic results for the Cauchy stresses are given by

$$
\frac{\tau_{11}}{\mu_1} = o(r^{1/2})
$$

$$
\frac{\tau_{22}}{\mu_1} = \frac{a_2^2}{4} r^{-1} h(\theta) - a_2 b_2 r^{-1/2} \sin(\theta/2)
\tag{3.3}
$$

$$
\frac{\tau_{12}}{\mu_1} = -\tfrac{1}{2} b_1 a_2 \sin(\theta/2),
$$

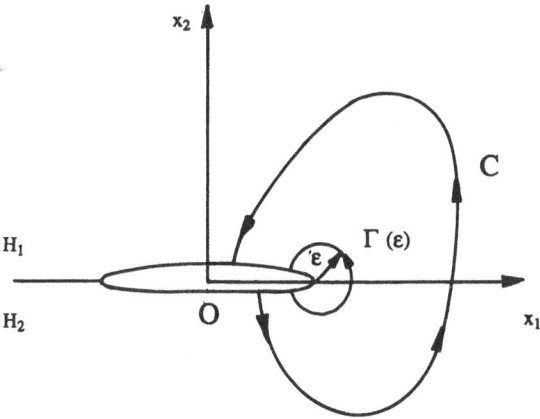

Fig. 2. Integration paths for calculating finite elastostatic *J*.

where

$$a_2 = [\hat{a}_1^2 + \hat{a}_2^2]^{1/2} \quad b_1 = 2A/\hat{a}_2 \quad b_2 = 2B/\hat{a}_2$$

$$A = (\hat{a}_2 \hat{b}_1 - \hat{a}_1 \hat{b}_2)/2 \quad B = (\hat{a}_1 \hat{b}_1 + \hat{a}_2 \hat{b}_2)/2.$$

It is seen that the asymptotic representation of the solution for the bimaterial crack problem for the Neo-Hookean material behavior does not show any interpenetration and the crack faces open monotonically, though with finite rotation as $r \to 0$.

The primary amplitude parameter a_2 can be expressed in terms of an appropriate path independent J integral in finite elastostatics [4]:

$$J = \int_C (U n_1 - \sigma_{\alpha\beta} n_\beta y_{\alpha,1}) \, \mathrm{d}S = \int_{\Gamma(\varepsilon)} (U n_1 - \sigma_{\alpha\beta} n_\beta u_{\alpha,1}) \, \mathrm{d}s. \tag{3.4}$$

The contour C and $\Gamma(\varepsilon)$ are shown in Fig. 2. One can evaluate the contour integral explicitly from the expressions (2.1) and (3.3) as

$$J = \frac{\pi}{8} \mu_1 (1 + s) a_2^2. \tag{3.5}$$

C is any simple curve issuing from an interior point of the crack, and terminating at such a point after surrounding the crack tip situated at $x_2 = l$.

4. Numerical procedure

A displacement based finite element is used to solve the full boundary value problem shown in Fig. 1. The material is modeled by Neo-Hookean behavior and has been extended to include the Mooney–Rivlin material model for modeling experiments. A detailed development of the finite element methodology for nonlinear continua may be found in Oden [7].

The variational form of the field equations is written as the energy functional

$$E(\phi) = \int_{B_0} U(\mathbf{F}) \, dV - \int_{S_{0t}} T_i \phi_i \, dS, \tag{4.1}$$

where B_0 is the reference or undeformed configuration and S_{0t} is the corresponding boundary where tractions T_i are prescribed. ϕ is the admissible variation of the displacement field which can be written in terms of the element interpolation functions N_a^e as

$$\phi_i(x) = \sum_e \sum_a x_{ia} N_a^e(x) \tag{4.2}$$

with x_{ia} as the i-th coordinate of the a-th node; e denotes the element. By making use of the virtual work principle, the stationarity of the functional $E(\phi)$ can be expressed as a system of nonlinear algebraic equations

$$\mathbf{M(x)} = \mathbf{f}, \tag{4.3}$$

where

$$M_i(\mathbf{x}) = \sum_e \int_{\Omega_0} \sigma_{ij}(F_h) N_{a,j}^e \, dV_0$$

$$f_i = \sum_e \int_{\Gamma_{0t}}^e T_i N_a^e \, dS$$

and σ denotes the Piola–Kirchoff stress tensor which in general is not symmetric. Hence the system of equations yield a nonsymmetric "stiffness" matrix and hence needs "additional" computational effort. This problem is overcome by subjecting the resulting equations to suitable transformations and making use of the principle of objectivity so that the system of equations can be rewritten as

$$Ku = f, \tag{4.4}$$

where K is the symmetric tangent stiffness matrix and u and f are the nodal displacement and force vectors. The tangent stiffness matrix can be written as

$$K = K^G + K^M, \tag{4.5}$$

where K^G and K^M denote the spatial geometric stiffness and spatial stiffness of the material. This stiffness matrix is given by

$$K = \int_{\Omega_e^i} \int_{\Omega_e^i} B^T D^T B \, dv, \tag{4.6}$$

and where D^T is the material matrix, $B = n_{a,j}^e$ with n_a^e the spatial interpolation function in the current configuration which is related to the spatial interpolation functions in the undeformed configuration by

$$n_a^e(x) = N_a^e F^{-1}. \tag{4.7}$$

The material matrix D^T can be expressed as

$$D^T_{ijkl} = \delta_{ik}\tau_{jl} + D_{ijkl}, \tag{4.8}$$

where

$$D_{ijkl} = \frac{\partial \sigma_{iq}}{\partial F_{ks}} F_{jq} F_{ls} \tag{4.9}$$

and δ_{ik} is the Kronecker delta. The symmetric system of nonlinear equations is solved by using the Newton–Raphson iterative procedure. At every iteration, the stiffness matrix K is updated in the current configuration.

The finite elastostatic J integral was computed using a domain integral form adopted for finite elastostatics. The line integral (3.4) is recast as an area integral, which is advantageous in numerical calculations. The domain integral formulation for obtaining the J integral for general material response has been outlined by Moran and Shih [8].

The finite element scheme described above is implemented in a modified version of the finite element program FEAP [9].

5. Computational model

To model the problem shown in Fig. 3a, only the right half of the body needs to be considered for finite element analysis since the problem possesses reflective symmetry about $x_1 = 0$. The half crack length is l and the semi-width of the region used in analysis is $100l$. This adequately models a crack centered in an infinite sheet.

The finite element discretization is shown in Fig. 3b for the right upper quarter of the problem, made up of four node quadrilateral elements. Wedge shaped four node elements surround the crack tip and have a radial length of $10^{-5}l$ as shown in Fig. 3c. Each decade of radial length is covered by four circular strips of elements and the region between $10^{-5}l$ and l is covered by 20 strips of elements generated on a logarithmic size scale. Within each strip, the angular distance $-\pi$ and π is spanned by 36 equally spaced elements. The mesh for the domain $r \leqslant l$ has 756 elements. The domain beyond $r > l$ and the remote boundaries is modeled by 320 elements. In the finite element model used there were 1108 nodes and 1076 elements. For the case $s = 1$ and $s = 0$ only the upper right quarter of the problem needs to be modeled in the finite element analysis. Otherwise the entire right half of the domain was discretized and analyzed. To test the adequacy of the mesh, computations were repeated in some cases using twice as many elements as in the model described above. The results obtained from the two computational models differed at most by 2 percent; results presented in this paper are for the coarser mesh described earlier in this section.

The smallest element used in the current calculations is $10^{-5}l$. The oscillatory singularities occur over distances typically of the order of $10^{-8}l$ in the linearized theory of elasticity. In a recent study Shih and Asaro [5] have conducted a finite element analysis for the elastic-plastic material behavior under small strain conditions, with a choice for the smallest element length of $10^{-15}l$. Disregarding the physical implication of such a "refined" grid, the reason one cannot choose such a fine element size in our current model problem is that in

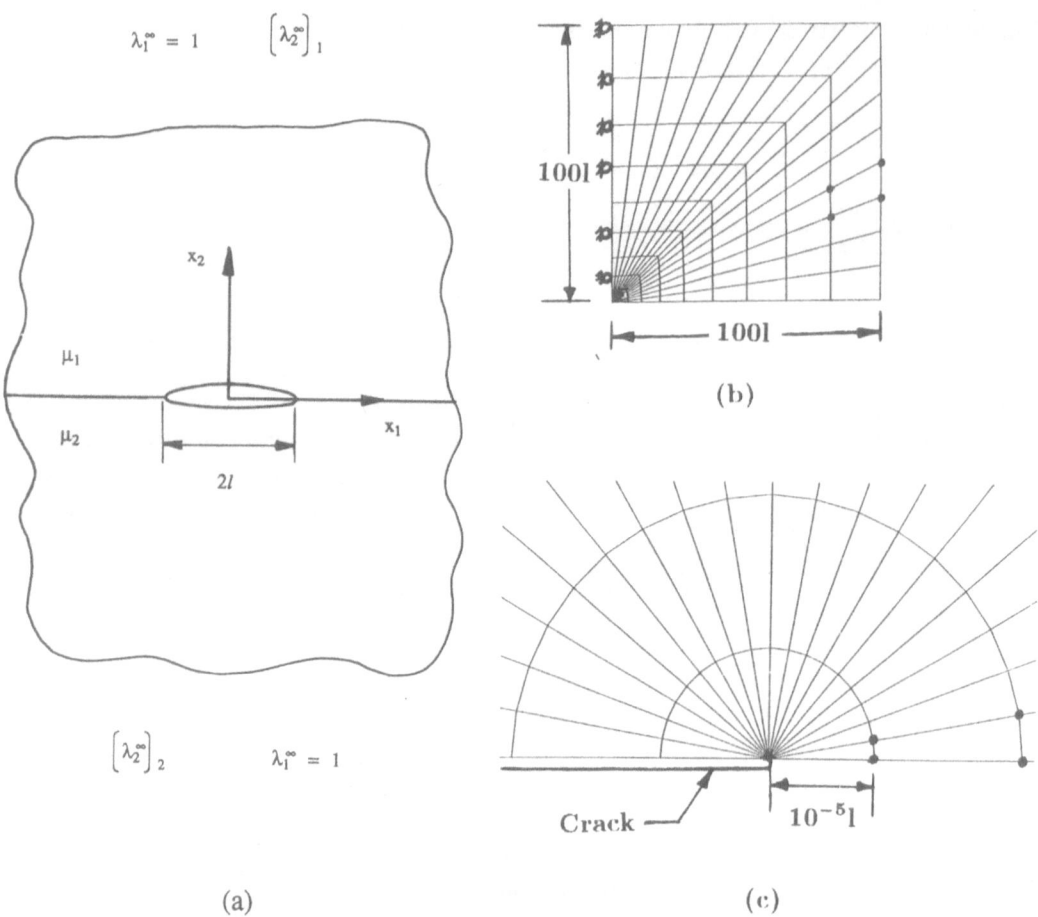

Fig. 3. (a) Far field loading on the cracked panel. (b) Finite element mesh for upper right half of the cracked panel. (c) Arrangement of elements at crack tip.

this class of finite deformation problems, the stiffness matrix used in the computations is formed with respect to the deformed configuration, which would become ill conditioned if the element size were extremely small. Also the increment in remote displacements is severely restricted for obtaining convergence through the Newton–Raphson iterative procedure. Preliminary studies show that the smallest increment in applied boundary displacement that does not pose any apparent numerical problems is on the order of 10^4 *times the smallest element length.* Thus the time required for computations would become truly excessive if one were to choose an element size on the oder of $10^{-15}l$ in the present situation. In this connection it should be recalled that the main purpose of this study is to examine the validity of the asymptotic fields presented by Knowles and Sternberg [4], the range of dominance of the asymptotic solution and to establish a relation between the far field loading conditions and the dominant term of the near-tip prevailing asymptotic field; our purpose is not to address the numerical resolution of the oscillatory fields over subatomic distances near the crack tip.

As a check on the consistency of the numerical solution the finite elastostatic J integral was calculated on five contours surrounding the crack tip. The value of the J integral on any given contour was within 1 percent of its mean value.

6. Results

The far field loading conditions for the boundary value problem shown in Fig. 1 are shown in Fig. 3a. The boundary of the square region is held fixed in the x_1 direction, so that the stretch is $\lambda_1^\infty = 1$. The boundary parallel to the crack faces of the lower half is held fixed in the x_2 direction. Uniform stretch is applied on the remote boundary parallel to the crack faces in the upper half. This loading simulates a homogeneous deformation in the far field characterized by the principal stretches $\lambda_1^\infty = 1$ and λ_2^∞ in the x_1 and x_2 directions respectively; such a loading satisfies the "bond" conditions required by Knowles and Sternberg [4] in order to satisfy the continuity of displacements and tractions across the interface. A useful measure of the applied loads in this configuration is the nominal normal stress at the remote boundary σ_{22}^∞,

$$\sigma_{22}^\infty = \mu_1 \left(\lambda_2^\infty - \frac{1}{(\lambda_2^\infty)^3} \right). \tag{6.1}$$

In all the cases studied here the normalized value of μ_1 was chosen to be $1/3$. Four different cases were analyzed for different ratios of the moduli namely, $s = 1, 0.5, 0.2$, and 0. $s = 1$ corresponds to the case where the material is homogeneous all around the interface. $s = 0$ corresponds to the case where the lower half plane H_2 is rigid.

In each case the sheets were stretched until $\lambda_2^\infty \approx 2$. The amplitude parameters $\hat{a}_1$ and $\hat{b}_1$ in (3.1) are determined from the displacement field surrounding the crack tip. $\hat{b}_2$ is determined from the interface displacements. The dominant amplitude parameter a_2 is determined from the finite elastostatic J integral. We find that $\hat{a}_1$ and $\hat{b}_1$ cannot be determined with a great deal of accuracy. For stretches λ_2^∞ greater than 1.3, $\hat{a}_1 \approx 0$ and $\hat{b}_1 \approx 1$. $\hat{b}_2$ can be determined accurately from the interface displacements. For $s = 1$ and $s = 0$ the amplitude parameter $\hat{b}_2 = 0$ and for $s = 0.5$ and 0.2 it is found that $\hat{b}_2 \approx 0$ for the loading and the geometry considered here for all applied loads. The dominant amplitude parameter a_2 is determined accurately with the aid of the computed finite elastostatic J integral.

The amplitude parameter a_2 calculated from the J integral is plotted against the normalized remote nominal normal stress σ_{22}^∞/μ_1 in Fig. 4. It was found that this curve is a universal curve for all moduli ratios under the far field loading conditions considered here. Under sufficiently small deformations the J integral computed from the linearized theory of elasticity, and denoted by j, and the finite elastostatic J are conceived to be the same [4]. This is termed the small load approximation. For an all around homogeneous incompressible material in the linear theory of elasticity the j integral has the well known expression

$$j = \frac{K_I^2}{3\mu_1}, \tag{6.2}$$

where K_I is the mode I stress intensity factor which for the loadings considered here, is given by

$$K_I = \sigma_{22}^\infty \sqrt{\pi l}. \tag{6.3}$$

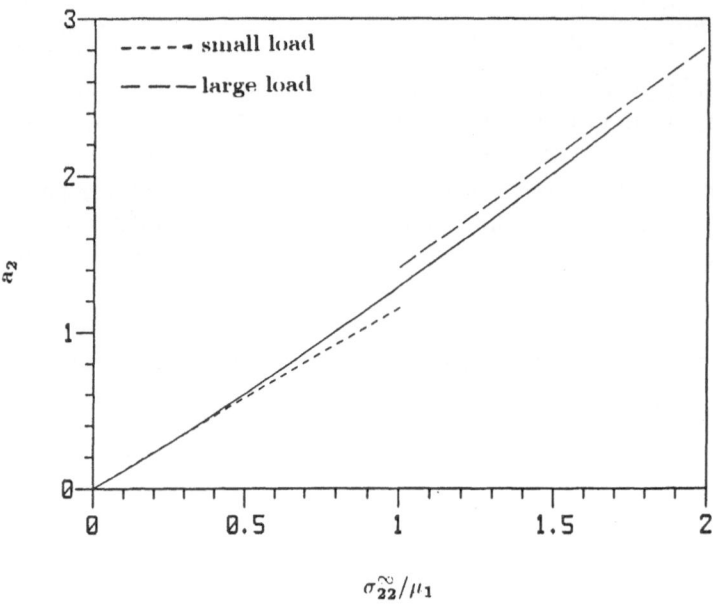

Fig. 4. Amplitude parameter a_2 versus normalized far field stress σ_{22}^∞/μ_1.

In finite elastostatics for $s = 1$, one has, from (3.5)

$$J = \frac{\pi}{4}\,\mu_1 a_2^2. \tag{6.4}$$

Under the small load approximation, we have

$$J = j \tag{6.5}$$

so that from expressions (6.2), (6.3) and (6.4) one obtains for this limiting case, the amplitude parameter a_2

$$a_2 = 2\sqrt{\frac{l}{3}}\,\frac{\sigma_{22}^\infty}{\mu_1}. \tag{6.6}$$

This approximation is plotted as a dotted line in Fig. 4. It is seen that the numerically computed a_2 shows excellent agreement with the small load approximation (6.6), for normalized nominal stress values up to $\sigma_{22}^\infty/\mu_1 \approx 0.5$. This corresponds to a normal stretch at the far field of $\lambda_2^\infty \approx 1.15$, in the cases where $s = 1$ and $s = 0$. Another useful result that is obtained for the homogeneous Neo-Hookean cracked solid relates the amplitude parameter a_2 and the normal stretch λ_2^∞ which can be found in Wong and Shield [10] (see also Knowles and Sternberg [4]). The amplitude parameter a_2 can then be expressed as

$$a_2 = \lambda_2^\infty\sqrt{2l}. \tag{6.7}$$

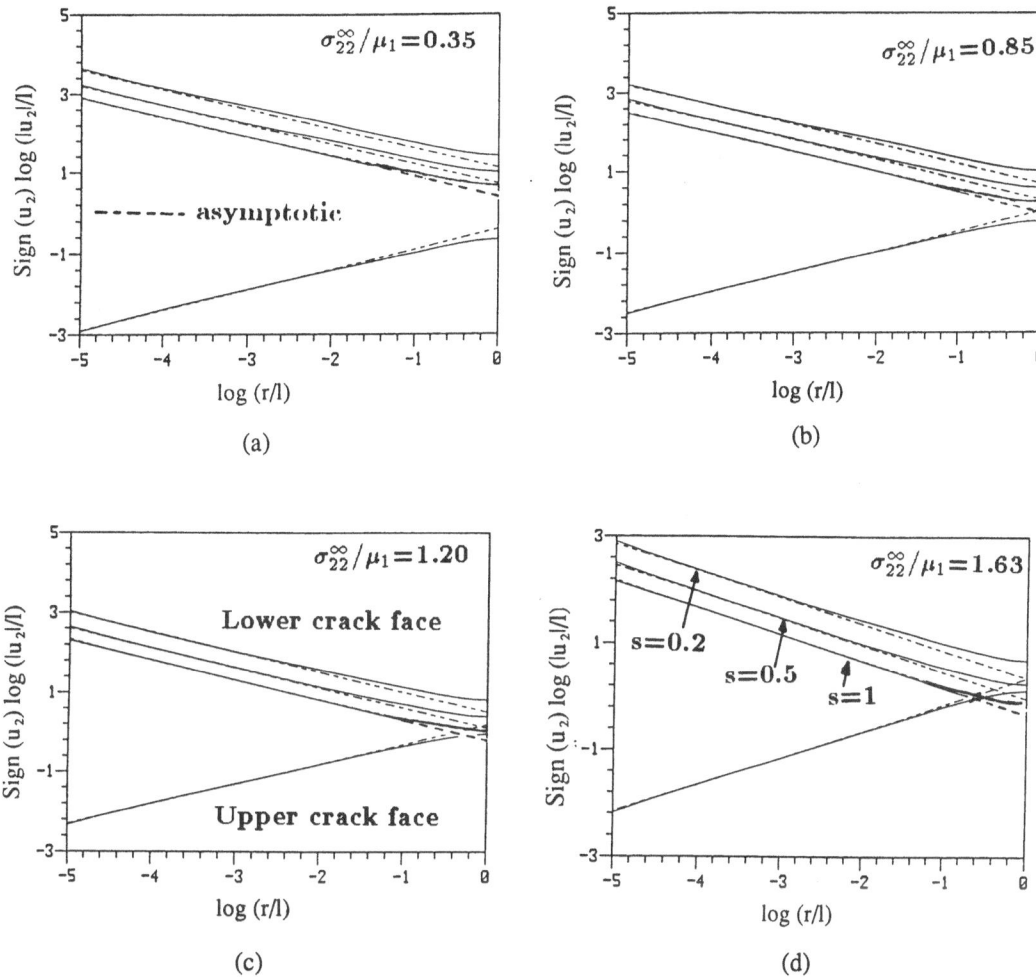

Fig. 5. Plot of log of normalized crack opening displacement versus log of normalized distance for modulus ratios $s = 1, 0.5, 0.2$, and 9 at normalized far field stress levels, (a) $\sigma_{22}^{\infty}/\mu_1 = 0.35$, (b) $\sigma_{22}^{x}/\mu_1 = 0.85$, (c) $\sigma_{22}^{x}/\mu_1 = 1.2$, and (d) $\sigma_{22}^{\infty}/\mu_1 = 1.63$.

This result is valid for large stretch ratios λ_2^{∞} and we call it the large load approximation. This result is shown as a dash-dot curve in Fig. 4. It is seen that this serves as an upper bound for the amplitude parameter a_2.

The small and large load approximations are illustrated for the case where the all around material is homogeneous. It is observed that although the amplitude parameter a_2 is independent of the modulus ratio s, the value of the finite elastostatic J does depend on s. There is no available result for the large load approximation in the bimaterial case. For the boundary value problem considered here, namely for the bimaterial interface problem, the large load approximation for the homogeneous material might be equally valid.

An important implication of this analysis is that for a given value of nominal stress at the remote boundary the opening profile of the crack face deformation in the upper half is independent of the modulus ratio s and the lower crack face shape is scaled by the factor s. It is also observed that the asymptotic results for the corresponding linear elastic problem have a similar feature where the normal displacement of the lower crack face is scaled by the

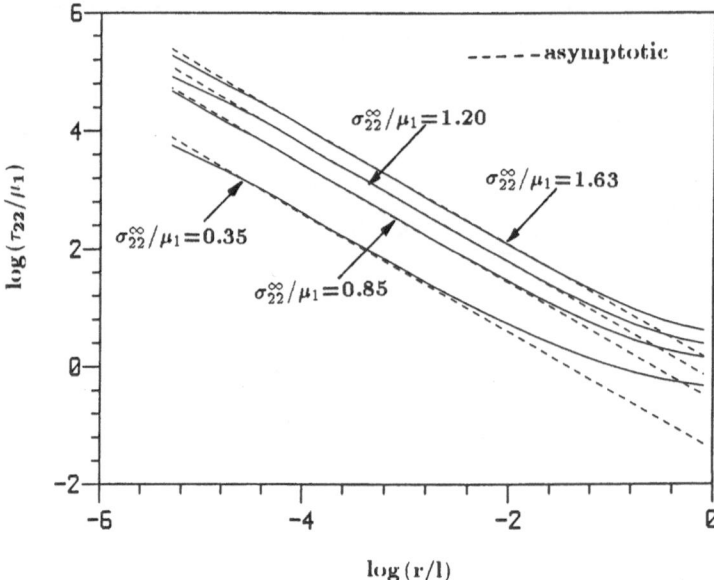

Fig. 6. Plot of log of normalized stress τ_{22}/μ_1 versus log of normalized distance ahead of the crack tip for normalized far field stress levels $\sigma_{22}^{\infty}/\mu_1 = 0.35, 0.85, 1.2,$ and 1.63.

factor s in comparison to the normal displacement of the upper crack face [2]. To establish the range of dominance of the two term asymptotic solution (3.1), the normalized opening profile of the crack faces is plotted against the normalized distance along the crack faces on a logarithmic scale. The results are plotted for four values of the normalized applied nominal stress $\sigma_{22}^{\infty}/\mu_1 = 0.35, 0.85, 1.20,$ and 1.63 in Fig. 5. In each of the Figs. 5, a through d, the results are plotted for the four modulus ratios $s = 1, 0.5, 0.2,$ and 0 considered in this study. The asymptotic prediction based on the results of Knowles and Sternberg [4] is also shown in Fig. 5. It is interesting to observe that the crack opening profile of the upper crack face is the same for all the four moduli ratios considered here, not just at distances where the asymptotic solution is valid but over the entire crack face. Note that the range of dominance of the asymptotic solution increases with increasing load level as can be seen in Figs. 5, a through d. At the same time the range of dominance in the lower half space does not increase in the same fashion as in the upper half space and the range of dominance appears to be smaller for a smaller ratio of moduli s for the same applied remote load level.

The normalized stress τ_{22}/μ_1 on $\theta = 5°$ is plotted against the normalized radial distance r/l ahead of the crack tip in the upper half on a logarithmic scale in Fig. 6. The asymptotic solution is shown on the figure as dotted lines for the four different load levels $\sigma_{22}^{\infty}/\mu_1 = 0.35,$ 0.85, 1.20, and 1.63. It is again seen that the range of dominance of the asymptotic solution is nearly the same as those found for the displacements.

To illustrate further the domain of dominance of the asymptotic solution, we show in Fig. 7a the discrepancy between the asymptotic and the numerical solution for the normal displacements as a percentage of the asymptotic solution against the radial distance for the normal displacements of the crack faces at different load levels at the remote boundaries. Figure 7b shows the increasing range of dominance as a function of the load level for the different levels of discrepancy (2, 5, and 10 percent) between the numerical and asymptotic solution in the normal displacement of the crack face.

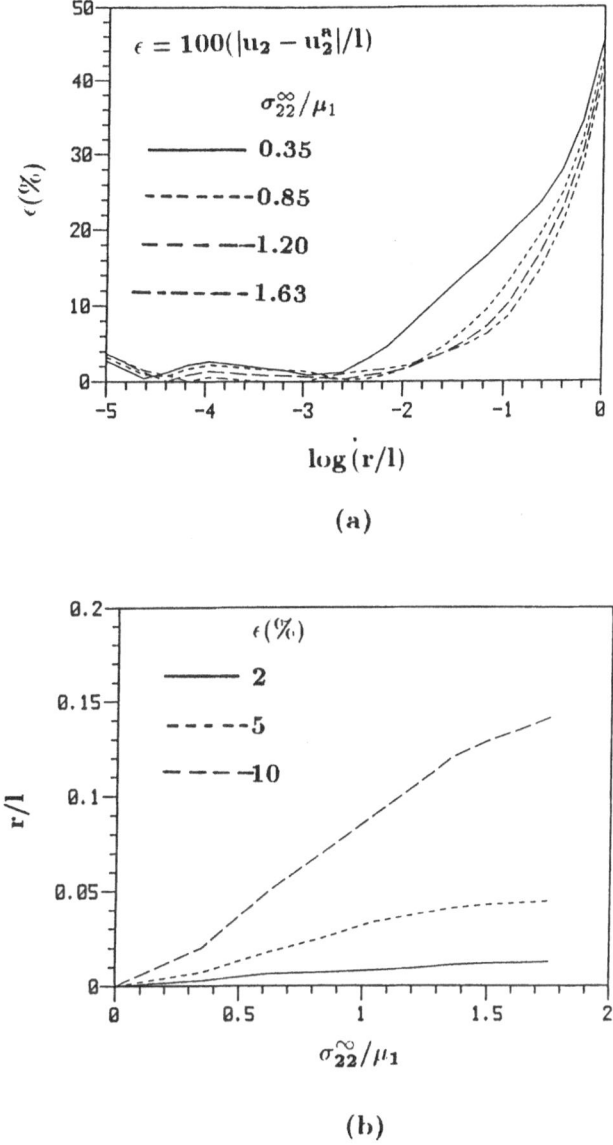

Fig. 7. (a) Plot of percent error in crack opening displacement versus log of normalized distance for normalized far field stress levels $\sigma_{22}^{\infty}/\mu_1 = 0.35, 0.85, 1.2$, and 1.63. (b) Range of dominance of the asymptotic solution: Plot of normalized radial distance versus the far field stress level $\sigma_{22}^{\infty}/\mu_1$ for $\varepsilon = 2, 5$ and 10 percent.

The Cauchy stresses ahead of the crack tip for $s = 1, 0.5, 0.2$, and 0 are plotted along the radial line ahead of the crack tip in Fig. 8. It should be observed that the stresses τ_{11} and τ_{12} exhibit a weaker singularity than τ_{22} (see Fig. 6) as expressed by the asymptotic analysis. The case $s = 0$ corresponds to the most severely constrained one in this class of problems. One might expect the shear stress τ_{12} to be quite dominant but it is seen that even in this case the normal stress τ_{22} is the dominant stress. Another interesting feature of the solution is that the stress component τ_{11} has a weaker singularity in comparison to that of τ_{22} as opposed to being the same as in the linear theory of elasticity.

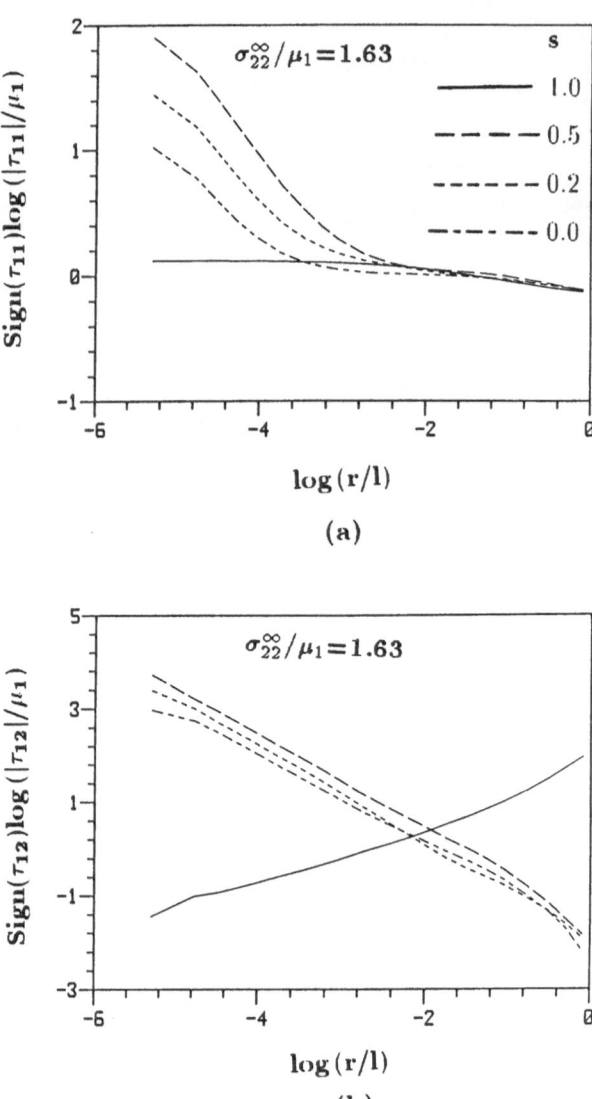

Fig. 8. (a) Plot of log of normalized stress τ_{11}/μ_1 versus log of normalized distance ahead of the crack tip at a normalized far field stress levels $\sigma_{22}^\infty/\mu_1 = 1.63$ for modulus ratios $s = 1, 0.5, 0.2$, and 0. (b) Plot of log of normalized stress τ_{12}/μ_1 versus log of normalized distance ahead of the crack tip at a normalized far field stress levels $\sigma_{22}^\infty/\mu_1 = 1.63$ for modulus ratios $s = 1, 0.5, 0.2$, and 0.

The angular distribution of the stress τ_{22} is plotted for $s = 0.5$ and $s = 0.2$ at a far field load level of $\sigma_{22}^\infty = 1.63$ along with the asymptotic predictions at radial distances $r/l = 5.5 \times 10^{-4}$ and $r/l = 5.5 \times 10^{-3}$ in Figs. 9a and b. It is found that there is good agreement between the asymptotic (3.4) and numerical solutions. Essentially τ_{22} is constant around the crack tip at a given radius and exhibits a jump across the interface. Though the tractions are continuous across the interface the stress component τ_{22} can experience a jump due to the finite geometry changes in the vicinity of the crack tip. Also it can be noted that the stress τ_{22} possesses a r^{-1} singularity.

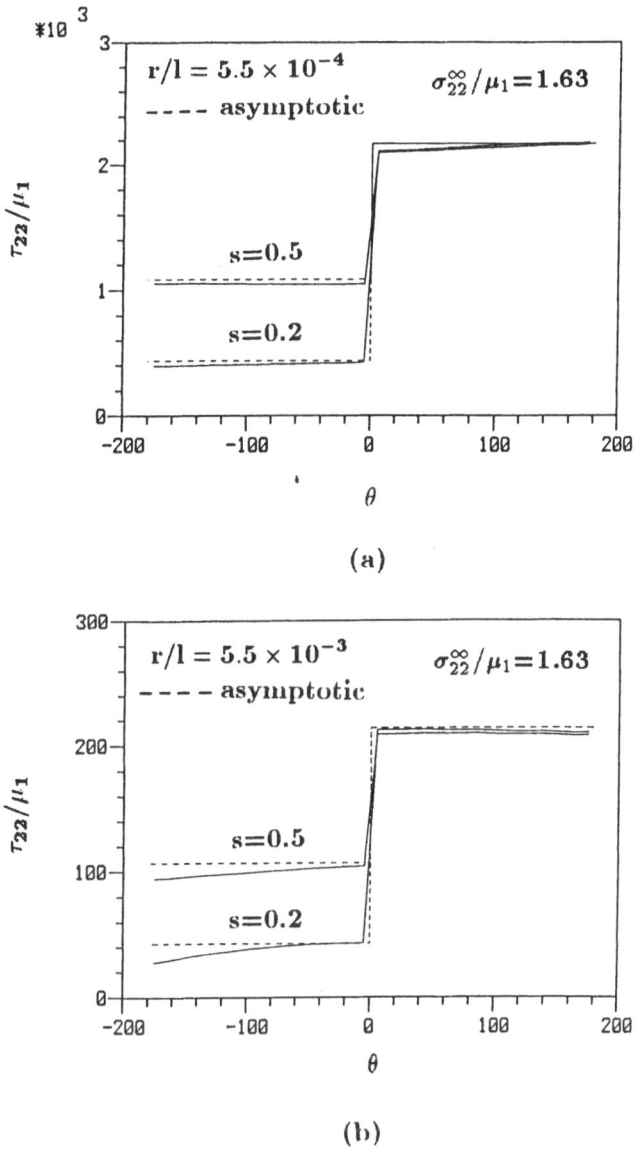

Fig. 9. Plots of angular variation of nomralized stress τ_{22}/μ_1 for modulus ratios $s = 0.5$ and 0.2 at normalized radial distances (a) $r/l = 5.5 \times 10^{-4}$, (b) $r/l = 5.5 \times 10^{-4}$ for $\sigma_{22}^{\infty}/\mu_1 = 1.63$.

7. An application

As an application of the analysis developed here a test geometry, the dimensions of which are shown in Fig. 10a, is analyzed. The specimen models a bimaterial interface crack between two Solithane plates of different composition. In this particular case the upper half of the specimen Solithane 50/50 contains 50 percent resin and 50 percent catalyst and is bonded to the lower half of the specimen which is made of Solithane 65/35 containing 65 percent resin and 35 percent catalyst. The plate is 12 inches long, 2 inches wide and 0.125 inches thick. A crack 3 inches long is introduced in the mid-plane at the interface. This specimen is modeled

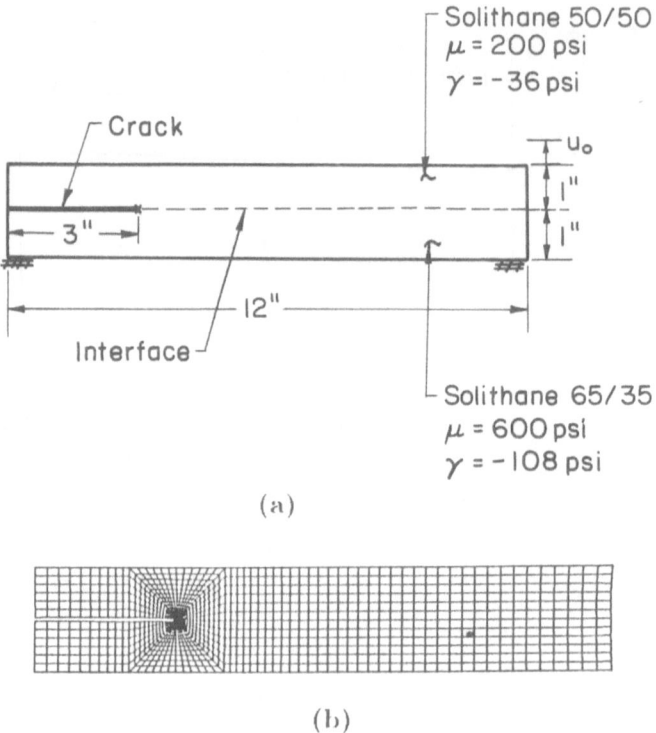

(a)

(b)

Fig. 10. (a) Specimen with an edge crack between two dissimilar Solithane plates. (b) Finite element mesh of the specimen (deformed configuration).

by using the mesh shown in Fig. 10b. The mesh consists of 1608 four noded isoparametric elements and the elements are concentrated near the crack tip to improve the resolution. The smallest element is 0.1 percent of the crack length. The material properties are obtained from [6]. This material is characterized by a Mooney–Rivlin material model with elastic potential expressed in the form

$$U = \tfrac{1}{2}[\mu(I_1 - 3) + \gamma(I_2 - 3)]. \tag{6.8}$$

The Mooney–Rivlin constants used for Solithane 50/50 are $\mu_1 = 200$ psi and $\gamma_1 = -36$ psi; as the softer material which forms the upper half of the specimen. The corresponding constants for Solithane 65/35 are $\mu_2 = 600$ psi and $\gamma_2 = -108$ psi (the stiffer material) and it describes the lower half of the specimen. The ratio of the moduli of the two materials is thus 3. The boundary conditions imposed represent fixed grip conditions at the top and bottom of the specimen. The bottom of the specimen is fixed and displacements are applied only to the top of the specimen. The displacements are applied in increments of 1 percent of the plate width (0.02 inches). From the field quantities, namely the out of plane deformation, synthetic caustic patterns are generated using the equations for the geometric caustics in the transmission mode (see Rosakis and Ravi-Chandar [11]). The caustic pattern from the analysis is shown in Fig. 11a for the simulated experimental conditions. The caustic obtained from analysis can be matched with those obtained from experiments by accounting for the retardation induced by the stress optic effect. Figure 11b shows the experimental

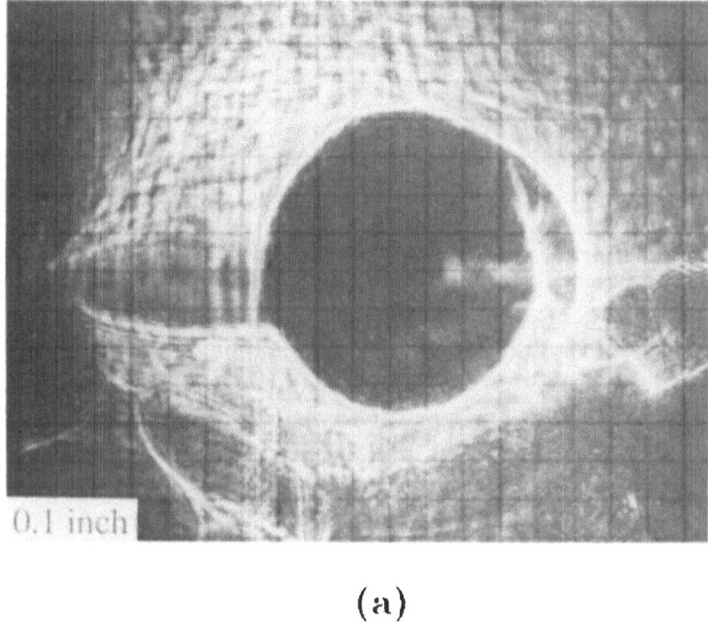

(a)

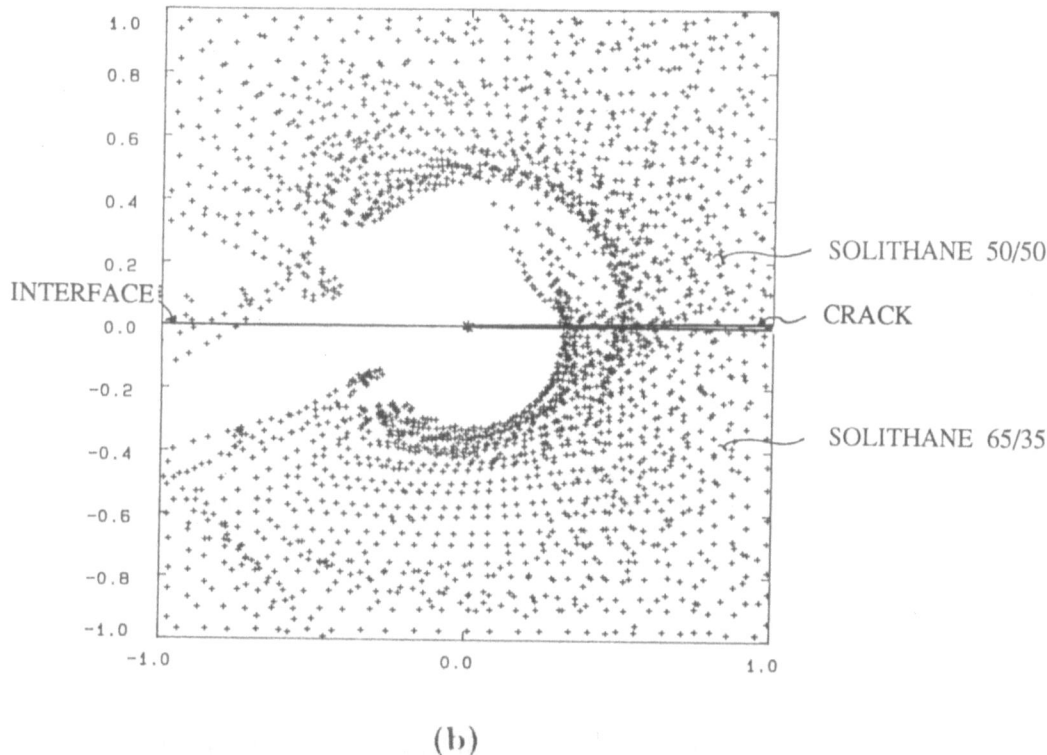

(b)

Fig. 11. (a) Caustics obtained from experiment. (b) Caustics generated from analysis.

caustic pattern obtained in the transmission mode for the same material and specimen geometry used in the analysis. The caustics pattern obtained from the experiment and analysis compares well in size and shape.

8. Discussion

A numerical procedure has been developed to analyze crack problems in bimaterial interfaces. The material model used is a finite elasticity model and the large deformations associated with the crack tip are handled using the finite element method. The results obtained from the numerical solution indicates the domain in which the asymptotic solution presented by Knowles and Sternberg [4] for the bimaterial interface crack problem is valid in the vicinity of the crack tip. The range of dominance increases approximately linearly with increasing load level. It is also found that on the scale of discretization used here there are no oscillations present either in stresses or displacements at all load levels and under the loading considered in this boundary value problem. Furthermore it is found that the relation between the asymptotic amplitude parameter a_2 and the applied nominal stress is independent of the modulus ratio s.

It is interesting to note that this dominant term agrees well with the linear approximation for small loads. For small stretches at the remote boundary ($\lambda_2^\infty \leqslant 1.15$) the region of dominance of the asymptotic solution is very small, about 0.1 percent of the crack length. It is conceivable that at the small load levels the linear theory approximates the behavior reasonably well at distances greater than 0.1 percent of the crack length. In commonly used materials the crack begins to propagate under sufficiently small loads. It might therefore be possible to interpret the behavior at the material interfaces with commonly used experimental methods such as caustics or photoelasticity.

Several questions remain to be answered within the frame work of this analysis. The precise nature of the approximate status of the linear solution with respect to the nonlinear problem needs to be addressed. This can be resolved only by resorting to very fine mesh sizes on the order of $10^{-15} l$. Such a study would address the mathematical rather than the physical character of the problem. Physically such dimensions are on the atomistic level where analysis based on continuum mechanics would not be valid. The exact nature of the mixity of the fields in the vicinity of the crack tip also needs to be examined.

The methodology presented in this paper can be extended to cover other material models with geometric nonlinearities taken into account. The problem of failure at interfaces and path of propagation of an interface crack needs to be addressed in the future. The asymptotic results and the numerical solution suggest that a stress based failure criterion (maximum principal stress) would imply that the interface crack would initiate along the interface line and cause the separation of interface. Currently further numerical and experimental work is underway to better understand the behavior of bimaterial interface fracture.

Acknowledgements

This work was conducted as part of a program in Interfacial Cracks at GALCIT funded by the Solid Rockets Division of the Aeronautics and Astronautics Laboratories at Edwards

Air Force Base, CA under the technical monitorship of Dr C.T. Liu. The computations were conducted on the Cray-XMP at the San Diego Super Computer facility. Access to SDSC was provided by the National Science Foundation through a grant MSM 8215438 to the California Institute of Technology. The authors are grateful to Professor J.K. Knowles for valuable discussions and Professor M. Ortiz of Brown University for providing the finite element program FEAP.

References

1. M.L. Williams, *Bulletin of the Seismological Society of America* 49(2) (1959) 199.
2. A.H. England, *Journal of Applied Mechanics* 32(2) (1965) 400.
3. J.R. Rice and G.C. Sih, *Journal of Applied Mechanics* 32(2) (1965) 418.
4. J.K. Knowles and E. Sternberg, *Journal of Elasticity* 13 (1983) 257.
5. C.F. Shih and R.J. Asaro, "Elastic-plastic analysis of cracks on bimaterial interfaces. Part I: Small scale yielding," *Brown University Report*, March (1987).
6. W.G. Knauss and H.K. Mueller, "The mechanical characterization of Solithane 113 in the swollen and unswollen state," *GALCIT SM 67-8*, California Institute of Technology, Pasadena, CA (1968).
7. J.T. Oden, *Finite Elements of Nonlinear Continua*, McGraw-Hill, New York (1972).
8. B. Moran and C.F. Shih, *Engineering Fracture Mechanics* 27(6) (1987) 615.
9. R.L. Taylor, in *The Finite Element Method* by O.C. Zienkiewicz, McGraw-Hill, London (1977).
10. F.S. Wong and R.T. Shield, *Zeitschrift für Angewandte Mathematik und Physik* 20(2) (1969) 176.
11. A.J. Rosakis and K. Ravi-Chandar, *International Journal of Solids and Structures* 22(2) (1986) 121.

Résumé. On met au point une méthode numérique pour l'étude d'un problème d'interfaces entre deux matériaux néo-Hookiens sollicités en état plan de tension. On compare les résultats avec les prédictions analytiques établies pour un champ asymptotique. On établit la gamme dans laquelle la solution asymptotique est dominante, à différents niveaux de charge, et on met en relation l'ampleur du champ asymptotique à l'extrémité de la fissure avec celle du champ de contraintes à une certaine distance. Le modèle numérique est étendu à l'analyse d'essais sur éprouvettes comportant une fissure de bord à l'interface de deux tôles de Solithane caractérisée par un comportement de matériau de Mooney–Rivlin.